IEIS 2020

Menggang Li · Gábor Bohács ·
Guowei Hua · Daqing Gong ·
Xiaopu Shang
Editors

IEIS 2020

Proceedings of the 7th International Conference on Industrial Economics Systems and Industrial Security Engineering

Springer

Editors
Menggang Li
School of Economics and Management
Beijing Jiaotong University
Beijing, China

Guowei Hua
School of Economics and Management
Beijing Jiaotong University
Beijing, China

Xiaopu Shang
School of Economics and Management
Beijing Jiaotong University
Beijing, China

Gábor Bohács
Budapest, Hungary

Daqing Gong
School of Economics and Management
Beijing Jiaotong University
Beijing, China

ISBN 978-981-33-4362-7 ISBN 978-981-33-4363-4 (eBook)
https://doi.org/10.1007/978-981-33-4363-4

This Springer imprint is published by the registered company Springer Nature Singapore Pte Ltd.
The registered company address is: 152 Beach Road, #21-01/04 Gateway East, Singapore 189721, Singapore

Contents

Network Centrality and Cross-Section of Stock Market Returns

Zhuo Xu, Zhen Li, and Tong Fang

Abstract We construct a network centrality portfolio strategy using Planar Maximally Filtered Graph, and investigate the relationships between centrality and cross-section of international stock markets. We find that the network centrality could significantly explain the cross-section of stock market returns. A subsample analysis shows that the strategy performs better before the crisis, which reveals that business cycles influence the performance of the strategy.

Keywords Stock market network · Portfolio strategy · Cross-section · PMFG

1 Introduction

Financial markets have been characterized as complex networks among the academics. With appropriate network filtering algorithms, the network approach provides a unique prospective for investigating the co-movement of financial markets, structures of market networks, and importance of the markets measured by network centrality [1, 2, 4, 5, 8].

The applications of network in asset pricing are quite limited, and there are only two exceptions. Tse et al. [9] use network analysis to select stocks and build a new stock index. Peralta and Zareei [7] employ the network to improve portfolio selection. However, these research only focus on the stocks within one market, and the cross-section of international stock market returns has not been found. These literature also ignore the vital role of centrality in financial market network.

Z. Xu (✉)
School of Economics and Management, Beijing Jiaotong University, Beijing, China
e-mail: 172912763@qq.com

Z. Li
Cultural Management Division, Beijing Capital Group, Beijing, China
e-mail: lizhencufe@163.com

T. Fang
School of Economics, Shandong University, Jinan, China
e-mail: fangtong1990@outlook.com

M. Li et al. (eds.), *IEIS 2020*, https://doi.org/10.1007/978-981-33-4363-4_1

Our contribution to the literature on financial market networks is fourfold. First, we investigate the relationships between network centrality and cross-section of stock market returns from an international perspective. Second, Planar Maximally Filtered Graph (PMFG) is employed for network construction, which reveals the same hierarchical structure to the methods mentioned in previous literature, but contains more valuable information. Third, we consider four measures of centrality and confirm the robustness of our results. Fourth, we further find that business cycle influences the performance of the portfolio strategy through a subsample analysis.

2 Network Approach and Centrality Portfolio Strategy

A network usually contains a collection of nodes which are connected by edges. For a stock market network, the nodes are stock market indices and the edges are the similarity between each pair of indices. Suppose $x^m_{t,i}$ and $x^m_{t,j}$ are daily returns of stock market index i and j at trading day $t = 1, 2, \ldots$ in month $m = 1, 2, \ldots$, respectively. We use correlation coefficient ρ^m_{ij} between index i and j as the proxy of similarity.

$$\rho^m_{ij} = \frac{\sum_{t,m}\left(x^m_{t,i} - \bar{x}^m_i\right)\left(x^m_{t,j} - \bar{x}^m_j\right)}{\sqrt{\sum_{t,m}\left(x^m_{t,i} - \bar{x}^m_i\right)^2}\sqrt{\sum_{t,m}\left(x^m_{t,j} - \bar{x}^m_j\right)^2}} \tag{1}$$

where $\bar{x}^m_i$ and $\bar{x}^m_j$ are the means of returns of $x^m_{t,i}$ and $x^m_{t,j}$ in month m. Following Mantegna [5], we transform the correlation coefficient to distance d^m_{ij} in Eq. (2). Small values of distance imply strong correlations, and vice versa.

$$d^m_{ij} = \sqrt{2\left(1 - \rho^m_{ij}\right)} \tag{2}$$

The adjacency matrix $\Omega^m = [d^m_{ij}]$ captures the pattern of co-movements across stock markets for month m, and leads to a fully connected network, which is too complex to analyze. We utilize PMFG to filter out this adjacent matrix to reveal the main structure of the stock market network. The filtering algorithm of PMFG proposed by Tumminello et al. [10] starts from considering the largest similarity. As we transform correlation coefficient to distance, and an algorithm starting from the smallest dissimilarity is employed.

We consider four measures of centrality that identify the most important nodes within a network. Degree centrality is defined as the number of edges that a node has. Betweenness centrality measures the number of times a node acts as a bridge along the shortest path between two other nodes. Eigenvector centrality is a measure of the influence of a node in the network. We also consider a hybrid centrality measured by the summation of normalized degree, betweenness and eigenvector centrality, similar to Pozzi et al. [8].

We construct the network centrality portfolio strategy as follows. For each month *m*, we sort stock market indices on the time-m centrality and allocate them to five portfolios, with 8 indices in one portfolio. Portfolio 1 includes the stock market indices with highest centrality, meanwhile Portfolio 5 corresponds to the market indices with lowest centrality. Returns of each portfolio are calculated as an equally weighted average of the stock market returns within the portfolio. As Pozzi et al. [8] show that stocks with lower centrality perform better, so we consider the difference between the returns of P_5 and P_1. It is a weak-minus-strong strategy that longs the stock market indices with lowest centrality and shorts the stock market indices with highest centrality.

3 Data

We collect 40 stock market index prices from Bloomberg, and each index represents one stock market. We obtain the stock market returns by taking natural logarithm of index prices. Our sample considers these indices: S&P500 (USA), GSPTSE (Canada), FTSE100 (UK), CAC40 (France), DAX (Germany), AEX (Netherlands), FTSE ITALIAN (Italy), IBEX35 (Spain), SMI (Switzerland), ISEQ (Ireland), BEL20 (Belgium), OMX20 (Denmark), OSEAX (Norway), OMXSPI (Sweden), TA100 (Israel), NIKKEI225 (Japan), AS51 (Australia), NZX50 (New Zealand), FTSE STI (Singapore), HSI (Hong Kong), IBOVESPA (Brazil), MERV (Argentina), IBVC (Venezuela), IPSA40 (Chile), BVLX (Portugal), ASE (Greece), RTS (Russia), ISE (Turkey), BUX (Hungary), SASEIDX (Saudi Arabia), DFMGI (United Arab Emirates), CASE30 (Egypt), KOSPI (South Korea), SENSEX (India), SETI (Thailand), JKSE (Indonesia), KLSE (Malaysia), PSI (Philippines), VNINDEX (Vietnam), Shanghai Composite (China). We use daily returns for network construction and monthly returns for portfolio allocations. The sample period is from January 2004 to October 2017, as the latest data we can find. The descriptive statistics of monthly returns are reported in Table 1.

4 Network Centrality and Cross-Section of Stock Market Returns

4.1 Cross-Sectional Evidence

Table 2 reports cross-sectional evidence. The average excess returns on the centrality portfolio strategy are statistically different from zero using Newey and West [6]

Table 1 Descriptive statistics

	USA	CAN	UK	FRA	GER	NED	ITA	ESP	CHE	IRE
Mean (%)	0.32	0.35	0.01	0.05	0.54	0.05	−0.15	0.07	0.09	−0.08
Std. (%)	3.54	3.21	3.39	4.25	4.76	4.42	4.96	5.07	3.36	5.38
Skew	−0.38	−0.50	−0.40	−0.79	−0.91	−1.04	−0.24	−0.36	−0.40	−0.87
Kurt	4.67	5.08	4.08	4.28	5.42	4.74	3.66	4.39	3.76	4.46
	BEL	DEN	NOR	SWE	ISR	JPN	AUS	NZL	SIN	HKG
Mean (%)	0.12	0.51	0.72	0.48	0.22	0.45	0.24	0.69	0.08	0.14
Std. (%)	4.38	4.47	5.05	4.29	4.11	5.25	3.83	3.19	4.54	5.76
Skew	−1.21	−0.49	−1.26	−0.73	−0.54	−0.69	−0.76	−0.75	−0.78	−0.68
Kurt	6.25	4.08	6.77	4.58	5.21	4.15	4.00	5.26	5.67	4.52
	BRA	ARG	VEN	CHI	POR	GRE	RUS	TUR	HUN	KSA
Mean (%)	0.18	1.65	5.55	0.58	0.06	−0.87	0.26	0.86	0.70	0.17
Std. (%)	5.73	8.40	14.80	3.96	4.47	8.18	8.84	7.25	6.18	7.53
Skew	−0.03	−0.30	2.17	−0.07	−0.70	−0.44	−0.43	−0.18	−0.72	−0.63
Kurt	2.87	4.31	10.35	2.81	4.19	3.77	3.69	3.53	6.01	4.58
	UAE	EGY	KOR	IND	THA	INA	MAS	PHI	VIE	CHN
Mean (%)	0.45	1.10	0.53	0.63	0.22	0.86	0.33	0.47	0.89	0.21
Std. (%)	9.53	9.59	4.89	6.15	5.38	5.34	3.21	5.03	8.68	8.11
Skew	−0.05	−0.50	−0.47	−0.62	−1.19	−1.04	−0.46	−0.82	0.21	−0.48
Kurt	5.64	5.05	5.11	4.07	7.12	7.65	6.03	6.36	4.61	4.42

Table 2 Centrality-sorted Portfolios

	Statistics	P_1	P_2	P_3	P_4	P_5	P_5–P_1
Degree	Mean (%)	0.26	0.10	0.46	0.55	1.09	0.83***
	Std. (%)	3.92	3.88	3.96	4.21	4.14	
Betweenness	Mean (%)	0.27	0.10	0.43	0.93	0.73	0.46**
	Std. (%)	3.99	3.95	3.62	4.20	4.48	
Eigenvector	Mean (%)	0.28	0.10	0.61	0.65	0.82	0.54**
	Std. (%)	4.03	3.96	3.85	4.12	4.22	
Hybrid	Mean (%)	0.26	0.11	0.42	0.45	1.22	0.96***
	Std. (%)	3.92	4.03	4.05	3.92	4.20	

***, **, * Indicate significance at 1%, 5% and 10% level, respectively

corrected standard errors. The average excess returns of degree, betweenness, eigenvector and hybrid centrality-sorted portfolios are 0.83%, 0.46%, 0.54% and 0.96%, significant at 1%, 5%, 5% and 1% level, which indicates the centrality could explain the cross-section of stock market returns. The hybrid-centrality portfolio strategy produces the highest excess returns. As a combination of the three measures of centrality, the hybrid-centrality performs the best in classifying the most peripheral and the most central stock market indices.

To examine whether centrality still matters controlling for risk factors, we regress the time series of excess returns of the centrality portfolio strategy on the international risk factors proposed by Fama and French [3]. The regression results reported in Table 3 confirm that there is a centrality return premium even after controlling for market, size, book-to-market, profitability, investment and momentum factors. The alpha in the model with hybrid centrality portfolio is the highest (1.20% per month), which is also consistent with the results in Table 2.

Table 3 Regression results with international risk factors

Portfolio	Intercept	Mkt-RF	SMB	HML	RMW	CMA	MOM
Degree	1.02*** (0.24)	−0.19** (0.06)	0.33* (0.16)	0.24 (0.18)	−0.27 (0.24)	−0.63 (0.23)	−0.02 (0.08)
Betweenness	0.52* (0.25)	−0.14* (0.07)	0.39* (0.16)	0.21 (0.19)	0.06 (0.25)	−0.66** (0.24)	−0.02 (0.08)
Eigenvector	0.61* (0.25)	−0.13 (0.07)	0.38* (0.16)	0.28 (0.19)	0.03 (0.25)	−0.68** (0.23)	−0.05 (0.08)
Hybird	1.20*** (0.24)	−0.24*** (0.06)	0.26 (0.16)	0.50** (0.18)	−0.35 (0.24)	−0.86*** (0.23)	0.01 (0.08)

***, **, * Indicate significance at 1%, 5% and 10% level, respectively

4.2 Portfolio Strategy Performance and Business Cycles

The stock markets are usually seen to be highly related to economic fundamentals, implying that the markets perform differently in expansions and recessions. So we consider a subsample analysis. The pre-crisis subsample starts from January 2004 to December 2007, corresponding to the expansion before the crisis. The crisis subsample starts from January 2008 to June 2009, corresponding to the recession. The last subsample starts from July 2009, representing the expansion.

Panel A of Table 4 presents the results for the pre-crisis subsample. The average excess returns of centrality strategy are all different from zero at 1% level. The degree centrality strategy has the highest average excess returns (1.51%), followed by hybrid centrality strategy (1.40%). In Panel B, the average excess returns are positive except for the betweenness-centrality-sorted portfolios. However, the average excess returns are not different from zero significantly at any level, implying the network centrality could not produce significant excess returns during the recession. In Panel C, the average excess returns become significantly different from zero during the expansion, except for the betweenness centrality. Comparing the average excess returns between the pre-crisis and post-crisis subsample, the centrality portfolio strategy is more efficient before the business cycle hits the peak. It is expected that the performance of the strategy would be better following the expansion of the world economy.

Table 4 Centrality-sorted Portfolios and Business Cycles

	P_1	P_2	P_3	P_4	P_5	P_5–P_1
Panel A: pre-crisis subsample						
Degree	0.83	0.95	1.09	1.19	2.35	1.51***
Betweenness	1.07	1.08	0.63	1.21	2.43	1.37***
Eigenvector	0.79	0.93	1.25	1.59	1.86	1.07***
Hybrid	0.78	1.09	1.17	1.19	2.18	1.40***
Panel B: crisis subsample						
Degree	−2.63	−2.88	−2.51	−2.42	−1.95	0.69
Betweenness	−2.78	−3.15	−2.00	−1.67	−2.79	−0.01
Eigenvector	−2.58	−2.75	−1.93	−2.42	−2.71	0.13
Hybrid	−2.68	−3.10	−2.37	−2.33	−1.91	0.77
Panel C: post-crisis subsample						
Degree	0.50	0.23	0.70	0.78	1.03	0.53**
Betweenness	0.44	0.21	0.78	1.26	0.55	0.12
Eigenvector	0.55	0.22	0.76	0.75	0.96	0.41*
Hybrid	0.54	0.22	0.57	0.59	1.32	0.78**

***, **, * Indicate significance at 1%, 5% and 10% level, respectively

5 Conclusions

We investigate the role of network centrality in explaining cross-section of international stock market returns using PMFG. We find that the centrality-sorted portfolio strategy could produce average excess returns that are significantly different from zero. The centrality premium remains when controlling for other risk factors. A subsample analysis reveals that the portfolios strategy performs better before the business cycle hits the peak, and centrality loses its efficiency during the recession. Business cycles could explain the different performances of the portfolio strategy.

References

1. Billio, M., Getmansky, M., Lo, A. W., & Pelizzon, L. (2012). Econometric measures of connectedness and systemic risk in the finance and insurance sectors. *Journal of Financial Economics, 104*(3), 535–559.
2. Diebold, F. X., & Yilmaz, K. (2014). On the network topology of variance decompositions: measuring the connectedness of financial firms. *Journal of Econometrics, 182*(1), 119–134.
3. Fama, E. F., & French, K. R. (2012). Size, value, and momentum in international stock returns. *Journal of Financial Economics, 105,* 457–472.
4. Firgo, M., Pennerstorfer, D., & Weiss, C. R. (2016). Network centrality and market prices: empirical evidence. *Economics Letters, 139,* 79–83.
5. Mantegna, R. N. (1999). Hierarchical structure in financial markets. *The European Physical Journal B-Condensed Matter and Complex Systems, 11*(1), 193–197.
6. Newey, W., & West, K. (1987). A simple, positive semi-definite, heteroskedasticity and autocorrelation consistent covariance matrix. *Econometrica, 55,* 703–708.
7. Peralta, G., & Zareei, A. (2016). A network approach to portfolio selection. *Journal of Empirical Finance, 38,* 157–180.
8. Pozzi, F., Di Matteo, T., & Aste, T. (2013). Spread of risk across financial markets: Better to invest in the peripheries. *Scientific Reports, 3,* 1665.
9. Tse, C. K., Liu, J., & Lau, F. C. M. (2010). A network perspective of the stock market. *Journal of Empirical Finance, 17,* 659–667.
10. Tumminello, M., Aste, T., Di Matteo, T., & Mantegna, R. N. (2005). A tool for filtering information in complex systems. *Proceedings of the National Academic of Science, 102*(30), 10421–10426.

Does the Non-penalty Regulation of CSRC Have Information Content? Evidence from the Accounting Risk Warning of Goodwill Impairment

Lei Zheng, Xuemeng Guo, and Kaiyuan Zhang

Abstract Using daily trading data and the event study method in Chinese capital market, we empirically test the information content of the "No. 8 Accounting Regulatory Risk Warning—Goodwill Impairment" issued by China Securities Regulatory Commission (hereinafter referred to as CSRC) in November 16th 2018. Our results show that, compared with companies that do not have goodwill, companies with higher proportion of goodwill experienced significantly lower cumulative abnormal returns (hereinafter refer to as CAR) around the event date. We also find that financial analyst coverage could aggregate the negative relationship between goodwill proportion and the company's CAR around the event date. All our results illustrate that investors do consider the accounting risk warning as a vital signal in the capital market, and the non-punitive supervisions of CSRC are important in improving the capital market efficiency, especially when potential risks are not commonly recognized by the public.

Keywords CSRC non-penalty regulation · Accounting risk warning · Goodwill impairment · Cumulative abnormal returns

1 Introduction

Although it is well acknowledged that corporate financial reporting and information disclosure are frequently regulated, Leuz and Wysocki [1] implies that researchers should conduct cost-benefit analyses of intended and past regulatory practices to

L. Zheng (✉) · X. Guo
School of Economics and Management, Beijing Jiaotong University, Beijing, People's Republic of China
e-mail: rayzhenglei@163.com

X. Guo
e-mail: xmguo@bjtu.edu.cn

K. Zhang
CRRC Industry Investment Co., Ltd., Beijing, People's Republic of China
e-mail: 13810436286@163.com

M. Li et al. (eds.), *IEIS 2020*, https://doi.org/10.1007/978-981-33-4363-4_2

see their real economic consequences. Unlike the U.S. capital market that is considered to be highly developed, Chinese capital market is still emerging at its initial stage. Various innovative regulatory practices, such as the inquiry letters issued by stock exchanges and the registration system for initial public offering (IPO), are set by the Chinese regulators in recent years to improve the market efficiency and to better protect investors in the capital market. However, not all regulatory practices achieved expected outcomes. Taking the circuit breaker system in 2016 as an example: to reduce individual investors' irrational trading activities in the capital market, the circuit breaker system was officially implemented in January 4th 2016. But this relatively innovative system did not achieve its initial goals, instead, it caused significant market shocks to both investors and regulators. Having proved that the circuit system may bring greater unstable factors to the market and the national economic system, Chinese regulators decided to close the circuit system in January 8th 2016. The implementation of the circuit system illustrated that market practices created in highly developed capital markets may not always suitable for the emerging capital market in China, and that not all regulators' practices in the capital market can receive the theoretically expected results. Thus, examining the real effects of different regulatory practices is an essential way to identify the suitability of such practices, and further to effectively protect investors as well as to improve the market efficiency.

Non-penalty regulations have been considered as both basic regulatory practices and important supplements to the formal penalties in the capital market. Contrary to the penalty regulations which give certain results to the market, the non-penalty regulations usually do not disclose the regulator's specific opinions to the public, but transit regulator's concerns about certain matters to the investors. Current researches have examined the economic consequences of various forms of non-penalty regulations practices. Drienko and Sault [2] finds that company's stock will experience about 3.3% decrease after the company receives an inquiry letter form the stock exchange. More specifically, Kubick et al. [3] examines the relationship between firms' tax avoidance behaviors and the comment letters issued by the U.S. SEC. Their results illustrate that, compared with companies not being the object of tax supervisions, companies that receive tax-related comment letters will significantly decrease their tax avoidance behaviors. Based on regulatory practices in Chinese capital market, the inquiry letters issued by Shanghai and Shenzhen stock exchanges and the case filing announcements issued by CSRC can cause effective market reactions soon after these files are released to the public, although these regulatory files do not necessarily mean certain penalties to related companies in the future [4, 5]. While most existing researches focused on the type of targeted "one-to-one" regulatory practices that reveal potential risks of some specific companies, the general risk warning practices implemented by the CSRC have long been neglected.

Focusing on one of the commonly used non-penalty regulatory practices in China, this paper aims to empirically examine the information content of the "No. 8 Accounting Regulatory Risk Warning—Goodwill impairment" issued by CSRC in November 16th 2018. The regulatory risk warning of goodwill impairment was not a punitive document aiming to fine some specific listed companies in Chinese capital market, but a general risk warning to remind investors about the potential impairment risks in the goodwill account of each public company. Different from other "Accounting Regulatory Risk Warning" documents, such as the "No. 6 Accounting Regulatory Risk Warning—Audit of NEEQ-Listed Companies" and the "No. 9 Accounting Regulatory Risk Warning—Occupation of Funds by Listed Companies' Controlling Shareholders and the Corresponding Audit" that mainly focused on the long-term business procedures, the "No. 8 Accounting Regulatory Risk Warning—Goodwill impairment" was directly focused on one specific account in the balance sheet of listed companies, making it easier to separate the immediate real effect of CSRC's non-penalty regulation from other long-term factors.

We find that: (1) the greater the proportion of the company's goodwill to its total assets, the more negative the cumulative abnormal return (CAR) of the company's stock experienced during the event period; (2) a higher level of financial analyst coverage aggravated the negative effect between goodwill proportion and company's CAR during the event period, suggesting that sell-side analysts failed to fully reveal the impairment risks related to company's goodwill account. All our results illustrate that the non-penalty regulatory practices of CSRC do provide effective decision-making information to the capital market and can be a useful tool to correct the market inefficiency when potential risks are commonly ignored.

Marginal contributions of this paper are as follows: (1) we empirically examine the validity of CSRC's accounting regulatory risk warning, one of the major non-penalty regulatory practices implemented by CSRC, and expand the current researches concerning the usefulness of official regulations on the financial reporting and information disclosure procedures; (2) we reveal that even professional analysts may fail to discover some potential risks in companies' accounting choices, making regulations from the market regulator a necessary way in improving market efficiency and offering protections to investors.

The remaining parts of this paper are as follows: Sect. 2 gives our hypothesis development; Sect. 3 introduces our empirical design. Empirical results are presented in Sect. 4. And finally Sect. 5 concludes our paper.

2 Hypothesis Development

The efficient market hypothesis believes that changes in stock prices are mainly caused by the arrival of new information related to the expected cash flows of listed companies [6]. Thus, any new information that are likely to change investors' expectations of companies' future cash flows will cause significant fluctuations to stock prices. The "No. 8 Accounting Regulatory Risk Warning—Goodwill impairment" issued by CSRC was such type of incremental regulatory information to Chinese capital market at the time it was released. Although this risk warning did not provide any new accounting rules that requested companies deal with their goodwill accounts in a totally different accounting method, issuing the risk warning itself fully revealed that regulators considered the risk of goodwill impairment significantly higher than it was estimated by investors. Thus, this risk-revealing signal further changed investors' expectations of companies' future cash flows, causing abnormal returns in the stock market.

When making investment decisions, although investors want to act rationally, they may be constrained by their limited rationality [7]. In addition, even investors want to be subjectively rational, they may lack professional knowledge in making the most rational decisions by themselves. Therefore, investors may choose to follow the professional opinions in the market to alter their believes about future situations and further make their current investment decisions. Among various views in the capital market, the opinions from regulators are always considered to be the most professional ones. Thus, if regulators give their opinions on specific risks in the market, investors are likely to consider regulators' opinions as value-carrying signals and thus to correct their extant optimistic expectations about future financial conditions of listed companies that they invest. And investors will further reflect their new believes in the stock prices, causing stock prices of companies warned by regulators experience negative fluctuations. Based on the analysis above, we give our first hypothesis:

H1a: other things being equal, companies with higher proportion of goodwill have more negative cumulative abnormal returns (CAR) during the CSRC's risk warning period.

However, investors might not unanimously act negatively to the stocks they hold since the CSRC's risk warning was not targeted at specific companies. In other words, investors could not directly determine the companies' future cash flows only based on the general warning issued by CSRC. According to newly implemented Chinese accounting standards, goodwill impairment decisions are highly depended on managers' subjective judgements about the real synergetic effects in previous M&A transactions. If previous M&A transactions achieved the expected synergetic effects, managers and investors do not need to concern potential risks of goodwill impairment, leading few or no abnormal fluctuations of company's stock price during the event period. Thus, having a large goodwill to total assets ratio itself may not have significant influence on the stock price during the event period. Based on the analysis above, hypothesis H1b in our paper is as follow:

H1b: the proportion of goodwill has no significant influence on the company's stock price fluctuations during the CSRC's risk warning period.

Sell-side analyst is an important information intermediary in the capital market. They collect information from various sources, track and evaluate the current performance of firms they follow, make recommendations to current and prospective investors, and forecast firm's prospects [8]. Thus, companies being covered by more analysts tend to have fewer information asymmetry problems between external investors and internal managers, making current stock prices fully reflect companies' future financial conditions. Under this circumstance, to the company that have a higher level of analysts coverage, CSRC's risk warning provides little new information to the market, causing few changes on stock prices. The above analysis leads to our hypothesis H2a:

H2a: higher level of analysts coverage will alleviate the potential negative relationship between the proportion of goodwill and company's CAR during the CSRC's risk warning period.

However, there is an alternative hypothesis on the moderate effect of analyst coverage. In practice, goodwill impairments and other one-time charges do not factor into analysts' released research reports and financial forecasts about the company's expected earnings [9, 10]. Thus, investors may not realize the potential risks of goodwill impairment in companies they invest, even they have acquired plenty of information from lots of analysts' reports. Under this practical situation, the more analysts following the specific company, the higher possibility that investors believe they have acknowledged every detailed risk of the stocks they hold, making the CSRC's risk warning an unexpected negative information to the investors. On the contrary, because of initially lacking external information about companies with few analyst followings, investors can be more cautious about those listed companies and realize the potential risks of goodwill impairment in advance. Under this situation, CSRC's risk warning provides no new information to investors holding stocks of companies with fewer analyst followings. Therefore, here we give our hypothesis H2b:

H2b: higher level of analyst coverage will strengthen the potential negative relationship between the goodwill proportion and company's CAR during the CSRC's risk warning period.

3 Research Design

3.1 Definitions of Key Variables

Dependent variable: Cumulative abnormal return (CAR) of individual stock in the Chinese capital market during the event window is the dependent variable in our paper. To our best knowledge, at the time when the CSRC published the "No. 8 Accounting Regulatory Risk Warning—Goodwill Impairment", the trading system in Chinese stock market was already closed. Under this circumstance, the stock market reactions to the risk warning will be included in stock price fluctuations in the next trading day [11]. Thus, we choose 19th November 2018 as the zero day in calculating individual stock's CAR during the event window.

Referring to Goodman [12], we employ the market model to calculate individual stock's CAR during the event window. We choose [−230, −31] as our estimation window and exclude samples which contain less than 120 trading records. The market model is as follow:

$$R_{i,t} = \alpha_i + \beta_i R_{m,t} + \varepsilon_{i,t} \tag{1}$$

where $R_{i,t}$ refers to individual stock's daily return and $R_{m,t}$ is the market return of the same trading day. After using the ordinary least squares (OLS) method to get the estimated β for each stock in the estimation period, we can calculate the predicted daily normal return of each stock in the event window. The difference between stock's real return and the predicted normal return is the abnormal return (AR) of that individual stock. The related model in estimating abnormal return (AR) of individual stock is as follow:

$$AR_{i,t} = R_{i,t} - \left(\hat{\alpha}_i + \hat{\beta}_i R_{m,t}\right) \tag{2}$$

As shown in Model (3), by adding all abnormal returns of each stock in the event period, we can finally get the cumulative abnormal return (CAR) of that stock.

$$CAR_{i,[t1,t2]} = \sum_{t1}^{t2} AR_{i,t} \tag{3}$$

Independent variable: As illustrated before, investors will response to the "No. 8 Accounting Regulatory Risk Warning—Goodwill Impairment" based on the company's financial status at the time when they receive the risk warning issued by CSRC. Thus, the status of company's goodwill account is the independent variable in our paper. We use the book value of goodwill account divided by the book value of firm's total assets to measure the magnitude of each firm's goodwill.

Moderate variable: We use the natural logarithm of the annual number of financial analysts covering the firm as the moderate variable in our research. As emphasized by He and Tian [13], a higher level of analyst coverage usually means fewer information asymmetry problems between company's insiders and outsiders.

Control variables: Referring to Chen et al. [4], Liu and Ye [11] and He and Luo [14], we select variables controlling firm size, leverage ratio, return on assets, Tobin's q, number of directors in the board, CEO duality, types of audit opinion in 2017 and the proportion of outstanding shares. In addition, industry effect is also controlled. All key variables used in this paper are presented in Table 1.

3.2 Data and Sample

We gather data of Chinese A-share listed companies from China Stock Market Accounting Research (CSMAR) database. All variables related to company's annual financial data are chose from financial reports ended on 31st December 2017 to ensure data reliability. We exclude companies that operated in financial industry and samples with missing data. Companies that were special treated are also excluded in our final sample. In order to alleviate the biases caused by extreme values, all the continuous variables are winsorized at 1% and 99% level. Our final sample contains 2752 firm-year observations.

Table 1 Definitions of key variables

Variable Name	Calculation Method
$CAR_{[0,1]}$	$CAR_{i,[0,1]} = \sum_0^1 AR_{i,t}$
GWratio	$\mathrm{GWratio_i = Goodwill_i/Total\ assets_i}$
Analyst	$\mathrm{Analyst_i = ln(number\ of\ analyst\ coverage_i + 1)}$
Size	$\mathrm{Size_i = ln(Total\ Assets_i)}$
Lev	$\mathrm{Lev_i = Total\ liability_i/Total\ assets_i}$
ROA	$\mathrm{ROA_i = Net\ income_i/Average\ total\ assets_i}$
Tobinq	$\mathrm{Tobinq_i = Market\ value_i/Total\ assets_i}$
BoardNum	Number of directors in company's board
Dual	Equals one if the CEO and the Chairman is the same person and zero otherwise
Aud	Equals one if the company received a clean audit opinion in the year of 2017 and zero otherwise
Liqratio	$\mathrm{Liqratio_i = Outstanding\ shares_i/Total\ shares_i}$

3.3 Models

To test our hypotheses, referring to He and Luo [13] focusing on Chinese stock market, we build two empirical models as follows:

$$\begin{aligned} CAR_{i,[0,1]} = {} & \alpha_0 + \alpha_1 GWratio_i + \alpha_2 Size_i + \alpha_3 Lev_i \\ & + \alpha_4 ROA_i + \alpha_5 Tobinq_i + \alpha_6 BoardNum_i + \alpha_7 Dual_i \\ & + \alpha_8 Aud + \alpha_9 Liqratio_i + \varepsilon \end{aligned} \tag{3}$$

$$\begin{aligned} CAR_{i,[0,1]} = {} & \alpha_0 + \alpha_1 GWratio_i + \alpha_2 Analyst_i \\ & + \alpha_3 GWratio_i \times Analyst_i + \alpha_4 Size_i + \alpha_5 Lev_i \\ & + \alpha_6 ROA_i + \alpha_7 Tobinq_i + \alpha_8 BoardNum_i \\ & + \alpha_9 Dual_i + \alpha_{10} Aud + \alpha_{11} Liqratio_i + \varepsilon \end{aligned} \tag{4}$$

Based on our hypotheses, coefficient α_1 in Model (3) and coefficient α_3 in Model (4) are what we are interested. According to our predictions, α_1 in Model (3) should be significantly negative or insignificant, and α_3 in Model(4) should be either significantly positive or significantly negative.

4 Descriptive Statistics

Table 2 shows the descriptive statistics of our regression sample. In average, the book value of Chinese listed companies' goodwill accounted for about 4.8% of companies'

Table 2 Descriptive statistics

Variable name	N	Mean	Std. Dev	Median	Max	Min
$CAR_{[0,1]}$	2752	−0.001	0.033	−0.004	0.136	−0.087
GWratio	2752	0.048	0.096	0.001	0.468	0.000
Analyst	2752	1.609	1.170	1.609	4.205	0.000
Size	2752	22.343	1.278	22.204	26.171	20.070
Lev	2752	0.415	0.197	0.408	0.879	0.063
ROA	2752	0.048	0.050	0.042	0.214	−0.121
Tobinq	2752	2.007	1.127	1.674	7.834	0.946
BoardNum	2752	8.483	1.693	9.000	18.000	0.000
Dual	2752	0.287	0.452	0.000	1.000	0.000
Aud	2752	0.023	0.148	0.000	1.000	0.000
Liqratio	2752	0.747	0.253	0.814	1.000	0.176

total assets in 2017. The biggest proportion of goodwill account was around 45.8% while the smallest proportion was 0. For the variable *Analyst*, each listed company in Chinese capital market attracted 5 analyst teams in average. The mean number of the variable $CAR_{[0,1]}$ is −0.1% with stand deviation of 3.3%. The highest cumulative abnormal return during the event period is 13.6%, while the lowest one is around −8.7%, illustrating that the "No. 8 Accounting Regulatory Risk Warning—Goodwill Impairment" issued by CSRC did cause fluctuations in Chinese stock market. As for the control variables, the average ROA of Chinese listed companies in 2017 was 4.8%. The scale of the company board was around 8 to 9 directors in average. 28.7% companies were governed by the same person who was both the CEO and chairman in the board and only 2.3% companies received a clean audit opinion in the fiscal year of 2017.

We next compare the mean of the variable $CAR_{[0,1]}$ between companies with goodwill and those without. The comparison results are presented in Table 3. It is clear that companies with goodwill account experienced significantly lower cumulative abnormal returns during the event period. Results in Table 3 preliminarily proves that our hypothesis H1a is correct.

Table 3 Comparison Between Groups

	Firms with goodwill		Firms without goodwill		Differences	
	N	Mean	N	Mean	Mean	t-value
$CAR_{[0,1]}$	1645	−0.003	1107	0.001	0.004	3.228***

*, **, ***Represent significance at 10%, 5%, and 1% (two-tails), respectively

5 Empirical Results

5.1 Baseline Empirical Results

Using statistics software Stata 14.0, we employ ordinary least squares (OLS) method to run Model (3) and Model (4) to test our hypothesis. Our baseline regression results are presented in Table 4.

It is clear in the regression results of Model (3) that the coefficient of our independent variable, GWratio, is -1.134% at 1% significant level, suggesting that 1%

Table 4 Baseline empirical results

Variable name	Model (3)	Model (4)
GWratio	−1.134*** (−2.868)	2.032*** (3.406)
Analyst		−0.021 (−0.325)
GWratio × *Analyst*		−1.699*** (−5.726)
Size	−0.049 (−0.687)	−0.018 (−0.242)
Lev	0.025 (0.059)	0.014 (0.034)
ROA	−5.654**** (−4.720)	−4.931*** (−4.589)
Tobinq	0.086* (1.814)	0.097** (2.082)
BoardNum	−0.000 (−0.022)	−0.000 (−0.018)
Dual	−0.246 (−1.535)	−0.238 (−1.493)
Aud	−0.661 (−1.082)	−0.713 (−1.165)
Liqratio	−0.449 (−1.540)	−0.472 (−1.595)
Intercept	1.558 (0.994)	0.854 (0.533)
Ind	Controlled	Controlled
N	2752	2752
R^2 (%)	0.94	1.31
Mean VIF	1.39	1.99

*, **, ***Represent significance at 10%, 5%, and 1% (two-tails), respectively
T-values are presented beneath the coefficient estimates in parentheses
The coefficient value of all of the independent variables is multiplied by 100

increase in the proportion of goodwill account to the total assets caused 1.1% decrease in the cumulative abnormal return of the company's stock price during the event period. The mean variance inflation factor (VIF) of Model (3) is 1.39, proving that there is no serious multicollinearity problem. This result gives solid empirical evidence to our hypothesis H1a that the accounting risk warning issued by CSRC does contain essential regulatory information to investors.

Empirical results in Model (4) examine the validity of our second hypothesis. The coefficient of the interaction term $GWratio \times Analyst$ is -1.699% with a significant level at 1%, suggesting that the regulatory effects are more pronounced in the companies with more analyst coverage. Examination of VIF shows that our results are not influenced by the multicollinearity problem. Thus, hypothesis H2b in our paper is proved. Results in Model (4) illustrate that analysts in the capital market did fail to fully reveal the potential risks behind the high book value of companies' goodwill account, making CSRC's non-penalty regulatory practices an essential mechanism to transfer risk-related information to the investors and to further improve the capital market efficiency.

5.2 *Robustness Tests*

Since CSRC's "No. 8 Accounting Regulatory Risk Warning—Goodwill Impairment" is considered as an exogenous event to both Chinese capital market and Chinese listed companies, there are few endogeneity concerns in our research design. Thus, we implement two commonly used methods to check the robustness of our empirical results, namely changing the estimation period of CAR during the event period and narrowing the regression sample, and we find our results are robust under different regression methods.

5.2.1 Change the Estimation Period in Calculating $CAR_{[0,1]}$

Empirical results may be diversified when using different estimation windows in calculating the dependent variable $CAR_{[0,1]}$, making our baseline results not solid enough to support our hypotheses. Thus, following Liu and Ye [11], we use [−182, −3] to re-estimate the dependent variable $CAR_{[0,1]}$ in our models. Regression results are presented in column 2 to column 3 in Table 5 below. We find no significant differences among robustness tests and original results when changing the calculation method of $CAR_{[0,1]}$. VIF tests show that it is no need to worry about the multicollinearity problem in our robustness tests of changing the estimation period in calculating $CAR_{[0,1]}$.

Table 5 Robustness tests

Variable Name	Change the estimation period in calculating $CAR_{[0,1]}$		Narrow the regression sample	
	[−182, −3]		Companies with goodwill	
	Model (3)	Model (4)	Model (3)	Model (4)
GWratio	−0.830** (−2.166)	2.310*** (3.228)	−0.788* (−1.878)	2.778*** (2.763)
Analyst		−0.000 (−0.011)		0.009 (0.088)
Gwratio × *Analyst*		−1.701*** (−4.742)		−1.920*** (−3.976)
Size	0.040 (0.555)	0.062 (0.835)	−0.024 (−0.336)	0.124 (0.159)
Lev	0.020 (0.051)	0.013 (0.032)	−0.951* (−1.737)	−0.977* (−1.679)
ROA	−4.556*** (−3.611)	−3.988*** (−3.649)	−5.693** (−2.502)	−4.559** (−2.168)
Tobinq	0.136*** (2.786)	0.144*** (3.016)	−0.015 (−0.159)	−0.012 (−0.128)
BoardNum	−0.008 (−0.251)	0.008 (−0.244)	0.022 (0.758)	0.022 (0.770)
Dual	−0.247 (−1.530)	−0.239 (−1.492)	−0.050 (−0.286)	−0.035 (−0.199)
Aud	−0.661 (−1.072)	−0.706 (−1.141)	−1.162* (−1.796)	−1.258** (−1.980)
Liqratio	−0.411* (−1.326)	−0.424 (−1.349)	−0.371 (−0.749)	−3.956 (−0.792)
Intercept	−0.564 (−0.355)	−1.074 (−0.680)	1.162 (0.747)	0.292 (0.193)
Ind	Controlled	Controlled	Controlled	Controlled
N	2752	2752	1645	1645
R^2 (%)	3.26	3.26	3.18	3.17
Mean VIF	1.39	1.99	1.39	2.03

*, **, ***Represent significance at 10%, 5%, and 1% (two-tails), respectively
T-values are presented beneath the coefficient estimates in parentheses
The coefficient value of all of the independent variables is multiplied by 100

5.2.2 Narrow the Regression Sample

Considering that there may be other essential differences, such as operating strategies and managers' horizons, between companies that have goodwill account and those do not, making our original results easier to be found. We restrict our regression sample to the companies with goodwill account to make our original results more convincing. Column 4 and column 5 in Table 5 provide empirical evidence to the narrowed sample regressions. As shown in Table 5, all results are similar with our original ones. Values of mean VIF are 1.39 and 2.03 respectively, suggesting that our results are not influenced by multicollinearities among variables.

5.3 *Additional Analysis*

Companies with a higher level of internal control are more likely to implement better corporate governance [15] , making it more difficult for managers to behave opportunistically through goodwill impairment decisions. Thus, we hypotheses that our baseline results are different in companies with different levels of internal control system.

Data measuring the level of internal control quality are from DIB database. After grouping the sample based on the median of the internal control variable, we rerun our Model (3) and Model (4) respectively. The empirical results of this additional analysis are presented in Table 6. It is clear that in the low internal control group, companies with goodwill account experience significantly negative abnormal returns than those in the high internal control group during the event period. However, the effect of analysts coverage is more negatively profound in the high internal control group. Our results suggest that financial analysts are less cautious in discovering and delivering goodwill impairment risks in companies with better internal control systems. Again, VIF test shows that multicollinearity problem is not severe in our OLS regressions.

6 Conclusions

Using the "No. 8 Accounting Regulatory Risk Warning—Goodwill Impairment" issued by CSRC as an exogenous shock in Chinese capital market, we empirically test the information content of this type of non-penalty regulation implemented by the CSRC. Our empirical results show that, compared with companies that do not have the goodwill account, companies with higher proportion of goodwill experienced significant negative abnormal returns during the CSRC's regulation event period, illustrating that investors in Chinese capital market do consider the non-penalty regulation of CSRC an effective warning signal. Also, higher level of analyst coverage can cause more negative abnormal returns during the event period, suggesting that financial analysts in Chinese capital market may fail to discover and transfer the potential risks related to goodwill impairment to the investors.

Our baseline results are robust in both robustness tests and additional analysis. We find that having a better internal control system can effectively alleviate the negative effect of CSRC's non-penalty regulation on companies' stock performance. However, a better corporate governance may make external analysts less likely to realize and transfer the potential risks in company's goodwill account.

Table 6 Additional analysis

Variable name	Companies with higher level of internal control		Companies with lower level of internal control	
	Model (3)	Model (4)	Model (3)	Model (4)
GWratio	−0.771 (−1.191)	5.350*** (2.626)	−1.472*** (−2.788)	0.378 (0.489)
Analyst		−0.220** (−2.028)		0.155** (2.246)
Gwratio × *Analyst*		−2.867*** (−3.190)		−1.182*** (−3.278)
Size	0.104 (1.173)	0.250*** (2.650)	−0.299*** (−4.273)	−0.342*** (−5.072)
Lev	−0.375 (−0.423)	−0.355 (−0.441)	0.429 (0.708)	0.445 (0.707)
ROA	−10.812*** (−6.701)	−8.156*** (−5.254)	−2.003 (−1.137)	−2.631* (−1.730)
Tobinq	0.219** (2.346)	0.289*** (3.486)	−0.030 (−0.481)	−0.039 (−0.623)
BoardNum	0.006 (0.150)	0.003 (0.075)	0.008 (0.186)	0.006 (0.143)
Dual	−0.314* (−1.689)	−0.307 (−1.630)	−0.249 (−0.939)	−0.239 (−0.897)
Aud	−1.643*** (−5.854)	−1.301*** (−5.631)	−0.436 (−0.740)	−0.410 (−0.690)
Liqratio	−0.838** (−2.189)	−0.932*** (−2.637)	−0.365 (−0.802)	−0.289 (−0.656)
Intercept	−1.361 (−0.756)	−4.468** (−2.243)	6.855*** (4.613)	7.605*** (5.335)
Ind	Controlled	Controlled	Controlled	Controlled
N	1335	1335	1335	1335
R^2 (%)	1.93	3.57	1.20	1.41
Mean VIF	1.43	2.15	1.37	1.91

*, **, ***Represent significance at 10%, 5%, and 1% (two-tails), respectively
T-values are presented beneath the coefficient estimates in parentheses
The coefficient value of all of the independent variables is multiplied by 100

Our research provide empirical evidence on the information validity of CSRC's non-penalty regulation in Chinese capital market. Our results suggest that, CSRC's non-penalty regulation is an effective method in Chinese capital market to improve the market efficiency, especially when professional financial analysts ignore the potential risks in companies that they are following. In addition, for listed companies themselves, implementing a high-quality internal control system can significantly reduce the regulatory risks they are facing and prevent their securities from fluctuating unexpectedly.

Acknowledgements This paper is supported by the general research program of National Natural Science Foundation of China: Research on Performance Evaluation System of urban rail transit PPP mode based on resource "passenger-value flow" (71973009).

References

1. Leuz, C., & Wysocki, P. D. (2016). The economics of disclosure and financial reporting regulation: Evidence and suggestions for future research. *Journal of Accounting Research, 54*(2), 525–622.
2. Drienko, J., & Sault, S. J. (2013). The intraday impact of company responses to exchange queries. *Journal of Banking & Finance, 37*(12), 4810–4819.
3. Kubick, T. R., Lynch, D. P., Mayberry, M. A., & Thomas, C. O. (2016). The effects of regulatory scrutiny on tax avoidance: An examination of SEC comment letters. *The Accounting Review, 91*(6), 1751–1780.
4. Chen, Y., Lu, Y., & Zhe, L. (2019). Effectiveness of the front-line regulation of the Chinese stock exchanges: Evidence from comment letters. *Management World, 3,* 169–185.
5. Wu, X., & Zhang, J. (2014). Stock market reaction to regulatory investigation announcements. *Accounting Research, 4,* 10–18.
6. Huberman, G., & Regev, T. (2001). Contagious speculation and a cure for cancer: A nonevent that made stock prices soar. *The Journal of Finance, 56*(1), 387–396.
7. Schijven, M., & Hitt, M. A. (2012). The vicarious wisdom of crowds: Toward a behavioral perspective on investor reactions to acquisition announcements. *Strategic Management Journal, 33*(11), 1247–1268.
8. Healy, P. M., & Palepu, K. G. (2001). Information asymmetry, corporate disclosure, and the capital markets: A review of the empirical disclosure literature. *Journal of Accounting and Economics, 31*(1–3), 405–440.
9. Bradshaw, M. T., & Sloan, R. G. (2002). GAAP versus the street: An empirical assessment of two alternative definitions of earnings. *Journal of Accounting Research, 40*(1), 41–66.
10. Ayres, D. R., Campbell, J. L., Chyz, J. A., & Shipman, J. E. (2019). Do financial analysts compel firms to make accounting decisions? Evidence from goodwill impairments. *Review of Accounting Studies, 24*(4), 1214–1251.
11. Liu, H., & Ye, K. (2018). The effect of value-added tax rate on firm value: Evidence from the reaction of stock market. *Management World, 11,* 12–35.
12. Goodman, T. H., Neamtiu, M., Shroff, N., & White, H. D. (2014). Management forecast quality and capital investment decisions. *The Accounting Review, 89*(1), 331–365.
13. He, J., & Tian, X. (2013). The dark side of analyst coverage: The case of innovation. *Journal of Financial Economics, 109*(3), 856–878.
14. He, D., & Luo, W. (2019). Reputation of venture capital firms and the contagion effect: Evidence from regulatory investigations of Chinese listed firms. *Journal of Financial Research, 9,* 169–187.
15. Chen, Z., & Fang, H. (2019). Can internal control fasten the fence of insiders' opportunistic sales? *Accounting Research, 7,* 82–89.

Risk Analysis of Major Financial Misstatement in Agricultural Enterprises

Yunjing Liang

Abstract In recent years, incidents such as Lvdi, Wanfushengke and Yinguangxia being punished by the Securities Regulatory Commission of agricultural listed companies have frequently occurred. Audit risk has once again become the focus of audit practice. This article takes Gongzhun Meat Food Co., Ltd. as an example to analyze the risk of major misstatement of financial statements from the macro-environmental level and the enterprise level. Among them, the enterprise level is divided into business risk analysis, internal control risk analysis and the use of financial indicators (profit, compensation, debt and operating). After analysis, this paper believes that there are problems with the lack of internal control, false financial data, and low audit effectiveness of related accounting firms in Gongzhan Shares. It is recommended that when internal control is missing, the auditor should focus on the financial data at the end of the audit period and expand the substantive procedures and audits. And for agricultural enterprises, identify risks in special fields such as related party transactions or events.

Keywords Audit risk · Material financial misstatement · Agricultural enterprises

1 Introduction

At this stage, certified public accountants generally use modern risk-oriented risk assessments for major misstatements. Major misstatement risk is key assessments of important subjects involving operating risks in target companies and design audit procedures based on them [1]. As the core content of modern risk-oriented audit, major misstatement risk assessment plays an important role in identifying major misstatement risks and determining risk levels. It can provide reference for the follow-up audit work and is an important guarantee for improving audit quality and efficiency [2].

Y. Liang (✉)
Beijing Jiaotong University, Beijing, China
e-mail: 19120659@bjtu.edu.cn

M. Li et al. (eds.), *IEIS 2020*, https://doi.org/10.1007/978-981-33-4363-4_3

According to the research, the reasons for the frequent occurrence of financial fraud cases in listed companies are not only the lack of basic professional ethics of certified public accountants, but also the fact that the certified public accountants in our country can not effectively identify and evaluate the major misstatement risks of the audited units effectively, which leads to the inability to design and implement targeted major misstatement risks based on the auditees and is difficult to ensure the effectiveness of the audit procedure and effectively control the audit risk at an acceptable low level [3, 4]. It shows that there is a conduction relationship between risk factors. External macroeconomic factors, industry and regulatory factors have an impact on risk consequences mainly through corporate strategy and corporate governance factors. Corporate strategy and corporate governance affect the risk consequences mainly through the company's operating conditions. Accounting policies have a direct impact on risk consequences. Financial distress has an impact on risk consequences through performance pressure [5]. According to the characteristics of real estate development enterprises, Ref. [6] constructs the risk evaluation index system of major misstatement, and obtains that the risk of major misstatement of real estate development enterprise statements is higher, among which the business cycle risk is the largest, followed by financing risk and investment risk. Through the understanding of the internal and external environment and internal control of the solid waste treatment company, combined with the analysis of audit data of the solid waste treatment company by J office over the years, it can be identified that the major misstatement risks of solid waste treatment enterprises mainly focus on four aspects: accounts receivable, fixed assets, intangible assets and operating income [7].

However, in recent years, incidents such as Lvdi, Wanfushengke and Yinguangxia being punished by the Securities Regulatory Commission have frequently occurred, and audit risks have once again become the focus of audit practice [8]. Agricultural listed companies such as Gongzhan shares generally have large cash transactions, scattered purchases and sales, and difficult inventory counting [9]. These have brought difficulties to auditors' auditing work, leading to the risk of auditing of such listed companies high. Therefore, before the audit institution officially starts the audit work, it should first evaluate the risk of material misstatement of the audited company to determine the follow-up audit procedures and audit priorities to reduce the audit risk.

This article uses Gongzhuan Meat Food Co., Ltd. as an example to analyze the risks of major misstatements in its financial statements, summarize the reasons for the risks, and put forward corresponding suggestions to help auditors better understand the audit risk points of agricultural companies and help identify them. And assess the risks of material misstatement of the audited company, and formulate corresponding audit procedures to improve audit quality and reduce audit risks. At the same time, it can also help to improve the supervision of the market, which is helpful for the supervision department to find the major and difficult points of supervision, improve the supervision system of the market, and ensure the healthy development of the market.

2 Case Study of Gongzhuan Meat Food Co., Ltd.

Gongzhun Meat Food Co., Ltd. (830916), formerly known as Heilongjiang Gongzhun Meat Food Co., Ltd., was established on December 30th, 2004. Approved by the State Administration for Industry and Commerce, Heilongjiang Gongzhun Meat Food Co., Ltd. changed its audited net assets into shares on August 23rd, 2011, and established a joint stock limited company, namely Gongzhun Meat Food Co., Ltd. The company's registered capital is 90.988 million yuan. Its main business has been the purchase, slaughter, refrigeration and sales of live pigs. The company's main products are cold fresh pork and frozen split meat.

2.1 Annual Report Disclosure Issues

On May 10th, 2017, the company issued an announcement. Due to the busy business of China Audit Asia-Pacific Certified Public Accountants (special general partnership) and the large number of audit operations, it was not possible to issue audit reports for the company on schedule. (Special General Partnership) changed to Asia Pacific (Group) Accounting Firm (Special General Partnership).

On June 22nd, 2017, Asia Pacific Certified Public Accountants issued a standard unqualified audit report on the annual report of Gongzhan Shares, but analyzed the data provided in the annual report of Gongzhan Shares and compared it with the indicators of other companies in the same industry. By comparison, it is found that there is a possibility of false financial data disclosure for public stocks.

On June 28th, 2017, as of April 30th, 2017, Gongzhun Meat Food Co., Ltd. did not prepare and disclose the annual report within four months from the end of the 2016 fiscal year, which violated the Article 11 of "the detailed rules for information disclosure of Listed Companies in the national small and medium sized enterprise share transfer system". Regarding the above-mentioned violations, the company's chairman and secretary of the board of directors failed to faithfully and diligently perform their duties, in violation of the relevant provisions of Article 1.5 of the "National Small and Medium-sized Enterprise Share Transfer System Business Rules (Trial)", Adopt self-regulatory measures to issue warning letters. Self-regulatory measures were issued to Yiwen Han, chairman of Gongzhun Meat Food Co., Ltd., and Chuanzhong Gong, secretary to the board of directors.

On August 11th, 2017, the stock transfer system issued an annual report inquiry letter to Gongzhun Shares and Asia Pacific Certified Public Accountants, asking Gongzhan Shares to explain the main business composition of 2016, the authenticity of monetary funds, the details of current assets, the dividend policy and Information on pledged loans; Asia Pacific Certified Public Accountants are required to explain the audit procedures and audit conclusions obtained for the authenticity of the fair share income and the authenticity of monetary funds.

On August 11th, 2017, Shanxi Securities Co., Ltd. announced its withdrawal from providing market quotation services for the stocks of Gongzhuan Meat Food Co., Ltd. On August 18th, 2017, BOCI Securities Co., Ltd. announced its withdrawal from Gongzhuan Meat Food Co., Ltd. Limited stocks provide market quotation services.

On August 17th, 2017, Asia Pacific Certified Public Accountants replied to the inquiry letter, stating that the audit of the revenue and monetary funds of the publicly-owned shares has been conducted with appropriate audit procedures and obtained reliable audit conclusions.

On August 21st, 2017, Gongzhun shares responded to the inquiry letter.

On October 26th, 2017, because the company and Han Yiwen were suspected of having leaked information and violated laws and regulations, according to the relevant provisions of the Securities Law of the People's Republic of China, the China Securities Regulatory Commission decided to file a case investigation into the company and its controlling shareholders and actual controllers.

On November 7th, 2017, Gongzhun's director and deputy general manager Ye Chao, Yunguo Sun, director Yiping Han and senior management were investigated by the China Securities Regulatory Commission.

On November 30th, 2017, the directors, supervisors and senior managers of Gongzhan shares were officially investigated by the China Securities Regulatory Commission. Including Yiwen Han, the controlling shareholder and actual controller of the company, director and deputy general manager Ye Chao, director and deputy general manager Yunguo Sun, chairman of the board of supervisors Qiuhui Wu, supervisor Ai Jing, former secretary of the board of directors and deputy general manager Chuanzhong Gong, director Zhiqiu Huo, Supervisor Dongsheng Gao, Former Chief Financial Officer Wenhua Sun, and Current Chief Financial Officer Lirong Tong, etc., the investigations were due to suspected information disclosure violations.

2.2 *The Problems of Gongzhuan Meat Food Co., Ltd. Pledge*

In August 2016, Yiwen Han, a shareholder of Gongzhun, pledged 42,118,060 shares, accounting for 18.13% of the company's total share capital. The pledge period was from August 3rd, 2016 to August 3rd, 2017.

On July 10th, 2017, the Gongzhun were reissued with a pledge announcement, stating that on April 2nd, 2017, the shareholder Ye Chao was a major shareholder and Yiwen Han, pledged 9,908,400 Gongzhun shares held by him, accounting for 4.27% of the company's total share capital The pledge period is from April 26th, 2017 to August 15th, 2017. In accordance with the requirements of the "New Third Board Listing Company Information Disclosure Details (Trial)", the company deemed that no announcement was required when pledged the shares, and after being supervised by the main board securities firm Huaan Securities Co., Ltd., it believed that it was not Han Yiwen's spouse and also owned the company's shares. The second pledge of shares is to supplement Han Yiwen's pledge. Therefore, in accordance with the rules of concerted action, the second pledge will be announced, so it is hereby announced.

On August 21st, 2017, Huaan Securities issued a risk alert, proposing that Yiwen Han, the controlling shareholder and actual controller of the quasi-shares, and directors, deputy general managers, and Yiwen Han's spouse, pledged a total of 22.40% of the company's total capital The maturity date of the loans was August 15th, 2017. As of August 21st, 2017, the loan had not been paid off. After that, on October 17th, 2017, Huaan Securities issued a risk alert again, stating that the loan had not been paid off as of October 17th, 2017.

On March 30th, 2018, Gongzhun shares issued a supplementary announcement of equity pledge. Yiwen Han pledged 110,000,000 shares on March 26th, 2018, accounting for 47.36% of the company's total share capital. The pledge period is from March 26th, 2018 to March 2021. Until the 25th the pledged shares are used to provide guarantee for the related party Harbin Gongzhun Economic and Trade Co., Ltd. loan. As of now, the total number of public stocks held by Yiwen Han is 168,222,240, of which 152,118,060 shares are pledged, accounting for 65.49% of the total share capital of public stocks.

2.3 Response of the Accounting Firm to the Annual Report Inquiry Letter

In response to the inquiry letter regarding the stock transfer system, Asia Pacific Certified Public Accountants responded on August 17th, 2017.

Aiming at the authenticity of the revenue of the quasi-shares, the auditing procedures performed by Asia Pacific Certified Public Accountants include collecting relevant evidence for revenue recognition, obtaining a detailed account of the revenue, performing internal control tests and substantive drafts, executing letter of credit procedures, and performing customer interview procedures. The company's income accounting strictly complies with the relevant provisions of the company's accounting policies, the income amount is measured correctly and there is no intertemporal phenomenon, and the amount can be confirmed after the conclusion.

Regarding the authenticity of monetary funds at the end of the period, the audit procedures performed by Asia Pacific CPAs include obtaining corporate credit reports, obtaining detailed accounts for opening bank accounts, detailed accounts for monetary funds, bank account statements and balance reconciliation statements for all accounts. Perform the letter verification procedure for the balance on the account cut-off date, conduct random checks on the original vouchers, conduct a reasonable analysis of the balance and balance of each account, and add specific verification procedures, including the calculation of interest, detailed testing of bank deposits, cut-off testing, etc. In the end, the firm concluded that the balance of monetary funds at the end of the period was true.

3 Risk Analysis of Major Misstatement of Gongzhun

In the analysis of the risk of material misstatement of Gongzhun, the concrete analysis is mainly carried out from the macro environment of Gongzhun, related operating risks of Gongzhun, internal control risks and financial risk analysis.

3.1 *Macro Environment Analysis*

Gongzhun Co., Ltd. is an agricultural company. The main business of Gongzhun Co., Ltd. is slaughtering business, which is mainly engaged in production and operation of pig slaughtering and processing. The industry is highly competitive. The factors affecting the industry status of the risk of material misstatement of publicly-owned shares mainly include two aspects. On the one hand, the slaughter industry has low entry barriers, a single product structure, strong replaceability, and fierce competition in the industry. The other is the risk of cyclical fluctuations in hog prices. The company's main raw material is hogs. The price of hogs is affected by various factors such as the growth cycle of pigs, stocks, feed prices, market supply and demand, etc., and they fluctuate greatly. Therefore, the company needs to adjust the inventory and meat product prices at any time according to market conditions. If the adjustment is not timely, it will affect the company's operating conditions.

In terms of legal environmental factors, according to the provisions of laws and regulations such as the "Enterprise Income Tax Law of the People's Republic of China" and the "Implementation Regulations of the Enterprise Income Tax Law of the People's Republic of China", the income from public stocks engaged in the raising of livestock and poultry and the initial processing of agricultural products may be exempt Levy corporate income tax. According to the provisions of the "Interim Regulations of the People's Republic of China on Value-added Tax" and the "Implementation Rules of the Interim Regulations of the People's Republic of China on Value-added Tax", the company's self-produced agricultural products such as pigs are exempt from value-added tax. Preferential tax policies have reduced the burden on the company, but at the same time, it has also made it more difficult to audit and verify the company's income.

Regarding the regulatory environment, on February 6th, 2016, the State Council issued the "Decision on Amending Certain Administrative Regulations", decided to amend some provisions of the "Regulations on the Management of Pig Slaughtering", strengthen the rectification of the slaughter industry, and continue to do well at designated slaughtering of pig Qualification review and clearance work. The Animal Husbandry and Veterinary Bureau of Heilongjiang Province also proposed in the "Key Points of Heilongjiang Province's Livestock Product Quality and Safety Supervision Work in 2016" to strengthen the supervision of pig slaughter in the province. State policies have put forward higher requirements for slaughter companies. Companies must improve their production quality and standardize production models to meet national policy requirements. In this context, the company has the possibility of using fictitious financial data to modify its financial statements so as to meet the requirements of the country.

3.2 *Operational Risk Analysis of Gongzhun*

The company's nature belongs to the processing of agricultural and sideline products. It is mainly engaged in the production and operation of pig slaughtering and processing. The product is single and the market competition is very fierce. In addition, the production mode of the Gongzhun is single, highly reproducible, and has low ability to withstand risks such as market competition, which will affect the sustainable operation of the Gongzhun. The supply sources and sales channels of agricultural companies generally have a single characteristic, which makes the supply and marketing chain of agricultural companies easy to break, and the supply and marketing parties of Gongzhun also have such characteristics.

According to the disclosure of the annual report of the suppliers and customers of Gongzhun, it is known that the top five suppliers and customers are natural persons, and the proportion of procurement and sales has reached 34.05% and 34.95%, respectively.

- Natural person operators are greatly affected by market fluctuations and have weak ability to resist risks. Therefore, the stability of the capital chain of public stocks is poor, the flexibility is high, and the company's production and operation risks are large, which affects the company's sustainable management ability.
- As a primary processing company of agricultural and sideline products, Gongzhun enjoys the preferential policies of national income tax exemption. Supervisory departments cannot understand the company's operating status through the company's income tax payment situation, and many of the company's economic business transactions are natural persons. After the transaction is completed, there will be a phenomenon that no corresponding bills are required. If the company uses this to whitewash its own operating results, it is difficult for the relevant regulatory authorities to verify the specific amount of the company's business transactions and the company's true profit. It makes the auditing work of accounting firms more difficult.
- The amount of the top four suppliers of Gongzhun all exceeded 80 million, the first place reached 89,228,456.90 yuan, and the sales of the top three customers of Gongzhun exceeded 100 million. This amount is for natural persons It is obviously too huge, indicating that the company's purchase amount and sales amount are fraudulent.

Table 1 Comparison of gross profit margin (10,000 yuan)

Year	Company	Operating revenue	Operating cost	Gross margin (%)
2014	Gongzhun	114,751.51	103,966.47	9.40
	Xinwufeng	130,249.39	127,129.56	2.40
2015	Gongzhun	119,641.75	108,353.17	9.44
	Xinwufeng	132,603.67	119,685.85	9.74
2016	Gongzhun	139,364.22	128,328.52	7.92
	Xinwufeng	169,137.43	139,863.53	17.31

3.3 *Risk Analysis of Internal Control of Gongzhun*

As of the end of the 2016 reporting period, the Company has not yet established the "Accountability System for Major Errors in Annual Reports". This system effectively restricted the accuracy and authenticity of the company's information disclosure, and played an important role in reducing the audit risk. As the first case of a company punished by the Securities and Futures Commission of the New Third Board, Shenxianyuan improved its internal control after accepting the penalty Among the measures adopted by the system is the establishment of the "Accountability System for Major Errors in Annual Reports." The lack of this system will make the company's fraudulent behavior in the preparation of the annual report without a system to restrict it, and subsequent accountability will be difficult, which will bring huge risks to the audit. Therefore, when conducting audit work, auditors should pay particular attention to the establishment and implementation of this system.

3.4 *Financial Risk Analysis of Gongzhun*

Although the Asia Pacific Certified Public Accountants certified the audit conclusion in the reply to the annual report inquiry letter, this cannot explain the problem that the annual report data of Gongzhun is too different from the indicators of other companies in the same industry. In order to objectively show the financial data of Gongzhun, it is compared with the financial data of Xinwufeng in the same industry. Gongzhun's main business is slaughtering business, and its products are chilled meat and frozen meat. Therefore, when comparing, only the similar operating business data of Xinwufeng was collected, which made the analysis results objective and credible.

3.4.1 Profitability Analysis

According to the data in Table 1, the gross margin of Xinwufeng from 2014 to 2016 has been in the growth stage, which may be related to the economic situation of the industry and its own products and strategies in the past three years. Or for companies that are smaller than Xinwufeng g, the three-year gross margin has not been affected by the environment, and is very stable. Even in 2014, the gross margin of Xinwufeng was 2.4%. The interest rate still reached a high of 9.4%, which indicates that the data of the Gongzhun is abnormal, and the amount of income of the Gongzhun may be fraudulent. At the same time, the auditors of the Asia-Pacific accounting firm may have not strictly implemented the audit procedures during the audit behavior.

3.4.2 Solvency Analysis

The debt-asset ratio is a comprehensive indicator of the company's debt level. It can reveal how much of the company's total funding sources are provided by creditors. From the data, it can be seen that the debt-asset ratio of Xinwufeng is in a downward trend for three years, but the asset-liability ratio is still at a double-digit level. Looking at the data of Gongzhun, we can find that the debt-asset ratio is similar to that of Xinwufeng. It is relatively low, and the increase in assets and the decrease in liabilities have occurred in the past three years. Although the debt-asset ratio alone cannot fully explain a company's ability to pay its debts, the indicators of Gongzhun shares are too different from those of other companies in the same industry. If the financial statement data provided by Gongzhun shares are true, then the debt settlement of Gongzhun shares is explained. The ability is not bad, even better than Xinwufeng, but its actual controller's equity pledge is not overdue, obviously not in line with the financial status reflected in the statement. Therefore, the authenticity of the data in the statement of public equity shares needs further verification (Table 2).

Table 2 Debt-Asset ratio comparison (10,000 yuan)

Year	Company	Debt	Asset	Debt-Asset ratio (%)
2014	Gongzhun	5281.19	50,884.48	10.38
	Xinwufeng	68,443.44	128,646.36	53.20
2015	Gongzhun	3988.88	75,367.24	5.29
	Xinwufeng	56,089.11	169,085.76	33.17
2016	Gongzhun	418.98	84,218.61	0.50
	Xinwufeng	61,736.60	191,483.52	32.24

Table 3 Comparison of fixed asset turnover (10,000 yuan)

Year	Company	Operating revenue	Net value of fixed assets	Fixed asset turnover
2014	Gongzhun	114,751.51	4899.68	23.42
	Xinwufeng	130,249.39	55,596.78	2.3
2015	Gongzhun	119,641.75	3848.24	31.08
	Xinwufeng	132,603.67	57,810.26	2.29
2016	Gongzhun	139,364.22	3521.44	39.57
	Xinwufeng	169,137.43	58,562.27	2.88

3.4.3 Operational Capacity Analysis

The fixed asset turnover reflects the company's utilization of fixed assets. The higher the ratio, the better the company's management level. In 2016, Xinwufeng was 2.88, and Gongzhun reached 39.57. After comprehensive consideration of the scale and market influence of the two companies, it can be concluded that the fixed asset turnover of Gongzhun shares is abnormal, and the differences between the data and the industry level are clear at a glance. Such differences should be easily audited during the audit. It was discovered by personnel, but it was not pointed out in the audit report, indicating that in the audit project for the fair shares, the auditor may fail to fully perform the audit procedures (Table 3).

When comparing the inventory turnover ratio, the difference between Gongzhun and Xinwufeng is more significant. From the data in 2014, it can be seen that the stock turnover rate of Gongzhun has reached 202.13, which is 54 times that of Xinwufeng. Although the stock turnover rate of Gongzhun has declined in 15 and 16 years, the 16-year data show The Gongzhun is still more than 25 times that of Xinwufeng. The inventory turnover rate reflects the company's inventory turnover rate. Generally, the higher the inventory turnover rate, the better the company's asset liquidity, but the excessive inventory turnover rate indicates that the company has problems in inventory management, such as frequent purchases, lower inventory quality, etc. (Table 4).

Based on the above analysis, we can clearly see that there is a problem in the disclosure of financial information of Gongzhun, and further understand the reasons

Table 4 Comparison of inventory turnover (10,000 yuan)

Year	Company	Operating cost	Average inventory balance	Inventory turnover
2014	Gongzhun	103,966.47	514.35	202.13
	Xinwufeng	127,129.56	34,116.64	3.72
2015	Gongzhun	108,353.17	1263.86	85.73
	Xinwufeng	119,685.85	33,607.56	3.56
2016	Gongzhun	128,328.52	1447.45	88.65
	Xinwufeng	139,863.53	40,984.41	3.41

for the Securities Regulatory Commission to conduct a case investigation of all directors, supervisors and senior executives of Gongzhun.

4 Questions and Suggestions

4.1 Questions

- The internal control of Gongzhun is lacking. On the issue of information disclosure, the company does not have an effective system to restrict it, which adds significant risk of misstatement to the audit.
- The financial data disclosed in the 2016 annual report of Gongzhun is fraudulent. For example, when Gongzhun were pledged in 2016, the liabilities did not increase but decreased, and the decline was large and there were obvious abnormalities.
- In the audit project for Gongzhun, the accounting firm has low effectiveness in performing audit procedures. Although the firm stated in its reply to the inquiry report of the annual report that it had performed the necessary audit procedures and reached reliable audit conclusions, it could not cover up the abnormal situation of the annual report data of Gongzhun.

4.2 Suggestions

- When internal control is missing, focus on the financial data at the end of the period and expand substantive procedures. When internal control is missing, firstly more audit procedures should be implemented at the end of the period rather than in the middle of the period. The financial data at the end of the period reflects the operating status of the enterprise throughout the year, and is closely related to the expected interests of the relevant interest groups. Enterprises tend to whitewash the statements at the end of the statement rather than in the middle. Therefore, auditors should implement more audit procedures at the end of the period to effectively control the risk of material misstatement. Secondly, through substantive procedures to obtain more extensive audit evidence and modify the nature of audit procedures to obtain more convincing audit evidence. When there are major defects in internal control, compliance test can not meet the requirements of risk control. Auditors must expand the scope and degree of substantive test to obtain satisfactory audit evidence to support the judgment of risk, so as to effectively reduce audit risk.
- In the audit of agricultural enterprises, focus on the identification of related party transactions or events and other special areas of risk. In view of the particularity of related party transactions, certified public accountants must keep a high degree of vigilance to the important related party transactions of the audited entity, and

achieve the audit objectives of focusing on identifying related party transactions and other areas that need special consideration. Agricultural listed companies have large trading volume, scattered terminals and cash settlement, which can not be verified by information system, and it is difficult to confirm with farmers. Therefore, auditors can not directly obtain transaction evidence, and the company has a large operating space [10]. From the analysis of public shares, it is very possible for the company to use related party transactions to falsify the financial statement data. Therefore, auditors should pay attention to the audit risk point, especially to understand whether there is an affiliated company with the same actual controller or shareholder although there is no related party transaction. Whether certified public accountants can correctly identify the risk of material misstatement in special areas such as related party transactions has actually become the key yardstick and decisive factor to test their audit quality.

References

1. Wu, G. B., & Li, M. Y. (2020). Construction and application analysis of audit material misstatement risk assessment model. *Accounting Communication, 5*, 134–137.
2. Liu, Z. H., & Zhang, C. J. (2020). Research on measurement and early warning of audit material misstatement risk-based on OWA operator and cloud model. *Accounting Communication, 03*, 129–133.
3. Liu, R., & Chen, F. L. (2015). Improving the effectiveness of identifying the risk of material misstatement by skillfully using analytical procedures—Taking the case of 'Wanfu Shengke' as an example. *Friends of accounting, 1*, 106–109.
4. Jiang, S. Y., & Lu, Z. N. (2013). Construction and application of audit material misstatement risk assessment system. *Science and Technology Management Research, 4*, 227–230.
5. Wang, L., Feng, Y. T., & Liu, H. F. (2015). Discussion on the formation and path effect of material misstatement risk in financial statements. *Chinese Certified Public Accountant, 9*, 43–50.
6. Wang, L., & Liu, H. F. (2011). Research on risk assessment of material misstatement of real estate development enterprises. *Chinese Certified Public Accountant, 9*, 85–90.
7. Liu, T. (2019). Study on risk assessment of material misstatement of D solid waste treatment company. *Rural Economy and Technology, 6*, 130–131.
8. Zhang, D. (2017). On the identification and evaluation of the risk of material misstatement in annual report audit—Taking the financial audit of Y company as an example. *China International Finance and Economics (Chinese and English), 11*, 55–56.
9. Yuan, J. H., & Liu, N. (2018). Study on risk assessment of material misstatement of Agricultural Listed Companies. *Business Accounting, 18*, 51–54.
10. Tian, G. J., & Yao, N. (2016). Risk identification and control of financial fraud audit of agricultural listed companies: A multi case analysis based on gone theory. *Financial and Accounting Communication, 19*, 93–95.

Research on the Driving Forces of Carbon Emissions in China's Manufacturing Industry: A Multi-sector Decomposition Analysis

Hua Fu, Yingying Shi, and Jiming Liu

Abstract As the largest emitter of CO_2 dioxide, the manufacturing industry shoulders the important responsibility of emissions reduction in China. Thus, it is important to explore the main driving force of emissions and propose appropriate emissions reduction measures. This paper implies temporal logarithmic mean divisia index (LMDI) method to analyze the driving forces of the CO_2 emissions change of manufacturing industry in 30 provinces and cities of China from perspectives of the industry and 28 subsectors from 2000 to 2017. The main conclusions reveal that industrial activity and energy intensity are the main factors that lead to the increase and mitigation of CO_2 emissions respectively. In the short term, economic development has a strong effect on carbon emissions, but in the long term, technology will become the main driving force of CO_2 emissions reduction.

Keywords CO_2 emissions · Manufacturing industry · LMDI · Decomposition analysis

1 Introduction

With the continuous economic growth, industrialization and urbanization, China surpassed the United States in 2007 as the world's largest emitter of carbon dioxide. According to the BP world energy statistics review, China's CO_2 emissions in 2016 reached 9.12 billion tons, accounting for 27.3% of the global CO_2 emissions. The

H. Fu (✉)
School of Government Management, Peking University, Beijing, China
e-mail: 1906385386@pku.edu.cn

Y. Shi
School of Management and Economics, Beijing Institute of Technology, Beijing, China
e-mail: shiyingying@bit.edu.cn

J. Liu
China Center for Industrial Security Research, Beijing Jiaotong University, Beijing, China
e-mail: 344397022@qq.com

M. Li et al. (eds.), *IEIS 2020*, https://doi.org/10.1007/978-981-33-4363-4_4

extreme climate phenomenon caused by the sharp increase of greenhouse gas concentration has brought great challenges to human survival and development. In order to undertake international responsibilities and obligations, in 2015, the Chinese government proposed to reduce CO_2 emissions per unit of GDP by 60–65% by 2030, returning to the level of 2005. As the pillar of China's economy, industry consumes 70% of the country's energy and contributes more than 50% of the country's CO_2 emissions. To achieve the goal of CO_2 emissions reduction while maintaining high-quality economic development, reduction of CO_2 emissions in manufacturing industry has become the top priority. In order to coordinate industrial development, energy consumption and CO_2 emissions reduction, it is necessary to identify the main driving factors of CO_2 emissions in manufacturing industry, and then explore effective path of CO_2 emissions reduction in manufacturing industry.

2 Research Status at Home and Abroad

As for the research of CO_2 emissions influencing factors, the academia usually uses decomposition technology to investigate the changes of energy and environment, and decomposes the emissions change (or intensity) into several effects. The main decomposition methods are IDA, SDA and PDA. Since 1980, IDA has been widely used to study energy consumption [1]. Considering the structure decomposition method (SDA) is based on input-output table, which is not suitable for cross period research, the index decomposition method (IDA) has gradually become the mainstream decomposition method. After 1995, Ang et al. improved LMDI decomposition method to realize complete decomposition and solve the problem of zero numerical impact calculation. Then LMDI method become widely used in research of country, region, and industry. Through decomposition analysis of CO_2 emissions of China's manufacturing industry, Chinese scholars found that the positive and negative contribution rate of economic output effect and energy intensity effect reduction to carbon emissions is the largest, and the output effect is far greater than the inhibition effect of energy intensity and energy structure on carbon emissions [2, 3]. Energy intensity can be further decomposed into technical factors, intermediate input, industrial output structure and other factors, but the effect of technological change on CO_2 emissions reduction is not obvious at this stage [4]. Lin and Du [5] further used the comprehensive decomposition framework to study the role of technological progress in China's energy intensity decline. In the long term, technological progress will be the key to achieve sustainable growth of energy productivity. In 2015, B. W. Ang et al. [6] expanded the LMDI decomposition method in the study of CO_2 emissions of manufacturing industry in 30 provinces of China. On this basis, domestic scholars further analyze the influencing factors of CO_2 emissions in 28 sub-industries of manufacturing industry in various provinces and cities of China [7].

3 Research Methods and Data Sources

3.1 Calculation Method

According to Kaya's equation, the carbon emissions of energy related manufacturing industry is mainly driven by five driving forces: carbon emissions coefficient, energy structure, energy intensity, economic activities and industrial scale, which can be expressed as follows:

$$C_i = \sum_j \frac{C_{ij}}{E_{ij}} \times \frac{E_{ij}}{E_i} \times \frac{E_i}{Y_i} \times \frac{Y_i}{P_i} \times P_i \tag{1}$$

On the right hand side of Eq. (1), the first component could be interpreted as the CO_2 emissions coefficient since the CO_2 emissions is deflated by the energy consumed in subsector i of manufacturing industry, where j repents the energy types (j = 1, 2, …, 18). The second component measures the energy structure since the CO_2 emissions is deflated by the of energy type consumed in the subsector. The third component could be interpreted as the energy intensity. An increase in energy usage technical efficiency will lead to a larger energy intensity change and therefore more of the change in E/Y. The fourth and fifth component measure output per capita and the number of employees in subsector i which represent industry activity and industry scale of the sub-industry respectively.

Using the notations described above, we can rewrite Eq. (1) as

$$C_i = \sum_j CE_{ij} + ES_{ij} + EI_i + IA_i + IS_i \tag{2}$$

According to the LMDI model given by the total change of carbon dioxide emissions related to energy between the base year 0 and target year t of China's manufacturing industry can be divided into five driving forces: carbon emissions coefficient effect, energy structure effect, energy intensity effect, economic activity effect and industry scale effect. Now that the change of CO_2 emissions from the period 0 to t is as follows:

$$\begin{aligned} \Delta C_i^{(t-0)} = C_i^t - C_i^0 &= \Delta C_{i,CE}^{(t-0)} + \Delta C_{i,ES}^{(t-0)} \\ &+ \Delta C_{i,EI}^{(t-0)} + \Delta C_{i,IA}^{(t-0)} + \Delta C_{i,IS}^{(t-0)} \end{aligned} \tag{3}$$

According to the LMDI model given by the total change of carbon dioxide emissions related to energy between the base year 0 and target year t of China's manufacturing industry can be divided into five driving forces: carbon emissions coefficient effect, energy structure effect, energy intensity effect, industry activity effect and industry scale effect, as follows:

The specific calculation is as follows:

$$\Delta CE = \sum_{ij} L(C_{ij}^t, C_{ij}^0) \ln\left(\frac{CE_i^t}{CE_i^0}\right) \tag{4}$$

$$\Delta ES = \sum_{ij} L(C_{ij}^t, C_{ij}^0) \ln\left(\frac{ES_i^t}{ES_i^0}\right) \tag{5}$$

$$\Delta EI = \sum_{ij} L(C_{ij}^t, C_{ij}^0) \ln\left(\frac{EI_i^t}{EI_i^0}\right) \tag{6}$$

$$\Delta IA = \sum_{ij} L(C_{ij}^t, C_{ij}^0) \ln\left(\frac{IA_i^t}{IA_i^0}\right) \tag{7}$$

$$\Delta IS = \sum_{ij} L(C_{ij}^t, C_{ij}^0) \ln\left(\frac{IS_i^t}{IS_i^0}\right) \tag{8}$$

where

$$L(C_{ij}^t, C_{ij}^0) = \begin{cases} \frac{C_{ij}^t - C_{ij}^0}{\ln C_{ij}^t - \ln C_{ij}^0} & \left(C_{ij}^t \neq C_{ij}^0\right) \\ C_{ij}^t & \left(C_{ij}^t = C_{ij}^0\right) \end{cases} \tag{9}$$

3.2 *Data Source*

According to the industrial classification of national economic activities (GB/T 4754-2011), manufacturing industry is divided into 31 sub sectors. Due to the data discontinuity of C42, C43 and C37, this study covers 28 subsectors of manufacturing industry. The output value and employment population of manufacturing subsector are from China Industrial Statistics Yearbook (2001–2019). The output value is calculated at constant price in 2000. Energy consumption and carbon dioxide emissions data of manufacturing industry are from carbon emissions inventory and energy inventory of CEADs. According to the revised China's carbon emissions factors published in nature (2015) by Liu et al. [8], the list was compiled with China's energy statistical yearbook data and widely used by scholars at home and abroad.

4 Calculation Process and Result Analysis

4.1 Relationship Between Energy Consumption and Carbon Emissions

In recent years, with the growth of manufacturing output value, the energy consumption and carbon emissions of manufacturing industry are also increasing. In 2000, 28 sub industries in the manufacturing industry consumed 6.05 million tons of standard coal, which increased to 1977 million tons of standard coal in 2017, an increase of 2.3 times in 18 years. In the same period, CO_2 emissions increased from 1365.1 tons to 4608.3 tons, an increase of 2.4 times. There is a positive correlation between energy consumption and CO_2 emissions as shown in Fig. 1. From 2000 to 2017, the carbon emissions and energy consumption of manufacturing industry can be roughly divided into four stages. In the first stage (2000–2005), growth rate of energy consumption and CO_2 emissions increased sharply from less than 9% to about 20%. In the second stage (2005–2010), the growth rates were basically controlled within 10%, once declined to 4% in 2008. In the third stage (2010–2014), the growth rates were both within 10%, and the growth rate of energy consumption even reached 0.85% in 2012. In the last stage (2015–2017), the growth rates were both negative.

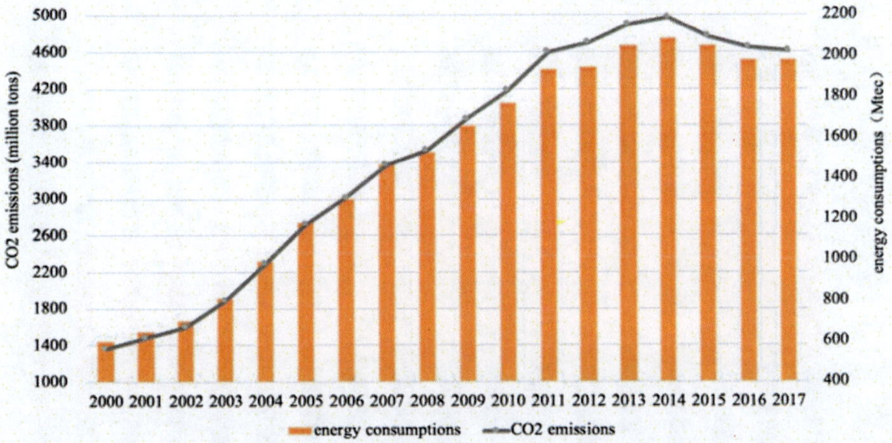

Fig. 1 Energy consumption and CO_2 emissions of manufacturing industry in 2010–2017

4.2 *Carbon Emissions Proportion of Different Energy Sources*

Coal has always been the main energy consumed by the manufacturing industry and the largest source of carbon dioxide emissions. Figure 2 reveals that the growth of CO_2 emissions is mainly due to the large use of coal after 2002. Ratio of coal-related CO_2 emissions has been declined to less than 50% since 2010. These 18 kinds of energy in the energy emissions list can be categorized into six main kinds: raw coal, coal products, oil, natural gas, electric energy and thermal energy, and process. It can be seen that the CO_2 emissions of raw coal reached the peak in 2004 and then gradually declined after 2005 while CO_2 emissions of coal products increased from 20.6% in 2000 to 29.3% in 2017. It reveals that energy structure of coal tends to be optimized.

Thermal power is the second largest source of CO_2 emissions of manufacturing energy. The share of thermal-related CO_2 emissions accounted for 33.4–39.3% in manufacturing industry during 2000–2017. Thermoelectric-related CO_2 emissions showed an inverted U-curve from 2004 to 2017, reaching a historical peak of 39.3% in 2017. The share of oil-related CO_2 emissions has declined since 2000 and then increased after 2013. The proportion of natural gas is the lowest among these energies and tends to rise as a whole.

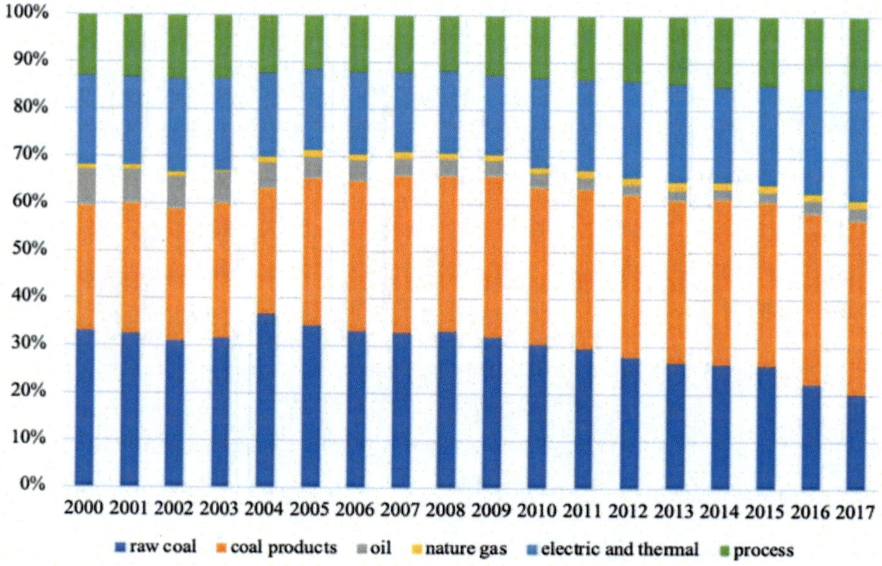

Fig. 2 Energy consumption of manufacturing industry in 2000–2017

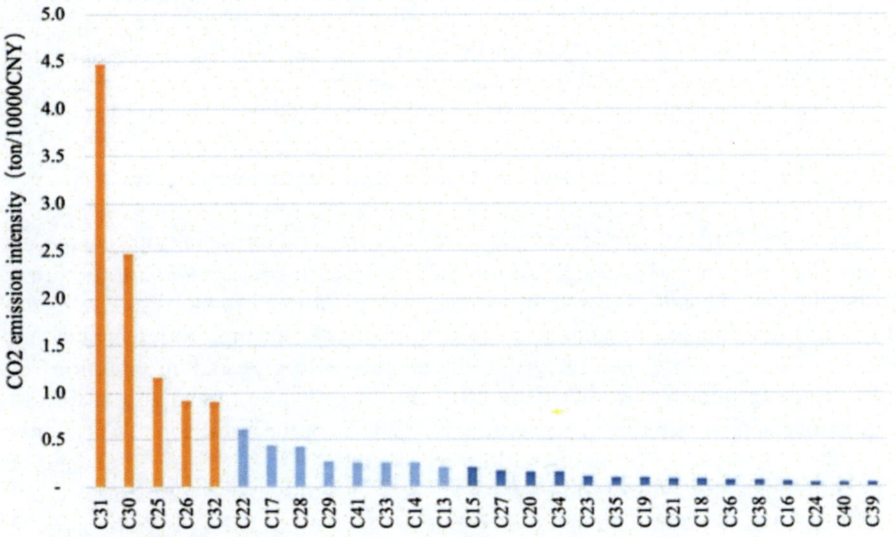

Fig. 3 Ranking of CO_2 emissions intensity of manufacturing sub-industries in 2017

4.3 *Carbon Emissions of Manufacturing Sub-industry*

Figure 3 shows the CO_2 emissions intensity of 28 manufacturing sub-industries in 2017. The industry with the highest CO_2 emissions intensity is C31 (447 tons/10000CNY), while the industry with the lowest CO_2 emissions intensity is C39 (0.04 tons/10000CNY). It's apparently that there are significant differences in CO_2 emissions across 28 subsectors. Therefore, according to the CO_2 emissions intensity, we divide 28 sub-industries into three categories: high emissions intensity industry (HEIs), medium emissions intensity industry (MEIs) and low emissions intensity industry (LEIs). HEIs include C31, C30, C25, C26, and C32. The CO_2 emissions of these five industries alone account for 82% of the total emissions of 28 industries. MEIs mainly include C22, C17, C28 and other traditional industries with high energy consumption. LEIs are mainly light industry, i.e. C39 and other high-tech industries. The average emissions intensity of HEIs is 1.99 tons/10000CNY, 5.9 times of MEIs (0.34 tons/10000CNY) and 20.6 times of LEIs (0.1 tons/10000CNY). Therefore, the efforts to reduce CO_2 emissions should be focused on HEIs and MEIs.

4.4 *Time Series Carbon Emissions Decomposition of Manufacturing Industry*

In this part, we use the LDMI decomposition model to analyze the main drivers of manufacturing CO_2 emission changes in 2000–2017. According to the trend of

CO_2 emissions in 2000–2017, we have conducted research in four stages, namely, 2000–2005, 2005–2010, 2010–2014 and 2014–2017.

4.4.1 Decomposition Results of the Whole Manufacturing Industry

During 2000–2017, the CO_2 emissions of China's manufacturing industry increased from 1.617 billion tons to 5.032 billion tons, with an increase of 211%. Figure 5 shows the decomposition results of manufacturing sub-sectors by using the factor decomposition method of time series. Compared with the time series analysis in Figs. 4, 5 shows more intuitively the influencing factors of carbon emissions of manufacturing industry from 2000 to 2017. EI effect led to a reduction of 582.83 billion tons of CO_2 emissions, accounting for 170.7% of the reduction of CO_2 emissions in the manufacturing industry. IS effect accounts for 12.8% of the change of CO_2 emissions in manufacturing industry. The CO_2 emissions coefficient of fossil energy is basically constant, and the change mainly comes from the technological

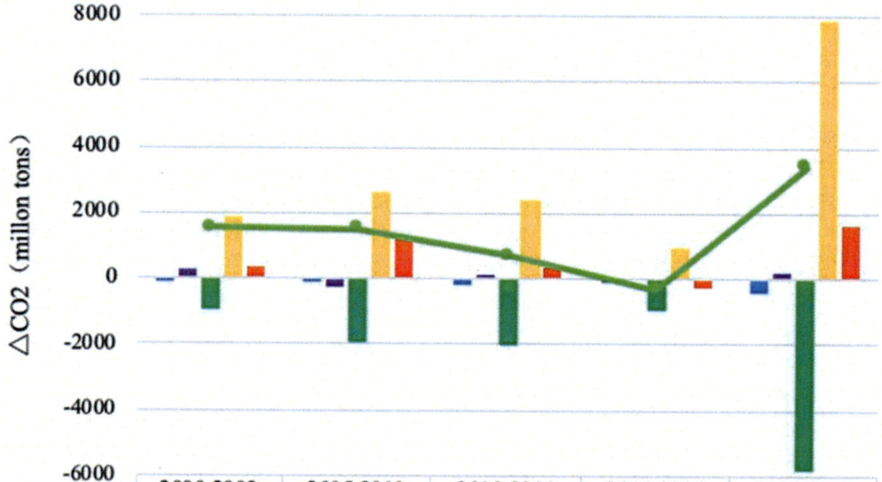

	2000-2005	2005-2010	2010-2014	2014-2016	2000-2017
CE	-5.07	-131.44	-190.75	-110.76	-438.02
ES	238.54	-237.27	138.15	23.30	162.71
EI	-913.21	-1955.78	-2017.79	-941.56	-5828.34
IA	1875.93	2596.93	2435.88	996.36	7905.10
IS	333.28	1226.64	341.44	-287.92	1613.44
ΔCO2	1529.46	1499.07	706.93	-320.57	3414.89

CE ES EI IA IS ΔCO2

Fig. 4 Decomposition result of CO_2 emission changes in menufactory industry progress and the improvement of production efficiency of electric heating production. In the four stages, CO_2 emissions increased much more rapidly than the follow years. Among the five influencing factors, IA effect is the most important force of CO_2 emissions.

4.4.2 Decomposition of Manufacturing Sub-industries

This part decomposes 28 subindustries of manufacturing industry to study the energy-related CO_2 emissions of these subsectors, so as to explore the main driving forces of CO_2 emissions changes in sub-industries. Although the cumulative contribution of the influencing factors of CO_2 emissions in each subsector is significantly different during 2000–2017, the decomposition results of 28 subsectors show that the economic activity effect is the main driving force for the increase of CO_2 emissions, and the energy intensity effect is the main contributor of CO_2 emissions reduction, as shown in Fig. 5.

From 2000 to 2017, the industries of HEIs, MEIs, LEIs show different characteristics in the effect of the influencing factors for CO_2 emissions.

With the acceleration of urbanization in China, the demand of raw materials for infrastructure construction is increasing, which leads to the rapid development of HEIs and CO_2 emissions. For HEIs, the cumulative contribution of IA effect in C31 is the largest in 28 sub-industries, accounting for 226.6% of CO_2 emissions changes. The expansion of industrial scale resulted in 738 million tons of CO_2 emissions of C31, followed by C30, C26, C25, and C32, with contribution of 49.2%, 49.9%, 33%, 44% and 13.3% respectively. Compared with the impact of industrial activities, the

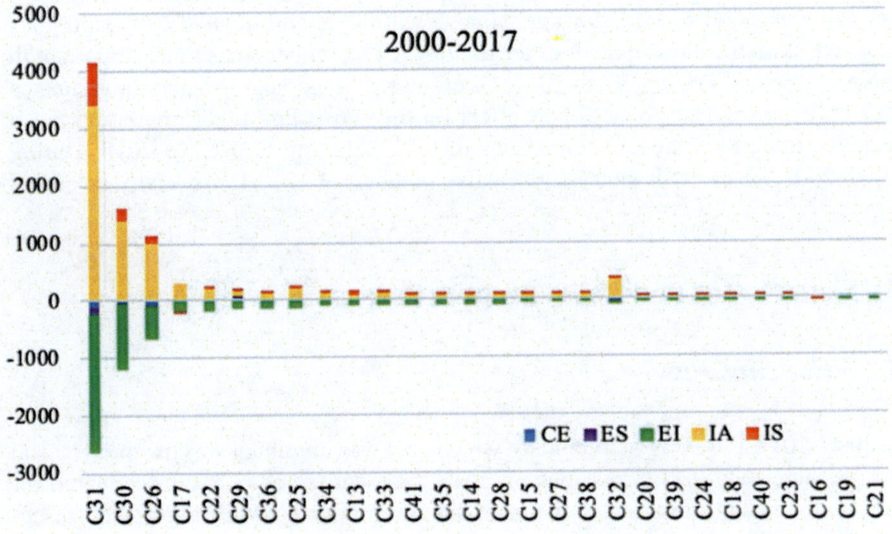

Fig. 5 Decomposition result of CO_2 emission changes in subsectors

EI effect mainly driven by technological progress is relatively limited. From 2000 to 2017, the EI effect of C31 reached −2417 million tons, followed by C30, C26, C25, and C32, respectively accounting for −161%, −261.1%, −127.9%, −175.5% and −14.8% of the total changes in CO_2 emissions. ES effect has a negative impact on the CO_2 emissions of C31, and positive impact on other HEIs, but the impact is not significant. CI effect has a positive impact on the CO_2 emissions of HEIs during 2000–2017.

During 2000–2017, the effects of economic activities and energy intensity are also key factors of CO_2 emissions for MEIs. The cumulative change of CO_2 emissions of C17 is the biggest among MEIs. One reason is IA effect contributed a lot, accounting for 301.3% of the changes of CO_2 emissions. EI effects of textile industry and paper industry reached −155 million tons and −151 million tons respectively, which greatly offset the increase of CO_2 emissions caused by economic activities. IS effect of C13 was the largest, followed by C37, C39 and C33. Industrial scale of C13, C37, C39 and C33 increased significantly, resulting in the increase of energy consumption and related CO_2 emissions. In contrast, CO_2 emissions reduced with downsizing of industrial scales in C17, C41 and C16. When CI effect reduces the CO_2 emissions of MEIs and LEIs from 2000 to 2017, ES effect leads to the increase of CO_2 emissions. CO_2 emissions increment of LEIs are relatively small, because the impact of IA effect and EI effect are relatively small.

Figure 6 shows the contribution of each driving force at different stages to the CO_2 emissions of the manufacturing subsectors. It can be seen that the impact of IA effect on CO_2 emissions in 2008–2011 is generally large. The main reason may be that rapid economic development, export and urbanization promote the rapid growth of industry and energy-related CO_2 emissions. During 2011–2015, China formulated a series of policies to promote the low-carbon development of manufacturing industry, and the expansion of HEIs became slow down and energy-related CO_2 emissions reduced. With the development of technology, EI effect becomes more important in most subsectors. During 2008–2011, CI effect remained negative and the influence was on the rise. The impact of IS effect on the subsectors is different. and has a negative impact on the CO_2 emissions of C17, C22 and C34 since 2010. During 2000–2005, ES effect is generally negative in most subsectors, especially in C31.

5 Conclusions and Policy Implications

5.1 Conclusions

In 2000–2017, the energy structure in which coal was dominant was the main reason for the increase of CO_2 emissions. In 2017, CO_2 emissions of HEIs accounted for 82% of total CO_2 emissions of the manufacturing industry, while high-tech subsectors such as C38, C39 and C40 contributed greater outputs with less CO_2 emissions. Industrial activity effect and energy intensity effect are main driving forces for the

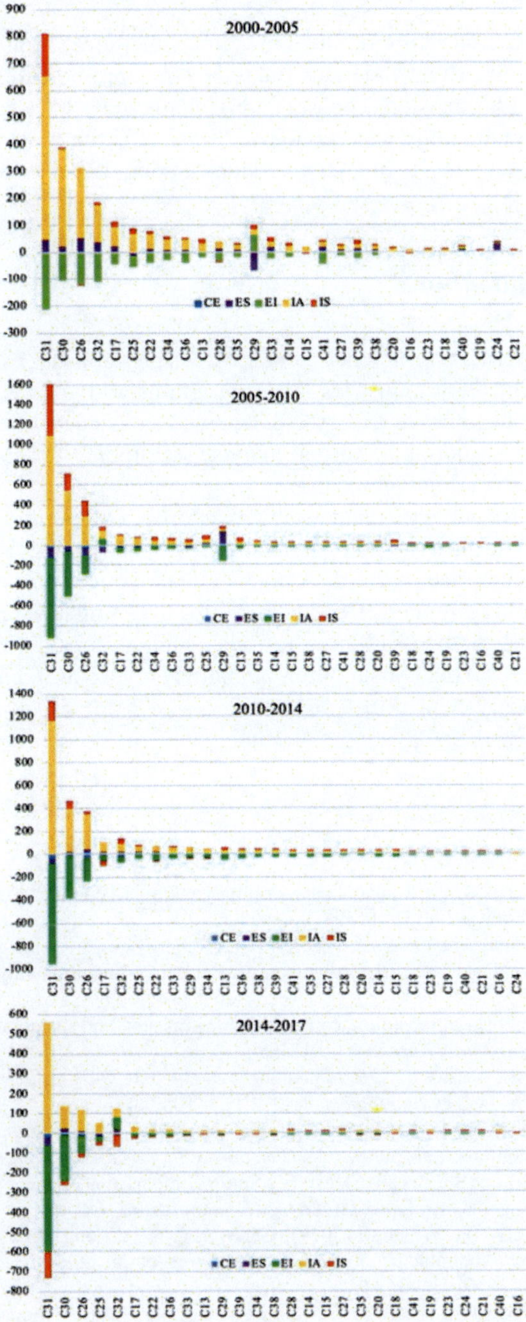

Fig. 6 Decomposition result of CO_2 emission changes in subsectors at different time intervals

increase and decrease of CO_2 emissions in China's manufacturing industry respectively. Energy structure effect and industrial scale effect promote the increase of CO_2 emissions, while CO_2 emissions coefficient effect alleviates CO_2 emissions. Due to the differences in technology and economy, EI effect is the most important factor that leads to the discrepancies of CO_2 emissions among 28 subsectors.

5.2 *Policy Implications*

Firstly, improve the energy efficiency of manufacturing industry, especially for C31 and other energy-intensity industries. More programs and actions should be initiated by authorities to improve energy efficiency, such as energy-saving projects in key industry, the construction of carbon market, subsidizing for energy-saving and energy-efficiency technology innovation.

Secondly, the government should establish an open, competitive and fair market mechanism, formulate effective policies and measures, and promote the development of manufacturing enterprises in the direction of energy conservation and low carbon.

Finally, we should formulate effective measures to optimize energy structure and encourage clean energy to substitution for coal, such as subsidizing production of renewable energy, or taxing consumption of coal energy, so as to change the energy structure dominated by coal.

References

1. Ang, B. W. (2004). Decomposition analysis for policymaking in energy: Which is the preferred method? *Energy Policy, 32,* 1131–1139.
2. Xu, J., Fleiter, T., Eichhammer, W., & Fan, Y. (2012). Energy consumption and CO2 emissions in China's cement industry: A perspective from LMDI decomposition analysis. *Energy Policy, 50,* 821–832.
3. Chen, S. (2009). Energy consumption, carbon dioxide emission and sustainable development of Chinese industry. *Economic Research, 4,* 41–55.
4. Zhou, P., & Ang, B. W. (2008). Decomposition of aggregate CO_2 emissions: A production-theoretical approach. *Energy Economics, 30*(3), 1054–1067.
5. Lin, B., & Du, K. (2014). Understanding the change of energy intensity in China: A comprehensive decomposition framework. *The Journal of World Economy, 4,* 69–87.
6. Ang, B. W., Xu, X. Y., & Su, B. (2015). Multi-country comparisons of energy performance: The index decomposition analysis approach. *Energy Economics, 47,* 68–76.
7. Shi, Y., Han, B., Zafar, M. W., et al. (2019). Uncovering the driving forces of carbon dioxide emissions in Chinese manufacturing industry: An intersectoral analysis. *Environ Science & Pollution Research, 26,* 31434–31448.
8. Liu, Z., Guan, D., Wei, W., et al. (2015). Reduced carbon emission estimates from fossil fuel combustion and cement production in china. *Nature, 524*, 335–338H.

The Impact of Urban Rail Transit on Financing Constraints of Small and Medium-Sized Enterprises

Rong Yang, Xuemeng Guo, and Minhua Song

Abstract In recent years, the continuous development of urban rail transit in China not only effectively alleviates urban traffic congestion, environmental pollution and other problems, but also drives the development of a series of upstream and downstream related industries. Based on the theory of industrial cluster, this paper analyzes the impact of urban rail transit operation on the financing constraints of small and medium-sized enterprises(SMEs) with the financial data of small and medium-sized listed companies in 2009–2018 as the research sample. It is found that the opening and operation of urban rail transit effectively eases the financing constraints of small and medium-sized listed enterprises, moreover, it has a more obvious role in easing the financing constraints of directly related industries and large-scale enterprises. The study of this paper provides some reference for alleviating the financing constraints of small and medium-sized enterprises, and also provides empirical evidence from the micro enterprise level for the economic consequences of urban rail transit construction and operation in China.

Keywords Urban rail transit · Small and medium-sized enterprises · Financing constraints

1 Introduction

Since the reform and opening up, small and medium-sized enterprises in China have also made great progress with the economic growth and technological progress, playing an important role in GDP growth, tax revenue and employment promotion. However, affected by a series of factors, SMEs are not only faced with the shortage of

R. Yang (✉) · X. Guo · M. Song
School of Economics and Management, Beijing Jiaotong University, Beijing, China
e-mail: 19113058@bjtu.edu.cn

X. Guo
e-mail: xmguo@bjtu.edu.cn

M. Song
e-mail: songmh009@163.com

M. Li et al. (eds.), *IEIS 2020*, https://doi.org/10.1007/978-981-33-4363-4_5

internal cash flow, but also with the difficulties of external financing. The lack of funds and the difficulty of financing have always been a common dilemma faced by SMEs in China, which to a large extent limits and restricts the survival and development of SMEs.

In recent years, urban rail transit construction in China has also achieved rapid development. According to the statistics of China Urban Rail Transit Association, as of December 31, 2019, 40 cities in mainland China have opened urban rail transit, with 6730.27 km of operating lines and 26 new lines. It has been confirmed that urban rail transit plays a positive role in promoting urban expansion [1] and regional economic development, as well as the site selection of residents and enterprises [2], and the housing price along the line. While urban rail transit drives urban development, it also provides a good opportunity for the upstream and downstream related industries, and gradually forms an industrial chain centered on urban rail transit. The proximity of industries in geographical location can produce agglomeration effect, which is conducive to ease the financing constraints faced by enterprises. According to this idea, based on the perspective of industrial clusters, this paper studies whether the opening and operation of urban rail transit is conducive to easing the financing constraints of small and medium-sized enterprises, and further test the difference of the impact effect according to different industry relevance and enterprise scale.

2 Theoretical Analysis and Research Hypothesis

The problems of "adverse selection" and "moral hazard" caused by the information asymmetry between the debtor and creditor, as well as the credit rationing problems caused by the "scale discrimination" of the fund providers to the small-scale enterprises, lead to the financing constraints of the small and medium-sized enterprises. Compared with the main board listed enterprises, small and medium-sized board listed enterprises are often faced with stronger financing constraints due to their small age and asset size. It is more practical to study the financing constraints of small and medium-sized board listed enterprises to solve the financing problems of SMEs.

Industrial cluster was first proposed by Michael Porter to explain the cluster phenomenon in real life. Specifically, the industrial cluster refers to the geographic gathering group composed of upstream and downstream enterprises, financial institutions and other relevant institutions with cooperation and competition within a certain space. The purpose of this paper is to study the economic consequences of the opening and operation of urban rail transit, that is, to study the industrial cluster formed with urban rail transit as the center. Promoting industrial cluster is conducive to regional industrial development and upgrading, and the impact of urban rail transit on corporate financing constraints may come from the value-added brought by industrial cluster.

Previous studies have shown that the external effects of industrial clusters have brought strong development impetus to SMEs in the cluster area. Compared with enterprises outside the cluster area, enterprises inside the industrial cluster have more advantages in obtaining external financing: First of all, the geographical location of the enterprises in the cluster area is close to that of the upstream and downstream enterprises, which makes the information of higher quality spread rapidly among enterprises, and is conducive to the formation of the business credit of the cluster enterprises [3]. The reduction of information asymmetry also reduces moral hazard and adverse selection [4]. Secondly, small and medium-sized enterprises in cluster area can apply for loans in the form of collective guarantee, which reduces the cost of credit and improves the controllability of bank credit risk [5]. Thirdly, industrial agglomeration is conducive to the specialization and cooperation between enterprises, and reduces the threshold of industry access to capital [6]. And the formation of commercial credit can make enterprises obtain price preference in the process of transaction. Rail transit industry cluster enterprises can also alleviate the difficulties caused by lack of funds through credit sale, inter-bank lending and other ways, and also reduce the financing constraints of small and medium-sized enterprises [7].

Based on this, the hypothesis 1 is put forward: the opening and operation of urban rail transit is conducive to easing the financing constraints of small and medium-sized enterprises.

Wassily Leontief, an American economist, put forward the theory of industrial association, which can be used to analyze the relationship between multiple industries. In the process of social reproduction, various production units will form a variety of complex technical and economic relations, which will not only have a direct connection, but also an indirect connection. The opening and operation of urban rail transit will definitely affect and drive the development of some upstream and downstream related industries, including construction industry, manufacturing industry, service industry, transportation industry, wholesale and retail industry and other related industries. However, there are some differences in the intensity of relevance between different industries and urban rail transit. Some scholars use the input-output table of urban rail transit to calculate the measurement coefficient of the relationship between urban rail transit and upstream and downstream industries. The larger the coefficient, the stronger the correlation between industries. In this paper, industries with strong relevance to urban rail transit are defined as directly related industries, and industries with weak relevance are defined as indirectly related industries. Generally speaking, the opening and operation of urban rail transit creates more convenient conditions and positive impact for the directly related industries, and brings more obvious industrial agglomeration effect, which can more effectively alleviate the financing constraints of enterprises.

Therefore, hypothesis 2 is put forward: the opening and operation of urban rail transit can significantly alleviate the financing constraints of enterprises in directly related industries, but not in indirectly related industries.

Although some studies have shown that industrial clusters can promote regional economic growth, the degree of industrial agglomeration and its promotion are not monotonous increasing, and some regions have experienced excessive or even decreasing agglomeration. Compared with state-owned enterprises and large-scale enterprises, private enterprises and small-scale enterprises are usually in a poor financing environment due to their internal and external factors, and the level of industrial agglomeration should have a more obvious role in easing their financing constraints. It shows that there will be some differences in the improvement effect of industrial clusters on the financing constraints of enterprises under different conditions. The higher the level of financing constraints enterprises face, the more significant the improvement effect of cluster effect on the financing constraints of enterprises. Generally speaking, compared with large-scale enterprises, small-scale enterprises face more serious financing difficulties.

Therefore, hypothesis 3 is proposed: The effect of the opening and operation of urban rail transit on the financing constraints of small-scale enterprises is more significant.

3 Research Design

3.1 Sample Selection and Data Sources

In recent years, the double difference in differences (DID) method is usually used to evaluate the implementation effect of public policies or projects. However, due to the different time of operating urban rail transit in each city, this paper adopts multi-phase DID, that is, there is no same policy implementation time, but different cities are allowed to have their own policy implementation year according to the actual situation. In view of the availability of relevant data, this paper selects the sample interval of 2009–2018, and takes the cities that have opened urban rail transit during this period as the processing group, and all cities that have not yet opened urban rail transit as the control group.

The financial data of small and medium-sized listed enterprises in this paper are from CSMAR database, and the relevant data of urban rail transit are mainly from the statistical data of China Urban Rail Transit Association.

3.2 Variables

- Dependent variable: Financing Constraint.

It is difficult to quantify the degree of financing constraints accurately, so in the early research, scholars have been trying to establish an appropriate measurement model to measure the level of financing constraints. The most representative measurement methods include classical investment cash flow sensitivity model, KZ model, cash-cash flow sensitivity model, WW index, SA index, etc. Although the calculation methods of each measurement index are different, the measurement results tend to be consistent. Among them, the investment cash flow sensitivity model was proposed by Fazzari et al., They think that the investment cash flow sensitivity of enterprises can be used to reflect the level of financing constraints. The greater the sensitivity coefficient of investment cash flow, the more serious the financing constraints faced by enterprises. In this paper, the investment cash flow sensitivity model is used to measure corporate financing constraints. Tobinq is used as the proxy variable of investment opportunities, and the following model is constructed to measure the degree of corporate financing constraints:

$$INVEST_{i,t}/ASSET_{i,t-1} = \alpha_0 + \alpha_1 CF_{i,t}/ASSET_{i,t-1} + \alpha_2 TOBINQ_{i,t-1} + \varepsilon_{i,t} \tag{1}$$

Among them, $INVEST_{i,t}$ represents the investment expenditure of enterprise i in period t, $ASSET_{i,t-1}$ represents the total assets of enterprise i at the beginning of period t, $CF_{i,t}$ is the net cash flow generated from operating activities in the cash flow statement of enterprise i in period t, $TOBINQ_{i,t-1}$ is the Tobin Q value of enterprise i in period $t - 1$. The coefficient α_1 in front of cash flow is the sensitivity coefficient of investment to cash flow. When $\alpha_1 > 0$, it is considered that there is a positive correlation between investment expenditure and internal cash flow of the enterprise, indicating that the enterprise is subject to external financing constraints.

- Independent variable: operation of Urban Rail Transit.

This paper uses DID (treat * after) as an independent variable. DID is a comprehensive virtual variable of two virtual variables: treated (whether urban rail transit has been put into operation) and after (whether after urban rail transit has been put into operation), which measure whether urban rail transit has been put into operation. When the urban rail transit is opened in the location of the enterprise and the time is later, the did is taken as 1, otherwise it is 0.

- Control variables.

In order to eliminate the influence of other potential factors and better study the impact of urban rail transit operation on the financing constraints of small and medium-sized enterprises, with reference to relevant research, the main control variables selected in this paper are: fluctuation of working capital, company size, corporate debt, company age and urban and annual fixed effects (Table 1).

3.3 Models

In this paper, the investment cash flow sensitivity model is used to measure the corporate financing constraints. Based on Formula (1), the control variables are added and Formula (2) is as follows:

$$INVEST_{i,t}/ASSET_{i,t-1} = \alpha_0 + \alpha_1 CF_{i,t}/ASSET_{i,t-1}$$

$$+\alpha_2 TOBINQ_{i,t-1} + \alpha_3 \Delta NWC_{i,t}/ASSET_{i,t-1}$$
$$+\alpha_4 SIZE_{i,t-1} + \alpha_5 LEV_{i,t-1} + \alpha_6 AGE_{i,t}$$
$$+\sum CIYT + \sum YEAR + \varepsilon_{i,t} \quad (2)$$

Expected result: the coefficient, α_1, is significantly positive.

In order to test the relationship between urban rail transit and corporate financing constraints, based on the model (2), this paper adds the $CF_{i,t}/ASSET_{i,t-1} \times DID_{i,t}$ to get Formula (3) as follows:

$$\begin{aligned} INVEST_{i,t}/ASSET_{i,t-1} = {} & \alpha_0 + \alpha_1 CF_{i,t}/ASSET_{i,t-1} + \alpha_2 DID_{i,t} \\ & + \alpha_3 CF_{i,t}/ASSET_{i,t-1} \times DID_{i,t} + \alpha_4 TOBINQ_{i,t-1} \\ & + \alpha_5 \Delta NWC_{i,t}/ASSET_{i,t-1} + \alpha_6 SIZE_{i,t-1} + \alpha_7 LEV_{i,t-1} \\ & + \alpha_8 AGE_{i,t} + \sum CITY + \sum YEAR + \varepsilon_{i,t} \end{aligned} \quad (3)$$

Table 1 Variable summary

Variables	Definitions
Whether the city has opened Urban Rail Transit (TREAT)	Virtual variable: cities with urban rail transit in operation from 2009 to 2018 are taken as 1; cities without urban rail transit in operation by the end of 2018 are taken as 0
Whether the time is after the opening of Urban Rail Transit (AFTER)	Virtual variable: when the time after the year of urban rail transit operation, it is taken as 1, otherwise it is taken as 0
Opening of Urban Rail Transit (DID)	DID = TREAT * AFTER
Operating mileage (MILEAGE)	Operating mileage of Urban Rail Transit
Investment expenditure of enterprises (INVEST)	"Cash paid for acquisition and construction of fixed assets, intangible assets and other long-term assets" in the cash flow statement
Total assets of the enterprise (ASSET)	Total assets at the end of the period
Cash flow of enterprises (CF)	Net cash flow from operating activities
Investment opportunity (TOBINQ)	Market value of enterprise stock/total book assets at the end of the period
Working capital (NWC)	Working capital = (current assets − current liabilities)
Size of the company (SIZE)	Natural logarithm of total assets
Liabilities of the company (LEV)	Asset liability ratio = total liabilities at the end of the period/total assets at the end of the period
Age of the company (AGE)	Year of sample observation − year of establishment
(CITY)	Fixed effect of city
(YEAR)	Fixed effect of the year

Expected results: the coefficient, α_1, is significantly negative, indicating that the operation of urban rail transit is conducive to easing the financing constraints of sample enterprises.

4 Empirical Analysis

4.1 Multiple Regression Analysis

The regression results of the relationship between the opening and operation of urban rail transit and the financing constraints of small and medium-sized enterprises are shown in Table 2.

Columns (1)–(3) show the regression results of the whole sample, the treatment group sample and the control group sample to the model (2), their investment cash flow sensitivity coefficient are significantly positive, indicating that the sample companies have financing constraints. In addition, the coefficient of the second row is significantly positive at the level of 5% (0.0529), and the coefficient of the third row is significantly positive at the level of 1% (0.112). It shows that the degree of financing constraint of the sample enterprises in the treatment group is less than that in the control group. This may be due to the relatively backward level of urban development and imperfect financial system of urban rail transit that has not yet been opened, leading enterprises to face more serious financing constraints. Column (4) shows the regression results of the whole sample to the model (3), that is, the regression analysis after adding the cross multiplication item $\mathrm{CF}_{i,t}/\mathrm{ASSET}_{i,t-1} \times DID_{i,t}$. The result shows that the coefficient of cross multiplication term is significantly negative (−0.0803) at the level of 5%. It shows that the opening of urban rail transit is conducive to ease the financing constraints faced by small and medium-sized enterprises, and the hypothesis 1 is valid.

In order to test hypothesis 2, to study whether the mitigation effect of urban rail transit operation on the financing constraints of small and medium-sized enterprises on different industries is different, based on the existing research, combined with the correlation strength between the industry and urban rail transit industry, this paper divides the whole sample of enterprises into two groups: directly related industries and indirectly related industries.

The regression analysis of the model (3) was carried out with the two groups of samples. The results are shown in columns (1) and (2) of Table 3. Column (1) shows the regression results of directly related industry samples to model (3), and the coefficient of cross multiplication item is significantly negative (−0.111) at the level of 1%, Column (2) is the regression result of indirect related industry samples to model (3), and the coefficient of cross multiplication term is not significant. That is to say, the opening of urban rail transit is conducive to alleviating the financing constraints faced by enterprises in directly related industries, while it has no significant effect on indirectly related enterprises. The results confirm hypothesis 2.

Table 2 Opening of urban rail transit and financing constraints of SMEs

	$INVEST_{i,t}/ASSET_{i,t-1}$	$INVEST_{i,t}/ASSET_{i,t-1}$	$INVEST_{i,t}/ASSET_{i,t-1}$	$INVEST_{i,t}/ASSET_{i,t-1}$
	Full sample	Treatment group	control group	Full sample
	(1)	(2)	(3)	(4)
$CF_{i,t}/ASSET_{i,t-1}$	0.0873***	0.0529**	0.112***	0.105***
	(5.294)	(2.009)	(5.367)	(5.687)
$CF_{i,t}/ASSET_{i,t-1}$*$DID_{i,t}$				−0.0803**
				(−2.121)
$DID_{i,t}$				0.0291***
				(5.751)
$TOBINQ_{i,t-1}$	0.00545***	0.00223	0.00805***	0.00565***
	(3.826)	(0.962)	(4.477)	(3.987)
$\Delta NWC_{i,t}/ASSET_{i,t-1}$	−0.0624***	−0.0677***	−0.0548***	−0.0622***
	(−7.275)	(−4.579)	(−5.252)	(−7.307)
$SIZE_{i,t-1}$	0.0103***	0.00954***	0.0109***	0.0104***
	(4.647)	(2.972)	(3.564)	(4.749)
$LEV_{i,t-1}$	−0.0216**	−0.0224*	−0.0282**	−0.0222**
	(−2.486)	(−1.712)	(−2.417)	(−2.574)
$AGE_{i,t-1}$	−0.000594*	−0.000454	−0.000644	−0.000580*
	(−1.717)	(−0.824)	(−1.455)	(−1.686)
City effects	Yes	Yes	Yes	Yes
Year effects	Yes	Yes	Yes	Yes
Constant	−0.113**	−0.117*	−0.126*	−0.113**

(continued)

Table 2 (continued)

	$INVEST_{i,t}/ASSET_{i,t-1}$	$INVEST_{i,t}/ASSET_{i,t-1}$	$INVEST_{i,t}/ASSET_{i,t-1}$	$INVEST_{i,t}/ASSET_{i,t-1}$
	Full sample	Treatment group	control group	Full sample
	(1)	(2)	(3)	(4)
	(−2.429)	(−1.775)	(−1.960)	(−2.448)
N	2,634	963	1671	2634
R^2	0.335	0.222	0.411	0.343
Adjusted R^2	0.292	0.190	0.360	0.301
F	7.93	6.93	8.05	8.14

***, **, *Indicate that the coefficients are significant at the level of 1%, 5%, and 10%, respectively. Same below

Table 3 Urban rail transit, industry relevance and financing constraints of SMEs

	$INVEST_{i,t}/ASSET_{i,t-1}$	$INVEST_{i,t}/ASSET_{i,t-1}$
	Directly related industries	Indirect related industries
	(1)	(2)
$CF_{i,t}/ASSET_{i,t-1}$	0.106***	0.0918**
	(4.876)	(2.478)
$CF_{i,t}/ASSET_{i,t-1}*DID_{i,t}$	−0.111***	0.0245
	(−2.609)	(0.291)
$DID_{i,t}$	0.0258***	0.0353***
	(4.370)	(3.582)
$TOBINQ_{i,t-1}$	0.00652***	0.00616**
	(3.746)	(2.342)
$\Delta NWC_{i,t}/ASSET_{i,t-1}$	−0.0508***	−0.0798***
	(−5.161)	(−4.839)
$SIZE_{i,t-1}$	0.00705***	0.0145***
	(2.736)	(3.135)
$LEV_{i,t-1}$	−0.0257**	−0.0200
	(−2.527)	(−1.100)
$AGE_{i,t-1}$	−0.00136***	−0.000523
	(−3.510)	(−0.473)
City effects	Yes	Yes
Year effects	Yes	Yes
Constant	−0.0421	−0.195**
	(−0.771)	(−2.038)
N	1870	764
R-squared	0.393	0.363
Adjusted R^2	0.345	0.287
F	8.31	4.74

To test hypothesis 3, based on the median size of enterprise size, the whole sample of enterprises is divided into two groups. Among them, the enterprises whose scale is less than or equal to the median level are divided into small-scale enterprises, and the rest are large-scale enterprises.

The results of regression analysis are shown in Table 4. Column (1) shows the regression results of large-scale enterprise samples to model (3), and the coefficient of cross multiplication item is significantly negative (−0.194) at the level of 1%, while in column (2), the coefficient of cross multiplication item is not significant. It shows that the opening of urban rail transit is conducive to alleviate the financing constraints faced by large-scale enterprises, but not significant for small-scale enterprises. This

Table 4 Urban rail transit, enterprise scale and financing constraints of SMEs

	$INVEST_{i,t}/ASSET_{i,t-1}$	$INVEST_{i,t}/ASSET_{i,t-1}$
	Larger scale	Smaller scale
	(1)	(2)
$CF_{i,t}/ASSET_{i,t-1}$	0.118***	0.0680***
	(3.366)	(3.086)
$CF_{i,t}/ASSET_{i,t-1}*DID_{i,t}$	−0.194***	0.0206
	(−2.991)	(0.440)
$DID_{i,t}$	0.0473***	0.0145**
	(5.538)	(2.338)
$TOBINQ_{i,t-1}$	0.0102***	0.00583***
	(3.283)	(3.624)
$\Delta NWC_{i,t}/ASSET_{i,t-1}$	−0.0630***	−0.0603***
	(−3.938)	(−6.093)
$SIZE_{i,t-1}$	0.00609	0.00436
	(1.166)	(1.259)
$LEV_{i,t-1}$	−0.0259	−0.0252**
	(−1.603)	(−2.378)
$AGE_{i,t-1}$	0.00110*	−0.0018***
	(1.702)	(−4.072)
City effects	Yes	Yes
Year effects	Yes	Yes
Constant	−0.0297	0.0155
	(−0.263)	(0.219)
N	874	1760
R-squared	0.361	0.415
Adjusted R^2	0.303	0.357
F	6.20	7.14

result is contrary to the expectation of hypothesis 3. The reasons for the failure of hypothesis 3 are as follows:

Although the opening and operation of urban rail transit will bring the agglomeration of related industries in geographical location and positive external benefits, it is precisely because the increase of the number of enterprises makes the competition between enterprises more intense, which may weaken the industrial cluster effect and the financing advantage caused by the opening and operation of Urban Rail Transit to some extent;

When the amount of capital supply is fixed, the proximity of geographical location makes the enterprises converge in financing behavior, which leads to the shortage of capital market. Compared with large-scale enterprises, small-scale enterprises are prone to credit discrimination, which aggravates the financing constraints of small-scale enterprises;

The sample selected in this paper is listed companies in small and medium-sized board. Too small sample size may affect the test results. In the follow-up study, we can expand the research sample size to test hypothesis 3.

4.2 Robustness Test

Because the capital market of our country is not perfect and the uncertainty of the market leads to heavy speculation, Tobin Q may not accurately reflect the growth opportunities of enterprises, so Tobin Q may lack the authenticity to reflect the investment opportunities of enterprises. Referring to the previous research, this paper uses the growth rate of business income to replace Tobin Q in the model to control the impact of growth opportunities, as an agent variable of investment opportunities to test the robustness of the conclusions.

Table 5 shows the regression results based on the growth rate of operating revenue in model (3). It is consistent with the results in Table 3. Among them, the sensitivity coefficients of investment cash flow in column (1)–(3) are significantly positive, indicating that the sample enterprises have financing constraints. Column (4) shows the regression analysis results based on the cross multiplication term, and the coefficient of the cross multiplication term is significantly negative (−0.0690) at the level of 10%. This shows that hypothesis 1 is robust.

In order to further verify the impact of the opening and operation of urban rail transit on the financing constraints of small and medium-sized enterprises, this paper also takes the cities that opened and operated urban rail transit from 2009 to 2018 as the research object. There are 25 cities in total: Chengdu, Shenyang, Foshan, Xi'an, Suzhou, Kunming, Hangzhou, Harbin, Zhengzhou, Changsha, Ningbo, Wuxi, Nanchang, Lanzhou, Qingdao, Huai'an, Fuzhou, Dongguan, Nanning, Hefei, Shijiazhuang, Guiyang, Xiamen, Zhuhai and Urumqi.

The operation mileage of urban rail transit, $MILEAGE_{i,t}$, (the data are from the China Urban Rail Transit Association) is used as the independent variable to replace the previous DID variable, and the regression analysis is conducted again, and the results are shown in Table 6.

Column (1) shows the regression results of model (2). The sensitivity coefficient of investment cash flow is significantly positive at the level of 10%, which indicates that the sample enterprises have financing constraints. Column (2) shows the regression results of model (3) on operating mileage, and its coefficient of cross multiplication is significantly negative (−0.00103) at the level of 10%. This result also shows that the opening and operation of urban rail transit is conducive to easing the financing constraints faced by sample enterprises. That is to say, with the continuous expansion of the operation scale of urban rail transit, it will drive the agglomeration and development of upstream and downstream industries. Through various ways such

Table 5 Test based on the growth rate of business income

	$INVEST_{i,t}/ASSET_{i,t-1}$	$INVEST_{i,t}/ASSET_{i,t-1}$	$INVEST_{i,t}/ASSET_{i,t-1}$	$INVEST_{i,t}/ASSET_{i,t-1}$
	Full sample	Treatment group	control group	Full sample
	(1)	(2)	(3)	(4)
$CF_{i,t}/ASSET_{i,t-1}$	0.0952***	0.0502**	0.125***	0.112***
	(6.041)	(2.081)	(6.024)	(6.214)
$CF_{i,t}/ASSET_{i,t-1}*DID_{i,t}$				−0.0690*
				(−1.926)
$DID_{i,t}$				0.0243***
				(4.891)
$GROWTH_{i,t-1}$	0.0183***	0.0157**	0.0198***	0.0179***
	(4.544)	(2.368)	(3.926)	(4.436)
$\Delta NWC_{i,t}/ASSET_{i,t-1}$	−0.0417***	−0.0493***	−0.0359***	−0.0421***
	(−5.551)	(−3.946)	(−3.840)	(−5.633)
$SIZE_{i,t-1}$	0.00645***	0.0085***	0.00423	0.00661***
	(3.093)	(2.855)	(1.439)	(3.178)
$LEV_{i,t-1}$	−0.0222***	−0.0205	−0.0291**	−0.0223***
	(−2.589)	(−1.629)	(−2.486)	(−2.605)
$AGE_{i,t-1}$	−0.000504	−6.74e−06	−0.000824*	−0.000479
	(−1.448)	(−0.0126)	(−1.801)	(−1.382)
City effects	Yes	Yes	Yes	Yes
Year effects	Yes	Yes	Yes	Yes
Constant	−0.0314	−0.0930	0.0182	−0.0322

(continued)

Table 5 (continued)

	$INVEST_{i,t}/ASSET_{i,t-1}$	$INVEST_{i,t}/ASSET_{i,t-1}$	$INVEST_{i,t}/ASSET_{i,t-1}$	$INVEST_{i,t}/ASSET_{i,t-1}$
	Full sample	Treatment group	control group	Full sample
	(1)	(2)	(3)	(4)
	(−0.721)	(−1.542)	(0.296)	(−0.741)
N	2834	1052	1782	2834
R^2	0.319	0.220	0.382	0.325
Adjusted R^2	0.278	0.191	0.332	0.284
F	7.97	7.53	7.65	8.08

Table 6 Inspection based on operating mileage of urban rail transits

	$INVEST_{i,t}/ASSET_{i,t-1}$	$INVEST_{i,t}/ASSET_{i,t-1}$
	(1)	(2)
$CF_{i,t}/ASSET_{i,t-1}$	0.0486*	0.0842**
	(1.838)	(2.573)
$CF_{i,t}/ASSET_{i,t-1}$ *$MILEAGE_{i,t}$		−0.00103*
		(−1.835)
$MILEAGE_{i,t}$		0.000141*
		(1.873)
$TOBINQ_{i,t-1}$	0.00258	0.00289
	(1.167)	(1.304)
$\Delta NWC_{i,t}/ASSET_{i,t-1}$	−0.0705***	−0.0692***
	(−4.750)	(−4.672)
$SIZE_{i,t-1}$	0.00945***	0.00940***
	(2.917)	(2.896)
$LEV_{i,t-1}$	−0.0291**	−0.0284**
	(−2.152)	(−2.108)
$AGE_{i,t-1}$	−0.000562	−0.000578
	(−1.006)	(−1.036)
City effects	Yes	Yes
Year effects	Yes	Yes
Constant	−0.106	−0.106
	(−1.597)	(−1.590)
N	986	986
R-squared	0.218	0.222
Adjusted R^2	0.186	0.189
F	6.93	6.74

as mutual learning, information dissemination, mutual guarantee, inter-bank lending and so on, enterprises effectively alleviate the financing constraints they are facing.

5 Conclusion

In this paper, the relationship between the opening and operation of urban rail transit and the financing constraints of small and medium-sized enterprises is empirically studied by multi period double difference method. The results show that the opening and operation of urban rail transit is conducive to easing the investment cash flow sensitivity of sample enterprises, that is, to easing the financing constraints faced by

SMEs. Further research on industry relevance and enterprise scale shows that the opening and operation of urban rail transit plays a more significant role in easing the financing constraints of directly related industries and large-scale enterprises. In addition, based on growth rate of business income and the operation mileage of urban rail transit, the robustness test is carried out, which further confirms the basic research idea of this paper.

Based on the research of this paper, we get the following enlightenment: As urban rail transit is conducive to easing the financing constraints of small and medium-sized enterprises, the government and relevant departments should give corresponding financial support and preferential policies to guide the formation of urban rail transit industry clusters, and give full play to the positive role of urban rail transit in easing the financing constraints of small and medium-sized enterprises and promoting the development of small and medium-sized enterprises; In addition, cities where conditions permit should vigorously promote the development of urban rail transit to form a perfect rail transit network and achieve the strategic goal of urban rail transit leading urban development.

Acknowledgements The National Natural Science Foundation of China subsidized project: "Research on the Performance Evaluation System of PPP Passenger Flow-Value Flow" in Urban Rail Transit (71973009).

References

1. Shen, Q., Chen, P., & Pan, H. (2016). Factors affecting car ownership and mode choice in rail transit-supported suburbs of a large Chinese city. *Transportation Research Part A, 94.*
2. Chatman, D. G., & Noland, R. B. (2011). Do public transport improvements increase agglomeration economies? A review of literature and an agenda for research. *Transport Reviews, 31*(6), 725–742.
3. Dei Ottati, G. (1994). Trust, interlinking transactions and credit in the industrial district. *Cambridge Journal of Economics, 18,* 529–546.
4. Rajan, R. G., Guiso, L., & Sapienza, P. (2004). The role of social capital in financial development. *The American Economic Review, 94,* 526–556.
5. Seidel, T., & Von Ehrlich, M. (2011). Agglomeration and credit constraints. Working Paper.
6. Fazzari, S. M., Hubbard, R. G., Petersen, B. C., et al. (1988). Financing constraints and corporate investment. *Brookings Papers on Economic Activity, 1988*(1), 141–206.
7. Badea, L., Ionescu, V., & Guzun, A. (2019). What is the causal relationship between Stoxx Europe 600 Sectors? But between large firms and small firms? *Economic Computation and Economic Cybernetics Studies and Research, 53*(3), 5–20.

Research on Service Capability Evaluation Index System of Non-track Operation Carrier Based on AHP

Na Dong, Yan Lu, and Jian chao Yan

Abstract Despite the logistics mode of Non-track Operation Carrier (NTOCC) has been proposed for a long time, a scientific framework that evaluates the service capabilities of NTOCC enterprises continues to elude us. Here we analyze the NTOCC's characteristics of China and develop a service capability evaluation index system based on questionnaire survey, field survey and copywriting survey. The evaluation system contains 7 primary indices and 22 secondary indices, which can evaluate NTOCC enterprises from both qualitative and quantitative aspects. Then we use improved AHP method to calculate the weight of each indices and take a NTOCC enterprise as research subject to verify our system's scientific nature. The evaluation result is consistent with the actual situation, which proves our system is scientific and reasonable. This paper considers the keys to measure the service ability of the NTOCC enterprises depends on standardized operation, security management and the in-depth application of advanced information technology.

Keywords Non-track operation carrier · Key factors · Model characteristics · Improved analytic hierarchy process · Evaluation indices

1 Introduction

Non-track Operation Carrier (NTOCC) refers to the efficient integration and allocation of social scattered logistics resources through deep mining and analysis of big data on the Internet platform, which can solve the problems of low transport efficiency and high cost caused by information asymmetry in the traditional freight

N. Dong · J. Yan
China Academy of Transportation Sciences, Beijing, China
e-mail: 122162518@qq.com

J. Yan
e-mail: jianchao_yan@126.com

Y. Lu (✉)
Beijing Jiaotong University, Beijing, China
e-mail: 18120851@bjtu.edu.cn

M. Li et al. (eds.), *IEIS 2020*, https://doi.org/10.1007/978-981-33-4363-4_6

industry. Promoting the standardized and healthy development of NTOCC is one of the important measures to improve the traditional logistics transformation under the "Internet +" strategy. Therefore, it is significant to study the service capability evaluation index system and guide the high-quality development of NTOCC for purifying the order of logistics market and improving the quality of logistics services.

Despite many researches on legal positioning and liability boundary, operation mode, existing problems and other aspects of NTOCC enterprises [1–5], there is still a lack of systematic researches on the key factors affecting their development and the measures to evaluate their service capacities scientifically. As new forms of logistics markets in China, the keys to promote the healthy development of NTOCC, whether the enterprises themselves or related industry management departments, are still not clear. In this paper, we develop a scientific, reasonable and operational service capability index system and study the key factors affecting NTOCC industry, which is significant for guiding its standardized development.

2 Characteristics of NTOCC

2.1 *Improving Logistics Service Quality Through Standardized Management*

NTOCC enterprises make full use of the advantages of their own information network and big data, integrate online and offline resources, and formulate the service rules that carriers should use unified brands, logos and standards. In terms of qualification verification, information release, online transaction, contract signing, loading, unloading, tracking in transit, freight settlement and insurance claim, the haphazard operations are avoided. Through online and offline standardized risk control system, NTOCC enterprises can effectively avoid transport and operation risks, and provide high-quality logistics services for shippers.

2.2 *Realizing Visible Supervision in the Whole Transport Process with Information Technologies*

NTOCC enterprises install video surveillances at loading and unloading sites to prevent the occurrences of mixed loading of dangerous cargos, inflammable and explosive cargos, contraband cargos by adopting network joint control technologies. During the transport process, NTOCC enterprises install video surveillances in the drivers' cab and use 4G active safety technologies to correct drivers' irregular operation behaviors in time. AS to the cargos with high added values, the trackers are installed to ensure the whole process monitoring and the cargos' safety, and provide the customers with the display and inquiry services of the dynamic information.

2.3 *Integrating Logistics Resources Through Accurate Calculation of Big Data*

NTOCC enterprises' intelligent matching systems analyze the valuable round-trip lines, triangle lines, island lines in the historical lines data through the accurate calculation for transport transactions, trajectory data, credit evaluation and other data. The intelligent matching systems give priorities to push the "to-be-returned" vehicles to consignors and push the "to-be-returned" cargo sources to drivers, which increases the vehicles utilization rate, reduces the vehicle idling rate, and cuts the waiting time of the drivers. According to the survey of typical enterprises, the freight reduction rate in long-distance transport is about 2–4%; in short-distance transport, it is 1.5–3.5%; the empty driving rate is effectively reduced by about 2–5%.

2.4 *Establishing Credit Evaluation System for Carriers Through Marketization*

NTOCC enterprises establish the credit evaluation system for carriers based on the vehicles and carriers qualification, the cargo delivery delay time, cargo loss and satisfaction levels, then publish the credit levels of carriers on the network platform and combine the credit levels with freight and the added value of cargos. By assigning high price cargos to carriers who has high credit levels, they are guided to standardize operations and operate with integrities [6–9].

3 Design of Service Capability Evaluation Index System of NTOCC

3.1 *The Principles of Index System Design*

Systematic principles. There are many factors that affect the service capabilities of NTOCC enterprises, including internal factors such as technology investments, talent teams and product innovations, and external factors such as market demands, service objects and policy supports, etc. Therefore, designing evaluation index system can not only consider a certain kind of factors, but consider all relevant factors in a comprehensive way so as to reflect the true situations of the service capabilities of NTOCC enterprises comprehensively.

Hierarchy principles. According to the characteristics of NTOCC operations, we evaluate the service capabilities of NTOCC enterprises from four dimensions: standardized operations, business scale, operating efficiencies and service qualities, each dimension subdivides several sub indicators. Through top-down decomposition and bottom-up synthesis, a two-hierarchies progressive evaluation index system is established.

Representative principles. The construction of the index system should consider the weak links of NTOCC operations and select the key elements that restrict its development to design targeted indicators that can accurately reflect the enterprises' service qualities.

Operational principles. First, the meaning and the concept of indicators should be clear, no ambiguity and conceptual confusion. Second, the index data should be obtained from the financial audit reports and the NTOCC monitoring systems as far as possible [10].

3.2 *Analysis of Key Elements*

Safety management. NTOCC is a kind of asset-lite strategies, that is, NTOCC enterprises can use social vehicles to engage in freight operations without owning any vehicles. The safety management of carriers and vehicles must be strengthened to ensure the legal qualifications and engaging in transport within the permitted business scope.

Service quality. Under the "Internet +" strategies, the new formats such as e-commerce and chain business are booming, constantly improving the consumers' logistics demands. NTOCC enterprises are required to provide on-time, efficient and customized logistics services, it is necessary to refine indicators such as delivery time and capacity demand satisfaction rate to guide NTOCC enterprises to improve logistics service qualities through standardized operation management.

Operation efficiency. The original intention of developing NTOCC is to promote the integration of logistics resources, reduce the vehicles empty driving rate and improve the efficiency of transport organization. Therefore, it is necessary to refine indicators in terms of vehicle utilization efficiency and car-cargo matching efficiency to guide NTOCC enterprises to improve online resources integration and offline operation efficiency through accurate calculation of big data.

Information technology. From the practical experiences, building a fully functional Internet platform which relies on information technologies is the basis for engaging in NTOCC business; from the legal relationship, NTOCC enterprises rely on the Internet platform to sign electronic contracts and allocate logistics resources for freight operations. The advanced technologies of the platform is a key factor in determining whether the carless operation is sustainable.

Business scale. NTOCC enterprises rely on the Internet platform to integrate social fragmented logistics resources, break the boundaries of geographical spaces and service areas and have a strong logistics resource integration capability. The indices reflecting its resource integration capability include the scale of integrated capacity, the number of completed orders, and the volume of freights completed.

Value-added service. In addition to monitoring the whole transport process, ensuring transport safety, improving service quality and logistics operation efficiency, NTOCC enterprises should also develop value-added services such as freight factoring, micro-credit, and online insurance claims in the fields of finance and insurance to provide more convenient services and increase the stickiness between the shippers and carriers.

3.3 Design of Index System

According to the key factors affecting the service capabilities of NTOCC and in combination with the suggestions of the enterprises under investigations, we develop the service capacity evaluation index system for NTOCC (Table 1).

The specific meanings of these indices are shown as follows, the research data are obtained from the National NTOCC Monitoring System.

1. *Business scale (A1)*
   (a) *Scale of integrated transport capability (A11)*: The total number of integrated freight vehicles of NTOCC enterprises (T).
   (b) *Number of customers served (A12)*: The total number of shippers that NTOCC enterprises service (H).
   (c) *Service network coverage (A13)*: The rate of the number of areas radiated by NTOCC enterprises (X') in the total number of areas in China (X):

$$A_{13} = \frac{X'}{X} \times 100\%$$

2. *Standardized operation (A2)*
   (a) *Vehicle qualification compliance rate (A_{21})*: The rate of the number of vehicles that obtain "Transport Certificate" (T') in the total number of vehicles owned by NTOCC enterprises (T):

$$A_{21} = \frac{T'}{T} \times 100\%$$

   (b) *Vehicle location matching rate (A_{22})*: Comparing the car number, beginning and ending time, administrative area code of the trucks and semi-trailer towing vehicles with a total mass of 12 tons and above (hereinafter referred to as heavy trucks) with the data in National Road Freight Monitoring and Servicing Platform, the rate of the number of matching waybills information

Table 1 Service capability evaluation index system

Primary index	Secondary index
Business scale (A_1)	Scale of integrated transport capability (A_{11})
	Number of customers served (A_{12})
	Service network coverage A_{13})
Standardized operation (A_2)	Vehicle qualification compliance rate(A_{21})
	Vehicle location matching rate (A_{22})
	Normal freight rate (A_{23})
Operation efficiency (A_3)	Mileage per month (A_{31})
	Vehicle actual loading rate (A_{32})
	Vehicle-cargo matching time (A_{33})
Service quality (A_4)	Capacity demand satisfaction rate (A_{41})
	Cargo loss rate (A_{42})
	Cargo delayed rate (A_{43})
Information technology (A_5)	Visual monitoring of loading and unloading and driving behavior with remote monitoring technology (A_{51})
	Cargo location-checking with more than two positioning methods (A_{52})
	Preventing fatigue driving, speeding and lane departure with driving assistance technology (A_{53})
	Number of computer software copyrights and patents (A_{54})
Safety management (A_6)	Safety inspection system of cargo delivery (A_{61})
	Safety training and education (A_{62})
	Information security management system (A_{63})
	Emergency plan (A_{64})
Value-added services (A_7)	Logistics financial services (A_{71})
	Insurance services (A_{72})

(D_1) in the total number of waybills completed by heavy trucks in NTOCC enterprises (D_h):

$$A_{22} = \frac{D_1}{D_h} \times 100\%$$

(c) *Normal freight rate (A_{23})*: Comparing the relevant information such as the consignment number with the waybills, the rate of the information matching bill number (D_2) in the total number of waybills (D):

$$A_{23} = \frac{D_2}{D} \times 100\%$$

3. *Operation efficiency (A_3)*
   (a) *Mileages per month (A_{31})*: The average mileages per vehicle in the effective integrated capacity every month, miles.
   (b) Vehicle actual loading rate *(A_{32})*: The rate of freight ton-kilometers completed by integrated vehicles in NTOCC enterprises (Z') in the total freight ton-kilometers (Z):

$$A_{32} = \frac{Z'}{Z} \times 100\%$$

   (c) *Vehicle-cargo matching time (A_{33})*: The time interval between two adjacent cargo transport tasks for the same vehicles.
4. *Service quality (A_4)*
   (a) *Capacity demand satisfaction rate (A_{41})*: The rate of capacity demand waybills published on NTOCC platform (D_d) in the actual completed waybills (D_c):

$$A_{41} = \frac{D_d}{D_c} \times 100\%$$

   (b) *Cargo loss rate (A_{42})*: The rate of waybills with cargo losing (D_l) in the total number of waybills (D):

$$A_{42} = \frac{D_l}{D} \times 100\%$$

   (c) *Cargo delayed rate (A_{43})*: The rate of waybills with cargo delaying (D_e) in the total number of waybills (D):

$$A_{43} = \frac{D_l}{D} \times 100\%$$

5. *Information technology (A_5)*
   (a) *Visual monitoring of loading and unloading and driving behavior with remote monitoring technology (A_{51})*: Qualitative.
   (b) *Cargo location-checking with more than two positioning methods (A_{52})*: Qualitative.
   (c) *Preventing fatigue driving, speeding and lane departure with driving assistance technology (A_{53})*: Qualitative.
   (d) *Number of computer software copyrights and patents (A_{54})*
6. *Safety management (A_6)*
   (a) *Safety inspection system of cargo delivery (A_{61})*: Each shipment must have a delivery security inspection record and the responsible person's signature.

(b) *Safety training and education (A_{62})*: Safety training and education for drivers must be organized regularly.
(c) *Information security management system (A_{63})*: Information security management system texts and operating procedures must be formed.
(d) *Emergency plan (A_{64})*: In response to possible traffic and safety accidents and emergencies caused by force majeure events, emergency plans should be formulated to minimize declaratens and property damage.

7. *Value-added services (A_7)*

(a) *Logistics financial services (A_{71})*: NTOCC enterprises provide low-cost and convenient microfinance services to shippers and actual carriers with financial institutions.
(b) *Insurance services (A_{72})*: In accordance with the characteristics of cargo categories and transport organization, NTOCC enterprises should innovate customized insurance products with insurance companies to reduce the transport at risks and improve their anti-risk abilities.

4 Service Capability Evaluation of NTOCC

4.1 Index Weight Models Based on Improved AHP

Analytic Hierarchy Process (AHP) is practical to deal with the quantitative and qualitative problems. However, the large proportion of qualitative components increases the subjective uncertainties, and it is difficult to achieve absolute consistencies when making pairwise comparisons, which makes the judgment matrix inaccurate, sometimes needs to be adjusted and retested until passing the consistency check. At the same time, the overly detailed scales of AHP (1-9) increases the difficulties of analysis when the relative importance of the two factors is not obvious. In this paper, we develop an improved AHP, which reduces the scales' number and takes into account human ambiguity and non-linearity. It can pass consistency check without reducing the process of adjusting the judgment matrix and has good transmission and rationality. The steps of the improved AHP are described below.

1. *Building the original comparison matrix*
   Supposing that there are n factors in a criterion layer, the original comparison matrix A can be defined as

$$A = (A_{ij})_{n \times n}, (1 \leq i, j \leq n)$$

$$A_{ij} = \begin{cases} 1, & i \text{ is more imortant than } j \\ -1, & j \text{ is more imortant than } i \\ 0, & i \text{ is as imortant as } j \end{cases}$$

2. *Calculating the transfer matrix corresponding to the original comparison matrix*

$$U = (U_{ij})_{n\times n}$$

$$U_{ij} = \frac{1}{n}\sum_{k=1}^{n}(A_{ik} + A_{kj})$$

3. *Calculating the judgement matrix*

$$R = (R_{ij})_{n\times n}$$

$$R_{ij} = e^{U_{ij}}$$

4. *Hierarchy sorting*

Calculating and normalizing the weight of the judgment matrix R by Asymptotic Normalization Coefficient (ANC), then the sorting weight of each factors can be obtained.

4.2 Calculation of Indicator Weights

According to the index system of NTOCC enterprises, Delphi method is used to compare elements at the same hierarchy to calculate the original comparison matrix. After scoring by 5 experts, we obtain the original comparison matrix.

The comparison matrix of primary indexes is

$$A = \begin{bmatrix} 0 & -1 & -1 & -1 & -1 & -1 & 0 \\ 1 & 0 & 1 & 0 & 0 & -1 & 1 \\ 1 & -1 & 0 & -1 & -1 & -1 & 1 \\ 1 & 0 & 1 & 0 & 0 & -1 & 1 \\ 1 & 0 & 1 & 0 & 0 & -1 & 1 \\ 1 & 1 & 1 & 1 & 1 & 0 & 1 \\ 0 & -1 & -1 & -1 & -1 & -1 & 0 \end{bmatrix},$$

the comparison matrixes of secondary indexes are

$$A_1 = \begin{bmatrix} 0 & -1 & -1 \\ 1 & 0 & 1 \\ 1 & -1 & 0 \end{bmatrix} \quad A_2 = \begin{bmatrix} 0 & -1 & -1 \\ 1 & 0 & -1 \\ 1 & 1 & 0 \end{bmatrix}$$

$$A_1 = \begin{bmatrix} 0 & -1 & -1 \\ 1 & 0 & 0 \\ 1 & 0 & 0 \end{bmatrix} \quad A_2 = \begin{bmatrix} 0 & -1 & -1 \\ 1 & 0 & 1 \\ 1 & -1 & 0 \end{bmatrix}$$

$$A_5 = \begin{bmatrix} 0 & -1 & 0 & 1 \\ 1 & 0 & 1 & 1 \\ 0 & -1 & 0 & 1 \\ -1 & -1 & -1 & 0 \end{bmatrix} \quad A_6 = \begin{bmatrix} 0 & -1 & -1 & 1 \\ -1 & 0 & 11 & 1 \\ 1 & 1 & 0 & 1 \\ -1 & -1 & -1 & 0 \end{bmatrix}$$

$$A_7 = \begin{bmatrix} 0 & 1 \\ -1 & 0 \end{bmatrix}$$

According to the original comparison matrix, the transfer matrix and the judgment matrix can be calculated [11–15]. After normalizing the weights of the judgment matrix, the sorting weights of all factors can be obtained (Table 2). It can be seen the first three factors that have a greater impact on the service capabilities of NTOCC enterprises are information security management systems, normal freight rate and cargo loss rate.

Table 2 Index weight sorting

Primary index weight		Secondary index weight		Weight	Sorting
A_1	0.061	A_{11}	0.148	0.009	22
		A_{12}	0.563	0.034	12
		A_{13}	0.289	0.018	19
A_2	0.165	A_{21}	0.148	0.024	17
		A_{22}	0.289	0.048	7
		A_{23}	0.563	0.093	2
A_3	0.093	A_{31}	0.269	0.025	16
		A_{32}	0.366	0.034	13
		A_{33}	0.366	0.034	13
A_4	0.165	A_{41}	0.148	0.024	17
		A_{42}	0.563	0.093	2
		A_{43}	0.289	0.048	7
A_5	0.165	A_{51}	0.218	0.036	10
		A_{52}	0.461	0.076	5
		A_{53}	0.218	0.036	10
		A_{54}	0.103	0.017	20
A_6	0.292	A_{61}	0.277	0.081	4
		A_{62}	0.164	0.048	6
		A_{63}	0.457	0.133	1
		A_{64}	0.102	0.030	15
A_7	0.061	A_{71}	0.731	0.044	9
		A_{72}	0.269	0.016	21

5 Empirical Analysis

5.1 Empirical Analysis

We select a NTOCC enterprises in Henan Province to test our index system. In terms of **business scale**, it has an effectively integrated transport capacity of more than 25,000 vehicles, which can complete an average of 52,000 waybills annually, a freight turnover of 220 million yuan and serve more than 495 shippers which covers 18 provinces and cities, including Beijing, Hebei, Liaoning and Hubei, and 1300 districts and counties.

In terms of **standardized operation**, the enterprise interfaces with public security department data and realizes driver face biometric recognition, ensuring the online identification is consistent with drivers providing services offline. According to the survey data, the enterprise's vehicle qualification compliance rate is 100%, vehicle location matching rate is 96%, normal freight rate is 100%.

In terms of **operation efficiency**, its vehicle-cargo matching time is 12–15 h, mileage per month is 15,000 km, vehicle actual loading rate is 75%..

In terms of **service quality**, its capacity demand satisfaction rate is 99.3%, cargo loss rate and cargo delayed rate are both controlled within 5%.

In terms of **information technology**, this enterprise uses 4G active security technology to monitor the whole transport process and reduce the transport risks. Through satellite dual positioning technology and extracting carriers' positioning information with time and geographic stamp by embedding SDK plug-ins in APPs, it achieves authenticity verification.

In terms of **safety management**, this enterprise installs video monitoring in loading and unloading site to prevent the occurrence of mixed loading of dangerous goods, flammable and explosive materials. In addition, it installs trackers for valuable cargo sources to ensure the whole process monitoring and provides shippers with the query services of dynamic information.

5.2 Service Capability Evaluation

Before service capability evaluation, we should first determine the evaluation set. Supposing that there are m possible comments, the evaluation set can be represented by $V = (v_1, v_2, \ldots, v_m)$. Our research adopts $V = (A, B, C, D)$ to represent the evaluation results of all levels of indices from good to bad, the score set is {8–10, 6–8, 4–6, 2–4}. Five experts score each evaluation index and take the three secondary indexes in business scale as an example, the evaluation result is shown in Table 3.

The final score can be seen in Table 4. According to the average score of each secondary index and its corresponding weight, the score of the primary index and the evaluation result can be obtained as shown in columns 1 and 3. According to the weight of primary indices $W = (0.061, 0.165, 0.093, 0.165, 0.165, 0.292, 0.061)$ and

Table 3 Evaluation results of secondary indexes in business scale

Primary index	Secondary index	Scores					Average score
A_1	A_{11}	8	7	9	8	7	7.8
	A_{12}	8	8	9	8	10	8.6
	A_{13}	7	9	9	6	7	7.6

Table 4 Final score

Evaluation result	Primary index and scores		Secondary index and weight		Secondary index scores
A	A_1	8.2	A_{11}	0.148	7.8
			A_{12}	0.563	8.6
			A_{13}	0.289	7.6
A	A_2	8.6	A_{21}	0.148	8.6
			A_{22}	0.289	9.4
			A_{23}	0.563	8.2
B	A_3	7.7	A_{31}	0.269	8.0
			A_{32}	0.366	7.4
			A_{33}	0.366	7.8
A	A_4	8.6	A_{41}	0.148	9.2
			A_{42}	0.563	8.8
			A_{43}	0.289	7.9
B	A_5	7.5	A_{51}	0.218	8.2
			A_{52}	0.461	7.4
			A_{53}	0.218	7.6
			A_{54}	0.103	6.4
A	A_6	8.5	A_{61}	0.277	8.2
			A_{62}	0.164	8.6
			A_{63}	0.457	8.8
			A_{64}	0.102	7.6
C	A_7	6.7	A_{71}	0.731	6.8
			A_{72}	0.269	6.4

the score vector $T = (8.2, 8.6, 7.7, 8.6, 7.5, 8.5, 6.7)$, the service capability score of this enterprise is 8.2.

Through the above analysis, it can be seen that the overall evaluation of the service capability is *A*, but there is still much room for improvement in three aspects of operation efficiency, information technology and value-added services. Especially in value-added services, we should innovate insurance products and financial services and utilize the advantages of massive data precipitation, further deepen cooperation with financial and insurance institutions. This helps to provide convenient value-added services enterprises in the supply chain and increase the stickiness of the platform with shippers and carriers, further, to lay the foundation for expanding services nationwide.

6 Conclusion

Through questionnaire surveys and analysis of model results, we conclude that key factors affecting the development of the NTOCC are mainly safety management, standardized operation and advanced technology, the three are interrelated and mutually reinforcing. Security management and standardized operations require the government to clarify corresponding requirements and standards, while the accelerated application of advanced technologies requires the active response of market entities. Only by the joint efforts of the government and the market can we promote healthy development of NTOCC.

The industry management department shall formulate accurate policies in terms of standards, supervisions, policy supports, etc. First, study the standards for the operations of NTOCC and make clear requirements in terms of safety management, information technology, and service quality; second, aim at the potential safety hazards to continuously upgrade the NTOCC monitoring systems and improve the information-based supervision abilities; third, select enterprises with high efficiency of resource utilization, good service quality and great innovation ability as the leading backbone enterprises to strengthen the promotion of typical models and mature experiences, and guide the high-quality development of NTOCC.

NTOCC enterprises should take safety, technology, and services as the core to continuously improve their market competitiveness. First, make full use of advanced information technology to strengthen the safety management and realize the whole process of visual tracking management; second, upgrade technology, promote the widespread application of blockchain and 5G technology and use big data analysis to optimize resource allocation schemes and improve resource utilization efficiency continuously; third, extend the service chains and fields, use the advantages of resource concentration in the aspects of centralized procurement of means of production, taxation, finance, insurance to reduce the operating costs and improve the market core competitiveness of NTOCC enterprises.

References

1. Dong, N., Jiang, C., Li, Y., & Feng, S. (2017). *Technical guidelines for pilot work of NTOCC* (pp. 1–10). China Communications Press.
2. Li, J. (2017). *Research on the development of NTOCC* (pp. 55–64). China: Nanjing University Press.

3. Xue, L., & Zhang, R. (2018). Research on legal characterization and liability guarantee of NTOCC. *Journal of Heilongjiang Administrative Cadre College of Politics And Law, 04,* 79–82.
4. Huang, S., & Li, H. (2016). Necessity and feasibility of NTOCC development in China. *Journal of Transport Management Institute Ministry of Transport, 26*(1), 17–20.
5. Zhang, B. (2017). Research on the problems and countermeasures of NTOCC development in China. *Journal of Guangdong Communication Polytechnic, 3*(1), 60–63.
6. Zhang, X. (2019). Analysis on the problems related to NTOCC. *Finance & Accounting for Communications, 08,* 66–68.
7. You, R. (2019). Development status and countermeasures of NTOCC logistics model. *Journal of Tianjin Sino-German University of Applied Sciences, 06,* 92–95.
8. Weiyu, H. (2019). Research on the current situation and development trend of the operation mode of NTOCC based on the 'Internet +'. *Henan Science and Technology, 17,* 103–105.
9. Xia, L. (2019). Important problems to be solved in the development process of NTOCC. *Management & Technology of SME, 11,* 82–83.
10. Wang, X. (2012). *Research on service quality and evaluation system of TPL.* Kunming University of Science and Technology.
11. Fang, H. (2017). *Research on the influencing factors of NTOCC platform's service quality.* Nanjing University.
12. Wang, Z. (2018). *Study on the classification and promotion strategy of truck broker based on the factors of the competitiveness of the company.* Jiangsu University of Science and Technology.
13. Wang, X. (2019). *Research on dynamic scheduling optimization of platform Non-Truck Operating Common Carrier Transport Carrier Capacity.* Chang'an University.
14. Shi, Y. (2011). *A study on components and evaluation of logistics capability in innovation product supply chain.* North China University of Technology.
15. Gu, P. (2018). *Research on the evaluation and cultivation of Jiangsu logistics enterprises' supply chain capability.* Nanjing University of Finance and Economics.

Research on Risk Identification and Evaluation of PPP in Traffic Infrastructure Construction—Take X City Rail Transit as an Example

Yuting Feng and Xuemeng Guo

Abstract Since the introduction of the PPP model in China at the end of the twentieth century, the PPP model has been widely used in China. As of the end of January 2019, 8735 new PPP projects with a total investment of 13.2 trillion yuan were included in the management database of the Ministry of Finance. In order to ensure the development and promotion of PPP projects, how to effectively manage the risks in PPP projects Identification, assessment and control have become the key to the development of the PPP model. This paper first reviews the relevant literature of the infrastructure PPP model through the method of literature research, introduces the classification and basic characteristics of the PPP model in China, and sorts out the existing literature from the aspects of risk identification, assessment and control. Secondly, this paper takes the X-city rail transit PPP project as the research object, identifies the risk factors based on the characteristics of the project's PPP mode of "separation of network and transportation", establishes a risk evaluation index system, and uses a fuzzy evaluation method based on AHP to evaluate the project. For risk evaluation, the SPC project company is used as the risk subject, and according to the different characteristics of different risks, its specific undertaking subjects are further subdivided. The risk sharing system for the PPP project of transportation infrastructure is designed, and suggestions are given.

Keywords PPP model · Transportation infrastructure · Urban rail transit · Risk identification · Risk assessment

1 Introduction

At the National People's Congress meeting held in Beijing on March 5, 2019, Premier Li Keqiang made a government work report, which mentioned that the capital ratio

Y. Feng (✉) · X. Guo
School of Economics and Management, Beijing Jiaotong University, Beijing, China
e-mail: 1036676465@qq.com

X. Guo
e-mail: 19113062@bjtu.edu.cn

M. Li et al. (eds.), *IEIS 2020*, https://doi.org/10.1007/978-981-33-4363-4_7

of projects such as infrastructure should be appropriately reduced, and innovative methods of project financing should be implemented; private investment support policies should be implemented, Orderly advance government and social capital cooperation (PPP). The reference to the PPP model under the direction of expanded investment means that the PPP model has once again received the attention and support of the government. After clearing up and rectifying last year, about 2557 non-compliant PPP projects were cleared from the PPP project database of the Ministry of Finance. The reason is that the PPP project has a long investment and construction period and involves multiple stakeholders, resulting in various risks in PPP projects. Therefore, the main research question of the text is: how to identify, evaluate and control the risks in the PPP mode of transportation infrastructure? What factors should be considered in risk control? What kind of risk control system can enable each participant in the PPP mode of transportation infrastructure to reasonably control the risks they faced?

The research idea adopted in this paper is to first review the literature on the infrastructure PPP financing model, introduce the classification and basic characteristics of China's PPP model, and summarize the literature from the aspects of risk identification, assessment and control. Subsequently, this article takes the X-city rail transit project as an example, identifies risk factors based on the characteristics of the project, and establishes a risk evaluation index system. A fuzzy evaluation method based on analytic hierarchy process was used to evaluate the project risk. Based on the analysis of various risk-bearing subjects, a risk sharing system for the PPP project of transportation infrastructure was designed based on the characteristics of different risks and specific bearing subjects. Against the recommendations.

The innovations studied in this article are reflected in the following aspects: First, the layered division and accurate identification of risk factors. Take the X-city rail transit project as an example to establish a comprehensive and complete risk index for the risks existing in the transportation infrastructure PPP project. The system. Second, based on the risk assessment results of X urban rail transit PPP projects, risk control strategies were given, and the risk control strategies were extended to general rail transit PPP projects, which enriched the risk control theory of urban rail transit PPP projects. The third is different from previous studies that only classified risks in a general way. In this paper, the SPC project company is taken as the main risk bearer, and according to the different characteristics of government and social sectors in the PPP project, risk sharing and control schemes are given.

2 Literature Review

2.1 Theoretical Research on PPP Financing Model

Foreign scholars took the lead in starting basic research around the PPP model around 1980, including the concept, basic theory, and characteristics of the PPP model. Grimsey and Lewis believe that the PPP model is a multi-party joint operation strategy adopted by government departments in order to achieve the purpose of improving economic benefits and involve the social private sector in social infrastructure construction projects [1]. The PPP model is to achieve the purpose of efficient construction and operation of public goods. A collaborative relationship established by multiple parties including operators, contractors, and government parties. The government party provides legal, policy, and economic Assurance, operators and contractors are responsible for the construction and subsequent operations of public goods. With the vigorous development of the PPP model in practice, the academic community has analyzed the factors affecting the success of PPP projects from a multidimensional perspective. Macroeconomic indicators such as GDP growth, capital market indicators and demographic factors, as well as macroeconomic indicators such as regional economic development level, system quality, marketization process, market demand and financialization index are fundamental factors that affect PPP projects [2–4].

2.2 *Research on the Risk of Urban Rail Transit PPP Mode*

2.2.1 Identification of Risks

First, risks can be classified into different levels. The most classic risk stratification study is that Li B divides project risks into macro-level risks, meso-level risks, and micro-level risks, and divides risks into project risks and external factors according to different sources of risks. Risks mainly include risks that may exist from project design to start-up construction and later contracted operations, which are at the meso level at the level of risks; while economic and political risks contained in the macro environment are external risks; due to multi-party responsibility and coordination. The risks caused by relationships are at the micro level [5].

Secondly, according to the theory of the whole life cycle, the risk can be divided according to the order of time, that is, the risk of the project. Foreign scholars started research in this area earlier. Based on the research on PPP projects constructed in Taiwan, Thomas divided the risks into the risks in the project life cycle, as well as the initial formation phase of the project and the construction phase of the project. And the risks of the operation phase when the project is put into use after completion [6]. The domestic research is relatively late, and it is based on the ideas of previous studies, so it has a similar risk score to study. Aiming at the example of the PPP model adopted in China's highway projects, the risks involved in the bidding process, decision-making process, and the same design, construction, and operation processes as those of previous studies were studied.

Thirdly, a risk classification method that is widely adopted in the existing research is that according to its nature, risks can be divided into political risks, financial risks, and technical risks. Afterwards, domestic scholar Fan Xiaojun took the domestic high-speed rail infrastructure PPP project as an example, and further refined the risk according to the nature and nature of the risks. The risks were divided into management risk, legal risk, operational risk, market risk, credit risk, and economic Risk, technical risk, natural risk, construction risk, etc. [7].

In addition, there are many different risk classification methods, such as dividing risks into systemic and non-systematic risks; external risks coordinating risks with internal organizations; and dividing risks into categories based on whether there is a strong correlation between risks and the PPP project itself. General risk and project risk; according to the breadth of risk, the risk is divided into basic risk and basic risk [8].

2.2.2 Evaluation of Risks

After identifying the risks of a PPP project, it is necessary to evaluate the risks that have been identified. In the current research, there are two main ideas, one is a qualitative evaluation of the risk, and the other is a quantitative evaluation of the risk. The earliest scholars, such as Fayek, used questionnaires to assess the risks in large-scale engineering projects. Questionnaires are a typical qualitative method for assessing risks. First, design questionnaires based on project characteristics. Experts who are related to the project or have professional knowledge and experience are selected as the survey objects, and the risk assessment is conducted by issuing questionnaires and collecting the results data. Professional literacy also affects the results of research [9].

In order to assess the risks more accurately, quantitative research methods are often used in the research, such as: Deng Xiaopeng can use factor analysis to refine macro risks, and use measurement software such as SPSS to further break down macro risk factors into contracts. Factors, economic factors, environmental protection factors, etc., and put forward corresponding risk control and response suggestions for different risk factors [10]; when Songer studied the risks of toll highway PPP projects, he identified eighteen types of Risk, using Monte Carlo simulation technology, research shows that cooperation risk is the risk that has the greatest impact on PPP projects [11].

2.2.3 Control of Risk

The purpose of risk control is to make the risk of the project be distributed among multiple participants in the project in a reasonable way, that is, the sharing of risk, in order to achieve the reasonable and effective risk management effect with the lowest management cost. In the PPP model, the principle of "risk sharing and benefit sharing" is promoted. This makes it necessary to pay attention to the establishment of a practical risk sharing mechanism, to control the cost with the lowest risk, reduce the possibility of risk occurrence, and reduce the project cost caused by risk The increase is minimized and all parties in the project have strong risk control capabilities [12].

Risk sharing should follow some basic risk sharing principles: First, the risk control ability possessed by multiple participants in the project is the basic factor that determines risk sharing, that is, the party with strong risk control should bear more responsibility accordingly. Peng Hua and Xiang Junyu summarized other principles involved in risk control as: the risk appetite principle, the fairness principle, the lowest risk cost principle, the risk-return equivalence principle, and the direct loss-taking principle, etc. [13, 14].

The research on risk control takes risk sharing as the main research idea. In addition to allocating risks among multiple participants, methods such as risk transfer, risk reduction and risk avoidance can also be adopted. The subject of risk bearing is not limited to the government department and the private sector, but also includes the departments that undertake the project contract design, construction contractors, relevant institutions responsible for the operation of the project after completion, and the departments that provide funds for PPP projects to share the risk.

2.3 Literature Review

Because the PPP project was applied earlier in the United Kingdom and the United States, and its development was relatively late in China, the focus of research carried out by domestic and foreign scholars according to the different development stages of domestic and foreign PPP projects has also been different. The research on basic theory has been relatively mature abroad, and the PPP model, characteristics, and methods have been discussed extensively. Now the focus of research is more on the further innovation and development of PPP projects and project performance and incentive mechanisms. China's PPP model is still in the stage of exploration and development, and foreign theories and research on PPP model may not adapt to the situation of China's PPP project. Therefore, in the study of PPP model, China's national conditions should be linked more realistically.

Existing literature has discussed the risk control of the PPP model to a certain extent, but according to the existing literature basis, the established risk control system is not complete enough, but only a general study of the risks common to the project is lacking. Complete risk indicator system and risk control mechanism, and exact analysis of specific projects.

3 Risk System Construction

3.1 Risk Assessment Subjects and Objectives

3.1.1 Research Subjects for Risk Assessment

This article takes the railway company jointly established by the X city government and social capital as an example to study the risk control in PPP projects in the field of transportation infrastructure. Therefore, when identifying and assessing risks, it is considered from the field of transportation infrastructure. The recommendations for control, sharing and sharing also correspond to the actual situation in the field of transportation.

3.1.2 Objectives of Risk Assessment

The construction of transportation infrastructure is a fundamental issue related to the people's lives in China. To ensure the smooth progress of the project, the core issue is to do a good job of controlling the risks. Therefore, the risk assessment must be objective and preliminary in the early or early stages of project construction. Identification, based on the preliminary identification of risks and the establishment of a reasonable and effective risk indicator system based on the actual situation of the project, selecting appropriate assessments of risks from existing risk assessment methods, conducting empirical research on cases, and finally putting forward according to the research conclusions Suggestions on risk control of X-city rail transit PPP projects provide a reference for China's transportation infrastructure PPP projects.

3.2 Choice of Risk Assessment Method

Based on the previous summary of risk identification and risk assessment methods, and analysis of the advantages and disadvantages of various methods, this article selects a combination of expert survey and qualitative and quantitative mathematical modeling when researching PPP projects in transportation infrastructure, and comprehensively uses expert surveys. Method, analytic hierarchy process and fuzzy comprehensive evaluation, it evaluates the risks in PPP projects of transportation infrastructure, and proposes risk prevention measures.

3.3 Risk System Design

3.3.1 Frequency Analysis of Risk Assessment Indicators

By reading 112 articles on the risks of PPP projects in the transportation field, the 54 documents that are most closely related to PPP projects in urban rail transit are summarized. Based on the Lib's risk hierarchy three classification method, the risk of force majeure has been increased and formed The four risk levels of the national level, market level, project level and force majeure level are analyzed, and the statistical analysis of the risk indicators appearing is as in Table 1.

3.3.2 Analysis of the Specific Risks of "Network Separation" Mode in X City

By analyzing the actual operation of the X city rail transit project and interviewing the experts of X city subway project companies and participating university professors, X city applied the PPP model of "separation of network traffic" to the PPP project of rail transit The "Separation of Network and Transportation" model is a PPP model that separates project construction from later operations. At different stages of the project, the city government of X has contributed to public tenders and introduced social capital with professional construction capabilities and PPP project operation experience. Fang, and then set up SPC project companies respectively, responsible for the construction period of the X-rail transit PPP project and the unified bundled operation period of the project. The "BLMT model" was applied to the design and construction phase of the project. In the later stage, the city's rail transit project was centralized and adopted the "IOT" model. Compared with the traditional PPP model, the application of the "separation of network and transportation" model in the city's urban rail transit projects has indeed brought core participation of social capital and shared risks and benefits; the introduction of a triple evaluation system for the project Performance evaluation; professional division of labor is more detailed, which is conducive to market operations and other advantages. However, due to the characteristics of its "separation of network traffic" model, it also has its own unique project risks.

First, the core participation of social capital in the PPP model of "separation of network and transportation" in X city. In the projects of line 1, 2, and 3, the shareholding ratio of the social capital side accounted for more than 75%, of which the social capital side of line 3 The proportion reached 98%. In the PPP project, the social capital party is responsible for the main daily operations, while the city government of X reserves only the right to decide and one vote of veto on matters that have a significant impact on the project. When the franchise operation of the X-city rail transit PPP project is over and the transfer is made to the government, the government may lack the experience of management and operation of the project, which may result in the risk of project transfer failure.

Table 1 Frequency analysis of risk indicators

Risk indicator	Frequency statistics	Risk indicator	Frequency statistics
Policy risk	18	Expropriation of assets	11
Risk of corruption	18	Political stability	11
Government intervention risk	7	Risk of government decision error	7
Government credit risk	7	Changes in industry laws	15
Interest rate risk	11	Upstream industry risk	5
Exchange rate risk	15	Environmental risks	11
Inflation risk	11	Social environmental risk	18
Changes to the tax system	15	Inadequate legal supervision	7
Influential economic events	5	Insufficient bidding competition	7
Industry orientation change	7	Impact of homogeneous projects	7
Price limit	5	Land acquisition policy	5
market competition	16	Price change	11
Change in demand	16	Financing feasibility	11
Financing costs are too high	15	Funds not available in time	6
Government limits profits	8	Poor financing environment	8
Supply risk	5	Currency exchange risk	5
Insufficient traffic	5	Infrastructure supporting risks	5
Design risk	20	Operation and maintenance risks	11
Engineering safety quality	20	Pricing risk	20
Construction cost	17	Technical risk	13
Land acquisition delay risk	9	Commitment risk	10
Project overdue risks	11	Survey risk	10
Risk of obstacles to cooperation	9	Contractor default	10
Handover risk	11	High cost of land demolition and compensation	9
Public-private sector risk	20	Residual value risk	9

(continued)

Table 1 (continued)

Risk indicator	Frequency statistics	Risk indicator	Frequency statistics
Improper distribution of responsibilities	11	Risk of insufficient project implementation experience	9

Secondly, X city adopts the operation mode of "module combination, regional property layering and back-feeding", that is, based on the "separation of network transportation" model, further development of land resources and resource development mode of superstructure properties. The starting point is The city government of X hopes to make full use of the subway and the resources along the line to create economic benefits for the project, but as there are few similar successful cases and experiences in the country that can be learned and used for reference, it will bring benefits to the project and increase construction technology. Risks and operational maintenance risks during operation.

Third, based on the PPP model, City X introduced the operation mode of "separation of network traffic", module combination, and super structured property, and began to use the current Internet of Things technology in 2017 to build "Internet + rail" intelligence. The travel application software platform, which makes the participants in the PPP project not only include the two parties of government and social capital, but also many participants of superstructure property development institutions, new media operation platforms, and Internet technology companies. There is a higher risk of improper allocation of responsibilities than traditional PPP projects.

3.3.3 Establishment of Risk Indicator System

Based on the above analysis, a risk indicator system for the X-city rail transit PPP project is established, and the project risk is divided into 9 first-level indicators and 27 second-level sub-indicators to comprehensively control the risks. Table 2 is a list of risks of PPP projects.

4 Empirical Analysis

4.1 Data Collection

In the calculation of risk assessment, this article adopts the method of expert scoring, and collects data by combining a face-to-face questionnaire with online questionnaire stars. According to the risk factors of 4 levels, 9 first-level risk indicators, and 27 second-level risk indicators established in the previous comparison, one by one comparison.

Table 2 Risk assessment and analysis system of X city rail transit PPP project

Main target	First-level indicators	Secondary indicators
Risk indicator system	B1 political risk	C1 policy risk
		C2 corruption risk
		C3 nationalization/requisition assets
		C4 political stability
	B2 economic risk	C5 interest rate risk
		C6 exchange rate risk
		C7 inflation risk
	B3 Legal risks	C8 tax system changes
		C9 industry law changes
	B4 Environmental risks	C10 Natural environmental risks
		C11 Social and environmental risks
	B5 market risk	C12 market competition
		C13 demand changes
		C14 price change
	B6 financing risk	C15 financing feasibility
		C16 financing costs are too high
	B7 project promotion risk	C17 Design Risk
		C18 Engineering Safety Quality
		C19 construction costs
		C20 Operation and Maintenance Risk
		C21 pricing risk
		C22 Technology Risk
		C23 handover risk
	B8 relationship risk	C24 public-private sector relationship risk
		C25 Improper distribution of responsibilities
	B9 force majeure risk	C26 Climate/natural disaster risk
		C27 Unforeseen risks

4.2 Risk Weight Calculation

On the basis of comprehensive analysis based on relevant literature, and using expert scoring methods, the evaluation and consistency of the judgment matrix are performed on the criterion layer and the index layer. We finally need to get the total weight of all elements, especially the sorting weight of the bottom-most index to the target, so as to choose a solution. The overall sorting weight is important to combine the weights of the layers from top to bottom (Table 3).

According to the results of the analytic hierarchy process, the risk weight vectors of the nine indicator layers B are [0.2542, 0.1206, 0.1953, 0.0517, 0.0926, 0.068,

Table 3 Total sort weight

Total target	Level 1 index	Level 1 weight	Level 2 index	Level 2 weight	Total WEIGHT
Risk indicator system	B1 political risk	0.2542	C1	0.4168	0.1060
			C2	0.1928	0.0490
			C3	0.2695	0.0685
			C4	0.1209	0.0307
	B3 legal risks	0.1206	C5	0.2605	0.0314
			C6	0.1062	0.0128
			C7	0.6333	0.0764
	B4 environmental risks	0.1953	C8	0.3333	0.0651
			C9	0.6667	0.1302
	B5 market risk	0.0517	C10	0.2000	0.0103
			C11	0.8000	0.0414
	B6 financing risk	0.0926	C12	0.1638	0.0152
			C13	0.5389	0.0499
			C14	0.2973	0.0275
		0.068	C15	0.25	0.0170
			C16	0.75	0.0510
	B7 project promotion risk	0.1469	C17	0.3054	0.0449
			C18	0.0563	0.0083
			C19	0.0774	0.0114
			C20	0.2419	0.0355
			C21	0.0369	0.0054
			C22	0.172	0.0253
			C23	0.1101	0.0162
	B8 relationship risk	0.0404	C24	0.6667	0.0269
			C25	0.3333	0.0135
	B9 force majeure risk	0.0303	C26	0.1667	0.0051
			C27	0.8333	0.0252

0.1469, 0.0404, 0.0303], which indicates that in the risk indicator system established for X cities The order of importance of various types of risks is political risk > legal risk > project promotion risk > economic risk > market risk > financing risk > environmental risk > relationship risk > force majeure risk. It can be seen that the largest proportion of this project is political risk, legal risk, and project promotion risk, because the PPP project on rail transit built in City X is based on the state's strategy of promoting rail transit to lead urban development. The promotion and development of the project have played a key role; the changes in the tax system related laws will directly affect the amount of project income, laws and regulations related to transportation and PPP are critical to the project risk; the risks of project promotion include the implementation of PPP For various risks in the whole process of the project, the "separation of network and transportation" model adopted by City X makes the project adopt a modular and professional operation method, so that the risk of project advancement occupies a higher proportion in the overall indicators. The economic risk, market risk, and financing risk are in the middle of the project risk system, indicating that changes in the economic situation, risks brought by market competition, and financing risks will have an important impact on the overall risk of the project. After that are environmental risks, relationship risks and force majeure risks.

4.3 *Risk Assessment*

In response to this evaluation of the risk indicator system, an indicator evaluation system questionnaire was designed, in which experts, front-line personnel and the general public were invited to evaluate the risk indicator. The questionnaire design was divided into three parts. The first is the X Urban Rail Transit Project., Followed by the main content of the questionnaire survey, asking respondents to evaluate the level of risk. Finally, the basic situation of the interviewees is to investigate which participant in the project the interviewee belongs to, and how many years are they engaged in PPP related projects or research (Table 4).

Based on the results of the fuzzy evaluation calculations, a summary of each risk indicator and overall risk evaluation is as in Table 5.

The overall risk evaluation of the X Urban Rail Transit PPP project is general. Among the 9 indicators, economic risk and project promotion risk are higher, political risk, legal risk, market risk, financing risk and relationship risk are general risks, while environmental risk and Force majeure risk is low. For economic risks with high risk evaluation and project promotion risks, important allocation and prevention measures should be carried out. It can be seen that the risks of inflation and interest rate changes are large in the X city PPP projects, and the projects existing in the process of project implementation The risk of advancement is also a risk factor that should be considered in the risk assessment.

This section analyzes the case of X-city rail transit projects, calculates the weight of the risk indicator system through the AHP method, and concludes that the importance of the risk indicator system in X-city rail transit projects is: political risk > legal risk > Project promotion risk > Economic risk > Market risk > Finance risk > Environmental risk > Relationship risk > Force majeure risk; then use matlab to make a final assessment of the risks in the PPP project and conclude that economic risk and project promotion risk are high, and political risk Legal risks, market risks, financing risks and relationship risks are general risks, while environmental risks and force majeure risks are lower. Based on the results of the risk assessment, opinions on risk control are further given.

Table 4 Index system and assessment form

Over all	Level 1	Level 2	Highest	Higher	Normal	Lower	Lowest
Risk indicator system	B1	C1	0.02	0.26	0.52	0.15	0.05
		C2	0.05	0.22	0.49	0.24	0.00
		C3	0.15	0.19	0.36	0.18	0.12
		C4	0.15	0.31	0.39	0.13	0.02
	B2	C5	0.28	0.44	0.19	0.06	0.03
		C6	0.26	0.46	0.20	0.03	0.05
		C7	0.02	0.35	0.38	0.17	0.08
	B3	C8	0.15	0.31	0.39	0.13	0.02
		C9	0.02	0.18	0.35	0.28	0.17
	B4	C10	0.03	0.06	0.19	0.44	0.28
		C11	0.05	0.03	0.2	0.46	0.26
	B5	C12	0.15	0.31	0.39	0.13	0.02
		C13	0.17	0.29	0.40	0.09	0.05
		C14	0.17	0.29	0.49	0.04	0.01
	B6	C15	0.05	0.19	0.41	0.34	0.01
		C16	0.03	0.04	0.43	0.33	0.17
	B7	C17	0.28	0.44	0.19	0.06	0.03
		C18	0.26	0.46	0.2	0.03	0.05
		C19	0.29	0.41	0.2	0.09	0.01
		C20	0.27	0.44	0.2	0.05	0.04
		C21	0.24	0.36	0.21	0.12	0.07
		C22	0.19	0.41	0.2	0.11	0.09
		C23	0.31	0.43	0.15	0.11	0.00
	B8	C24	0.03	0.18	0.44	0.32	0.03
		C25	0.09	0.04	0.46	0.36	0.05
	B9	C26	0.00	0.03	0.28	0.47	0.22
		C27	0.03	0.05	0.26	0.46	0.20

5 Risk Control of X Urban Rail Transit PPP Project

According to the results of the previous risk assessment of risks, a proposal for centralized risk control is proposed. First, it is necessary to distinguish the risk bearers. Different risks should have the main bearer or be shared by both parties. Second, a comprehensive risk management system should be established, Analyze the risks layer by layer, and propose solutions accordingly; at the end, design key risk control measures for the high-risk areas calculated in the risk assessment.

Table 5 Risk Assessment Results

Risk indicator	Risk assessment level
Political risk	General
Economic risk	Higher
Legal risks	General
Environmental risk	Lower
Market risk	General
Financing risk	General
Project advance risk	Higher
Relationship risk	General
Force majeure risk	Lower
General comment	General

5.1 Division of Risk-Bearing Entities

By consulting the literature and combining the analysis of the characteristics of the X-city rail transit PPP project, the 27 main risks identified in the risk indicator system are distinguished from the main risk bearers.

Risk level	Risk factor	Risk bearer
B1	C1 policy risk	Government department
	C2 corruption expropriation	Government department
	C3 nationalization/asset risk	Government department
	C4 political stability	Government department
B2	C5 interest rate risk	Share
	C6 exchange rate risk	Share
	C7 inflation risk	Share
B3	C8 tax system changes	Share
	C9 industry law changes	Government department
B4	C10 natural environmental risks	Social capital
	C11 social and environmental risks	Social capital
B5	C12 market competition	Share
	C13 demand changes	Share
	C14 price change	Share
B6	C15 financing feasibility	Social capital
	C16 financing costs are too high	Social capital
B7	C17 Design Risk	Social capital
	C18 engineering safety quality	Social capital
	C19 construction costs	Social capital
	C20 operation and maintenance risk	Social capital

(continued)

(continued)

Risk level	Risk factor	Risk bearer
	C21 pricing risk	Social capital
	C22 technology risk	Social capital
	C23 handover risk	Social capital
B8	C24 public-private sector relationship risk	Share
	C25 Improper distribution of responsibilities	Share
B9	C26 Climate/natural disaster risk	Share
	C27 Unforeseen risks	Share

The X-city rail transit PPP project is invested and constructed and operated in accordance with the "Provincial X-city Comprehensive Transport Plan" and "J-X-city Urban Master Plan" formulated by the provincial government. The risk of related legal changes is mainly borne by the X city government. Since X city rail transit adopts the PPP mode of "separation of network and transportation", during the project design and construction period and operation service period, the X city government recruits social capital parties through public bidding, and jointly establishes the SPC project company responsible for the project The operation, and in different periods, the social capital side of the SPC company to carry out specific project design, construction and operation work, so the financing risk and project promotion risk are mainly borne by the private capital side. The risks caused by changes in the national economic environment, the risk of force majeure, and the risks of relations between the two parties in the X-city rail transit PPP project are shared by the public and private parties represented by SPC.

5.2 Key Risk Management Solutions

According to the results of the risk assessment of the X-city rail transit PPP project in the previous study, it was determined that the core risk control areas were key risk areas such as economic risk, project advancement risk, relationship risk, and market risk. Provide specific risk response strategies for these core risk areas and other risks.

5.2.1 Political Risk

The political risks in the X-city rail transit PPP project are mainly affected by the state's policy on PPP "Notice on Promoting the Use of Government and Social Capital Cooperation Models", Notice ", the local policy of the X city government," Comprehensive Transportation Plan for the X City of J Province "and the" General Plan of the X City of J Province "and the possibility of expropriation related to the corruption of the X City government. The overall evaluation of political risk is a general risk, which is greatly affected by policy risks and nationalization risks. Therefore, the following strategies should be adopted when preventing political risks:

First of all, according to the different social capital parties during the construction and operation periods, do a good job of communication and negotiation before signing the contract with the city government of X, so that relevant government units such as the Development and Reform Commission, the Finance Bureau, the Land and Resources Bureau, the Local Taxation Bureau, and Business Administration Departments, etc. make good prior regulations on local finances, operating permits, and non-constructing competitive projects during the construction period of X-city rail transit projects, and make written statements on the specific content.

Secondly, in order to prevent the risk of nationalization from being expropriated by the government before the expiration of the franchise period in the X urban rail transit project, the social capital party during the operation and maintenance period should sign a franchise agreement with the city government of X and specify in the franchise period The city government of X can terminate the PPP project before the end of the period.

Third, the SPC company of the X-rail transit PPP project can share the occurrence of political risks by purchasing international guarantees, and purchase policy risk insurance from international guarantee insurance institutions before the project starts construction. When changes in political risk arise, you can obtain financial compensation from insurance institutions to subsidize losses caused by political risk.

Finally, political risk is a systemic risk that is borne by government departments, and the risk responders are private sector SPC project companies and lenders. In the event of political risk, if the contract allows or agrees to the suspension clause, the SPC company may choose to terminate the contract and claim loss of capital and estimated profits from the government. When the contract is terminated or the claim is unsuccessful, the sponsor of the SPC company may lose capital. When political risk arises, the lender can request a loan from the government or sign a new contract instead of the old one. When the political risk generated violates the terms of the financing contract, the bank can choose to take effect the guarantee/mortgage contract. The principal and interest of the loan issued.

5.2.2 Economic Risks

Economic risk is the risk of interest rate changes, inflation, and exchange rate changes caused by changes in the economic environment when constructing a rail transit PPP project in City X. The objective attributes of economic risk are related to the country's macroeconomic situation and directly affect all stages of a PPP project.

According to the existing policies of the PPP model and the capital structure characteristics of PPP projects, changes in exchange rates often have a smaller impact on PPP projects, but the risks of inflation and interest rate changes have a significant impact on the project. The mechanism for bearing economic risks is shared by government departments and social capital parties according to the life cycle of the project. The main risks undertaken by social capital parties are during the construction and franchise period of the project, and other risks during the operation period It is borne by government departments.

When faced with the risks of inflation and interest rate fluctuations, SPC companies can respond by taking backup loans and raising fees or exploring other incomes. At the same time, when interest rate fluctuations occur, the issuing bank will also test the project company to increase it. Fees increase revenue to reduce the risk of bad loans on loans. According to the demand-price curve, an increase in fees may result in a decrease in market demand, which in turn will lead to a reduction in total revenue. The remaining risk at this time is that if the financing contract is violated, after the agreed correction period, the bank can replace the project company and the effective guarantee/mortgage contract. At this time, the bank may lose the principal and interest of the loan; if it violates the concession contract The government may suspend the contract after the agreed corrections, at which point the project sponsor may lose capital.

5.2.3 Environmental Risks

The level of risk assessment for environmental risks faced by X-city rail transit PPP projects is low, including both natural and social environment risks. According to the different phases of the project, corresponding strategies for dealing with environmental risks are developed. Before the project is designed and constructed, the environmental and cultural conditions of City X must be fully investigated. It faces Lianyungang in the east and Liuzhou in the west. The terrain is mainly plain, and it belongs to a temperate monsoon climate. The terrain is flat and the climate is mild. At the same time, it is necessary to make an investigation report on the living habits and wages of residents along the subway in X city, and provide sufficient information for the response to environmental risks in the project construction in the form of "X City Social Environment Investigation Report". In addition, during the operation phase of the project, the city government of X can promote the residents 'priority in using public transport to promote the social habits of residents in X city, so as to cope with the risks caused by the changes in residents' travel modes.

5.2.4 Market Risk

As the X city applies the PPP model to the field of urban rail transit construction, the construction projects have quasi-public goods and non-profit characteristics. For example, the toll of the subway is determined by the government to hold a hearing. It is not possible to adjust internally as the market changes. Therefore, the market risk prevention in the PPP project of city rail transit in city X is mainly at the stage of preliminary research and project feasibility analysis. Government departments and social capital should coordinate and make a good forecast of market demand and future market changes, and establish reasonable prices And income system. For subway transportation infrastructure, the passenger flow is mainly determined by the route selection route, which is the rigid demand of residents of X city. Therefore, the investigation and route selection during the project planning period are of vital importance. At the same time, market competition is also an important factor that causes market risk. For the method of competitive risk, the city government guarantee method can be adopted. When the social capital party signs an agreement with the government department, it sets up anti-competitive provisions to enable the government department. Make a guarantee that in the future, the local area will not invest in similar competitive projects.

5.2.5 Project Promotion Risks

X city urban rail transit PPP project promotion risk is a risk that is mainly borne by the social capital party during the implementation of the project. According to the theory of full life cycle, combined with the PPP model of "city separation" in X city, the whole process of project promotion can be promoted. Divided into different phases, corresponding risk management programs should be formulated for different phases. First of all, in the project design phase, a comprehensive and systematic study of the feasibility of the project should be conducted, and the actual situation of the project implementation area should be fully investigated. Market demand, project feasibility, technological maturity, and the choice of PPP model and traditional government investment model The evaluation is worthwhile, and it is necessary to fully identify the various risks that may exist in the PPP project and design an operable implementation plan. Then during the construction phase of the project, it is necessary to strictly control the engineering quality and safety issues of the project, and make reasonable expectations for the possibility of rising project construction costs.

5.2.6 Relationship Risk

In the X-city urban rail transit PPP project, the relationship between government departments and various social capital parties, such as project investors, contracted construction companies, project companies, and post-contracted operating companies, is the basis of the PPP model because of the many participants It also became a possibility of risk. In order to prevent the occurrence of interrelationship risks, all-round control before and after the risks should occur. When establishing a multi-party cooperation relationship, it is necessary to sign a cooperation contract, which clearly defines each party's responsibilities, their respective scope of work, and a reasonable allocation of risk control schemes; in the process of project implementation, they must strictly follow the rules established in advance All parties should fulfill their rights and obligations. For deviations from pre-established standards, even if corrective measures are adopted to ensure the achievement of project objectives, breaches of contract can be prohibited by fines and other acts. Strengthen communication between various partners, and timely communicate and solve problems in the implementation of PPP projects, so as to avoid greater risks.

5.2.7 Force Majeure Risk

Force majeure risks cannot usually be circumvented through technology, experience, prejudgment, or care. To cope with force majeure, first of all, it is clear that all parties must share and different parties must agree on the definition of "force majeure". Second, the force majeure risks that can be transferred should be transferred as much as possible, such as purchasing insurance; the non-transferable parts should be risk controlled to reduce risks Such as preparation for unforeseen fees.

6 Research Conclusions and Recommendations

6.1 Risk Response of Urban Rail Transit PPP Project

Risk response can be negotiated to establish contract terms that should include risk management mechanisms, define rights and responsibilities, and pricing and price adjustment mechanisms; reserves should include specific measures such as unforeseen fees, reserve capital, and reserve loans. In the risk response process, both the government and the private sector should abandon misconceptions. For example, government officials have the wrong idea of transferring as much risk as possible to the enterprise, or even not taking any risks. Wrong idea returned. The risks should be based on the premise that the project goals are achieved, so that the project participants can achieve the goal of mutual benefit and win-win.

Risk response strategies commonly used in PPP investment and financing include avoiding risks, transferring risks, and mitigating risks. Avoidance of risks, that is, project participants do not take certain actions to avoid risks, such as not bidding on high-risk projects. By adopting this strategy, you may also lose the opportunity to obtain high returns. Transfer risk, reduce risk by transferring risk to other units. For example, purchase engineering insurance; for uninsurable risks, adopt contract terms to transfer to other participants, such as subcontracting according to the "fixed period fixed price" contract to transfer and reduce the risk of delays and cost overruns; change the organizational structure and Shared by other participants. Adopting this strategy will bring additional financial expenditures, such as paying insurance premiums, or bear new risks, such as in the new organizational structure, where the other party fails to implement effective risk management. Risk mitigation includes two aspects: one is to reduce the possibility of risk and minimize the probability of adverse consequences; the other is to minimize the loss after the risk occurs, such as reducing inventory or purchasing engineering insurance.

6.2 Establish a Risk Management System

In the risk management of PPP projects, it is very important to establish a comprehensive risk management system based on the identification and evaluation of risk elements.

First, do a good job in the feasibility study of PPP projects. Before the investment and construction of a PPP project, the government should cooperate with the social capital party to conduct a comprehensive and systematic study on the feasibility of the project. According to the national policy and economic environment, fully investigate the actual situation of the project implementation area, and the market demand and project Feasibility, technology maturity, and value-for-money evaluation of the choice of PPP model and traditional government investment model. It is necessary to fully identify various risks that may exist in PPP projects and work out corresponding risk response plans.

Second, identifying risks involves distinguishing between levels and subjects. The general identification in risk identification has no practical application value. Therefore, in the risk identification of PPP projects, it is necessary to pay attention to the hierarchical division of risks. The identification of layer-by-layer may affect the risks of project implementation, and for the identified risks, look for risks The undertaking body helps to further develop risk prevention measures. In order to make a clear judgment of the risks in practical applications, comprehensive classification and evaluation of the possibility of the identified risks and the magnitude of the losses should be performed. The first-class risks are those with high probability and serious losses, and the second-class risks. The risk is the risk that the possibility is small and the loss caused is serious or the possibility is large and the loss caused is normal, otherwise it is a third-class risk. Therefore, the implementation of PPP projects should focus on the prevention of first-class risks, and pay attention to the emergence of second-class risks.

Third, formulate a comprehensive risk control plan. According to the characteristics of risks of different levels and different natures, formulate corresponding risk control programs, and reasonably share the risks among different entities, minimize the possibility of risk occurrence and reduce the losses caused by risks, and give full play to risk transfer. Methods. And you must have a sense of risk and anxiety. You cannot wait until the risk has occurred before taking remedial measures. At that time, the losses that have occurred cannot be recovered. You must take precautions against risks before they occur, and implement effective control measures during the PPP project, And make a timely response after the risk occurs.

Fourth, risk management is implemented in the responsible department. On the basis of a comprehensive and accurate identification of risks, the management of risks should be implemented to specific responsible departments, and even in the actual operation, the responsibility should be reached to the people, so that there is supervision and monitoring from time to time, such as project As an example, the risk of engineering safety and quality risks in the promotion of risk should be the construction body of social capital, and the engineering safety and quality risks should be detailed into specific engineering areas, and the responsibility should be implemented in practice.

Fifth, in the PPP project, it is necessary to establish a sound risk protection mechanism, improve the professional and technical capabilities of personnel responsible for key control points, and cultivate the awareness of risk prevention of relevant personnel; quantify core risk indicators and set standards for risk early warning; The management of the company is implemented on the job and in person, and the specific work scope and responsibilities of all relevant personnel are clarified; and a regular research mechanism for key risks is established, and efficiency and effectiveness are emphasized in risk management.

6.3 In Conclusion

In the context of the country's supply-side structural reforms and the new year's government report advocating the promotion of government-social capital cooperation (PPP model), the construction of PPP projects for transportation infrastructure construction not only reduces the government's financial burden and broadens social capital The investment method has also enabled citizens to enjoy better public services. In order to further promote the construction of national PPP projects, it is necessary to focus on the implementation of risk control. This article takes X cities as the research object and conducts risk research on PPP projects in the field of transportation infrastructure. The conclusions are as follows:

1. This article takes the X-city rail transit PPP project as a case, and based on the project's full life cycle theory and risk hierarchical theory, it has initially and accurately identified the risks in the project, and established 9 types of risks, 27 Risk list for specific risk indicators.
2. For the 27 risks identified, use the AHP method to calculate the weight of each risk indicator, the results show that the risks with a larger weight of risk are political risk, legal risk, project promotion risk, and economic risk.
3. According to the established risk index evaluation system, use fuzzy comprehensive evaluation method to conduct risk assessment. The risk assessment shows that in the X city rail transit PPP project, the core risk control areas are identified as economic risks, project promotion risks, Relationship risk and market risk.
4. Formulate specific risk control plan. According to the characteristics of different levels and different natures of risk, we should formulate corresponding risk control plans, share the risk reasonably among different subjects, reduce the possibility of risk occurrence and the loss caused by risk as much as possible, and give full play to the method of risk transfer.

Acknowledgements "Study on the Performance Evaluation System of PPP Passenger Flow-Value Flow" in the PPP Model of Urban Rail Transit supported by the National Natural Science Foundation of China (71973009).

References

1. Grimsey, D., & Lewis, M. K. (2002). Evaluating the risks of public private partnerships for infrastructure projects. *International Journal of Project Management, 20*(2), 107–118.
2. Ma, N. T., & Li, X. (2019). Research on the implementation rate of PPP projects in my country and its influencing factors. *Economic and Management Review, 35*(02), 32–43.
3. Liu, H., Chen, S. J., & Chen, C. F. (2018). Analysis of the effectiveness of the national infrastructure PPP projects along the Belt and Road. *Journal of National Institute of Administration, 05*, 57–63 + 188–189.
4. Luo Yu, Y., Wang, F., & Chen, X. (2017). Institutional quality and how international financial institutions affect the effectiveness of PPP projects: A study based on the Belt and Road 46 countries' empirical data. *Finance Research, 04*, 61–77.
5. Li, B., Akintoye, A., Edwards, P. J., & Hardcastle, C. (2005). The allocation of risk in PPP/PFI construction projects in the UK. *International Journal of Project Management, 23*(1), 25–35.
6. Thomas, A. V., Kali, S. N., & Ganesh, L. S. (2006). Model ling and assessment of critical risks in BOT road Projects. *Construction Management and Economics, 24*(4), 407–424.
7. Fan, X. J., Zhao, Y., & Zhong, G. Y. (2007). Research on the proportion of financing risk of basic projects. *Journal of Management Engineering, 01*, 98–101.
8. Yan, L. J. (2009). Identification and allocation of PPP contract risks. *Construction Management Modernization, 23*(03), 271–274.
9. Fayek, A., Young, D. M., & Duffield, C. F. (1998). A survey of tendering practices in the Australian construction industry. *Engineering Management Journal, 10*(4), 29–34.
10. Deng, X. P., Li, Q. M., Xiong, W., & Yuan, J. F. (2009). Research on the key risks of PPP projects in urban infrastructure construction. *Modern Management Science, 12*, 55–59.
11. Songer, A. D., Diekmann, J., & Pecsko, R. S. (1997). Risk analysis for revenue dependent infrastructure projects. *Construction Management and Economics, 15*(4), 377–382.
12. Chen, B., & Li, Y. F. (2008). Risk allocation principles and allocation methods of infrastructure financing PPP model. *Railway Engineering Cost Management, 03*, 5–7.
13. Peng, H., Xiang, J. Y. (200). The main risks and prevention of private enterprises in the PPP model. *Journal of Guangdong Technical Teachers College, 06*, 41–43.
14. Coser, A., Maer-Matei, M., & Albu, C. (2019). Predictive models for loan default risk assessment. *Economic Computation And Economic Cybernetics Studies And Research, 53*(2), 149–165.

A Visualization Analysis of Environmental Accounting Research Based on CiteSpace

Yixuan Gao and Meijun Ning

Abstract To investigate the development course, research hotspots and latest trends on environmental accounting, this paper regarded 1191 articles in the WoS Core Collection database as the research objects. Besides, through descriptive statistics, keyword analysis, and citation analysis based on CiteSpace V, the research results are as follows. (1) The research has gone through three stages, focusing on environmental accounting concept, environmental accounting system, and national policies. (2) The research paths of environmental accounting mainly include biodiverse environmental planting, Western Europe, decision support, long-term forest inventory data, green cost, using graph search algorithm, international flow, Austrias livestock system, and carbon accounting tool. Meanwhile, the research frontiers mainly included carbon sequence, aboveground bioma, climate, and life cycle assessment. (3) References that attract scholars' attention include Daily et al. (Science 289:395–396, 2000 [1]), Solomon et al. (Climate change, vol 18, pp 95–123, 2007 [2]), Costanza et al. (World Environ 387:253–260, 1997 [3]), Luyssaert and Schulze (Nature 455:213–215, 2008 [4]), and Maler (Environ Resour Econ 1:1–15, 1991 [5]). Our findings provide insights into the promotion of the research on environmental accounting.

Keywords Environmental accounting · Sustainable development · Visualization analysis

1 Introduction

In 1987, the United Nations proposed the concept of sustainable development for the first time, and defined it as "development that meets the needs of present without compromising the ability of future generations to meet their own needs." It requires that economic and social development must be based on natural resources and the

Y. Gao · M. Ning (✉)
School of Economics and Management, North China University of Technology, Beijing, China
e-mail: nmj@ncut.edu.cn

Y. Gao
e-mail: 884603977@qq.com

M. Li et al. (eds.), *IEIS 2020*, https://doi.org/10.1007/978-981-33-4363-4_8

environment. With ecological resources depletes, a tension has emerged between human society and the ecological environment. Worries have multiplied about how to coordinate environmental protection and economic development. Only relying on the market can not effectively allocate resources. Therefore, it is necessary to use modern accounting systems and property rights to clarify the company's rights and responsibilities, so as to make effective use of environmental resources.

In the 1970s, the concept of environmental accounting was firstly proposed. It officially started the research of environmental accounting. So far, scholars have explored environmental accounting theory from various perspectives, including environmental accounting concepts, research content, environmental accounting systems, information disclosure, and related policies. In this context, this paper regarded 1191 articles in the WoS Core Collection database as the research objects. Through descriptive statistics, keyword analysis, and citation analysis by using CiteSpace V, this paper intends to summarize the development course, research hotspots, and latest trends in the field of environmental accounting. Besides, the paper makes some suggestions for subsequent research in this area.

2 Data Sources and Methodology

2.1 *Data Sources*

This study regarded the WoS Core Collection as the data collection platform. The search strategy can be described as the following. (1) Time span = 1987–2019. (2) TS = ("environmental accounting" OR "green accounting" OR "carbon accounting"). (3) Journal source = (SCI-E, SSCI, A&HCI, CPCI-S, CPCI-SSH, ESCI). (4) Literature type = Article, and language = English. Through preliminary search, our research obtained 1191 articles, and then CiteSpace V was used to reduce repetition. In the end, 1189 articles were obtained, with a repetition removal rate of 99.83%.

2.2 *Methodology*

CiteSpace is a Java-based visualization software developed by Professor Chen, C. It takes citation network analysis as its core. It can achieve scientific network analysis, such as collaborative network analysis, co-keyword analysis, and full-text mining. More specifically, the software uses multiple, diachronic, and dynamic visualization techniques to form knowledge maps that represent the citation age by color rings and clusters by timeline [6]. In this context, this paper uses CiteSpace V to draw the mapping knowledge domains about environmental accounting. Through keyword analysis and citation analysis, the paper tends to explore its evolution path and make some suggestions for subsequent research on environmental accounting.

3 Descriptive Statistical Analysis

The distribution of environmental accounting articles in the time dimension is shown as Fig. 1.

According to Fig. 1, it can be seen that the number of articles issued on environmental accounting from 1987 to 2019 was generally on the rise. On the whole, the research of environmental accounting can be divided into three stages:

1. 1987–1997: The number of articles was small. In the early 1990s, International Standards of Accounting and Reporting repeatedly discussed environmental accounting issues. It proposed that environmental accounting standards should be established to provide guidance for environmental accounting practices in many countries. At the same time, the revision of the national economic account system with SNA as the core took resource and environmental factors into account. China has also responded positively. The support and promotion of the government were linked to the beginning of environmental accounting research. However, during this stage, the annual publication volume was below 10.
2. 1998–2008: The annual change in the number of articles was relatively fluctuated. In 1999, the Japan Environment Agency issued an interim announcement entitled "The Principles of Understanding and Disclosure of Environmental Protection Costs", encouraging various industries to evaluate and disclose environmental costs. Meanwhile, the Environmental Protection Branch of Canada issued the Introduction Guide to Environmental Accounting. It aims to make environmental managers realize the concepts of environmental accounting and improve the economic benefits of pollution prevention. The German and Austrian

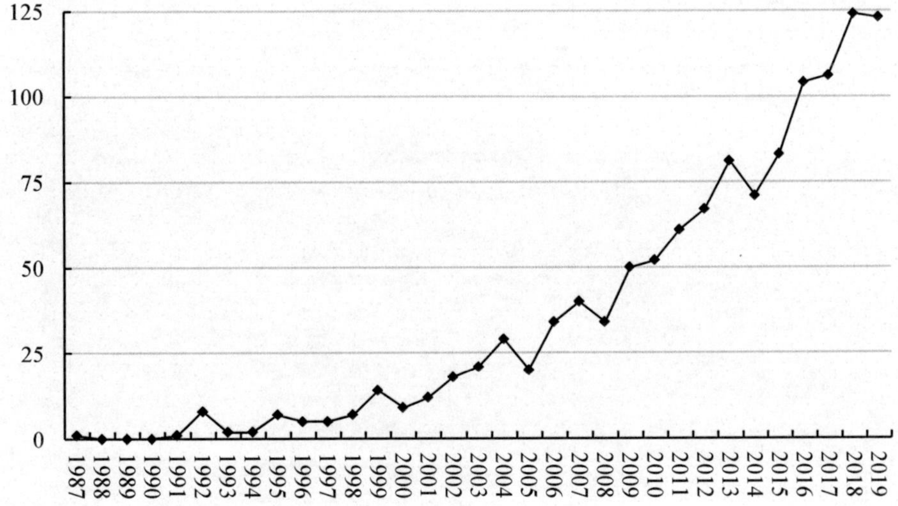

Fig. 1 Distribution of environmental accounting articles

governments also cooperated with the Commission on Sustainable Development to develop guidelines suitable for national conditions. On the whole, as society's calls for sustainable development gradually increases, environmental accounting research is also constantly deepening.

3. 2009–2019: The number of annual articles has increased rapidly. The 2009 United Nations Climate Change Conference firstly proposed that the world face climate issues together. Moreover, it advocated establishing the national carbon emission standards and reducing global carbon emissions. In 2012, the United Nations Climate Conference in Doha issued the Kyoto Protocol, emphasizing the necessity and importance of environmental accounting research. In 2015, the Environmental Protection Law revised by China also strengthened restrictions on sewage treatment companies and clarified the specific research directions of corporate environmental accounting. In this context, the number of articles has increased significantly during this period, and reached a peak in 2018, with 124 articles published.

4 Keyword Analysis

4.1 Keyword Co-occurrence Analysis

The keyword can express the main content of the article. And it can fully reflect the focus of environmental accounting in the field of accounting. By extracting the distribution of keywords, the research trends and research hotspots are clearly found [6]. Based on these, using CiteSpace V, the keyword map of environmental accounting research is shown in Fig. 2.

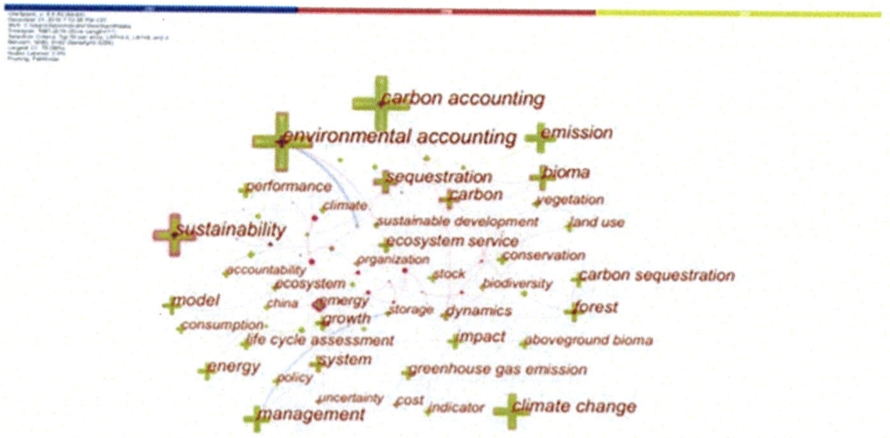

Fig. 2 Keyword map of environmental accounting research

Table 1 High frequency keyword and high centrality keyword (top 10)

Sort by frequency			Sort by centrality		
Freq.	Centr.	Keywords	Freq.	Centr.	Keywords
187	0.21	Environmental accounting	95	0.25	Management
169	0.18	Carbon accounting	187	0.21	Environmental accounting
139	0.13	Sustainability	85	0.2	Sequestration
110	0.06	Climate change	169	0.18	Carbon accounting
95	0.25	Management	87	0.15	Bioma
92	0.07	Emission	139	0.13	Sustainability
87	0.15	Bioma	55	0.12	Growth
85	0.2	Sequestration	68	0.11	Energy
72	0.05	Forest	26	0.09	Land use change
69	0.08	Carbon	69	0.08	Carbon

By analyzing Fig. 2, the top 10 keywords sorted by frequency and centrality are shown in Table 1.

The larger the cross, the higher the frequency. As can be seen from Fig. 2, the crosses of environmental accounting, carbon accounting, sustainability, climate change, and management were larger. Among them, keyword with the highest frequency of occurrence was environmental accounting. In another respect, keyword nodes such as management, environmental accounting, sequestration, carbon accounting, and bioma had purple rings around the outer rim, indicating that the centrality was above 0.1. For centrality, management ranked the highest (0.25). It means that environmental accounting management played a vital role in promoting the development of environmental accounting.

4.2 *Keyword Cluster Analysis*

The timeline view focuses on expressing the relationship between clusters and the historical span of literatures in a cluster. Using CiteSpace V, the keywords cluster map of environmental accounting research is shown in Fig. 3. There are 9 clusters in the network. Chen and Chen [7] pointed out that CiteSpace V provides two indicators to evaluate the accuracy of map based on the network structure and the clarity of clustering: Modularity value (Q) and Silhouette value (S). In this part, the Q = 0.78 (>0.3), and the S = 0.77 (>0.5), indicating that the clustering structure is prominent and clusters are rational.

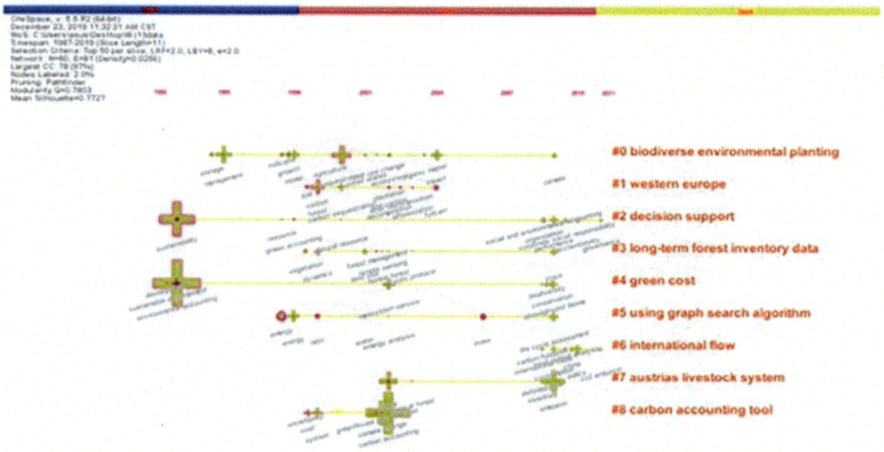

Fig. 3 Keyword cluster map of environmental accounting research

It is difficult to obtain detailed information from the cluster names shown in Fig. 3. Therefore, in order to analyze the specific content contained in each cluster in depth, sub-clusters need to be combined. Through a comprehensive analysis of Fig. 3 and related articles, it can be seen that Cluster #0 biodiverse environmental planting, the largest cluster, contained a total of 10 articles. The main research included integrated farm forestry and forest cover trend. Cluster #1 Western Europe contained 9 articles. Through analysis, we found this cluster mainly studied increasing carbon stock and forest soil. In addition, Cluster #1 had a strong connection with Cluster #2 and Cluster #5. Cluster #2 decision support, the third largest cluster, had 10 articles. This cluster focused on sustainability accounting and environmental accounting research. Cluster #3 long-term forest inventory data contained 10 articles. This cluster mostly illustrated carbon intensity, evaluation, and low-carbon sustainable precinct. Cluster #4 green cost contained 10 articles. It is also found that this cluster focused on biodiversity valuation, discount rate problem, and managing forest. Dandois and Ellis [8] introduced a new aerial remote sensing system to observe dynamics in ecosystem structure and spectral traits. Edens and Hein [9] identified the challenges in developing ecosystem accounts and put forward some suggestions for addressing the challenges of developing ecosystem accounts by analyzing the different perspectives on these challenges. Cluster #5 using graph search algorithm contained 10 articles. It mainly studied energy algebra rule and rigorous application. Cluster #6 international flow contained 7 articles. Research hotspots included conservation viewpoint and forest biomass. Cluster #7 Austrias livestock system mainly studied managing forest, carbon matter, and integrating energy. Cluster #8 carbon accounting tool contained 7 articles. It mainly studied Full CAM model, evaluation, and low-carbon sustainable precinct.

4.3 Keyword Burst Analysis

Using CiteSpace V, the keyword burst of environmental accounting research are shown in Table 2. Among them, the burst value represents the degree of increased attention to the keyword, and the cited column represents the rapid growth of the citation.

Combining with Table 2, we found the research frontier of each period on environmental accounting, which was divided into the following three stages:

1. Initial stage: 1987–1997. As mentioned above, the study of environmental accounting was just in its infancy. Evolving from the fields of environmental protection and sustainable development, the basic concept of environmental accounting was put forward. Among them, the burst value of green accounting was 10.49, and the duration was 14 years. Maunders and Roger [10] studied the relationship between accounting and ecology with economics as an intermediate variable. Serafy [11] pointed out that the traditional national economic accounting system could not reflect the environmental change, and pollution information should be included.
2. Growth stage: 1998–2008. Environmental accounting research began to emerge a large number of burst keywords in this period. It indicated that environmental accounting has become a hot topic of academic research, as well as a diverse research center. At this stage, the research focused on the construction of environmental accounting system. The burst keywords sorted by the burst value included forest, index, resource, afforestation, ecosystem, natural resource, Kyoto Protocol, boreal forest, growth, ration, and organic matter. According to the research of Brack [12], afforestation had important economic value for energy conservation, emission reduction, pollution mitigation, and carbon sequestration. In another respect, Gray [13] provided suggestions for the sustainable development of ecological environment by analyzing the significance and internal contradictions of sustainable development.
3. Outbreak stage: 2009–2019. At this stage, the research expanded to the national level. It focused on the formulation of international bills and national policies. Among them, United States and China had high value of burst. As known, China has entered the "new normal" of economic development. That is to say, green development has become the main keynote of the Thirteen Five Plan. It requires enterprises to realize green transformation as soon as possible. On the other hand, the government encourages enterprises to improve the environmental evaluation system and environmental information disclosure. Therefore, stakeholders pay more attention to environmental accounting research. On the whole, scholars have improved the research of environmental accounting in China by studying the relevant research results of environmental accounting in the United States and other developed countries. In addition, carbon sequence, aboveground bioma, climate, and life cycle assessment have become new fields of environmental accounting research.

Table 2 Keyword burst of environmental accounting research

Period	Keyword	Burst	1987–2019
1987–1997	Environmental accounting	4.86	
	Green accounting	10.49	
1998–2008	Resource	4.63	
	Ratio	3.64	
	Natural resource	3.89	
	Forest	6.54	
	Boreal forest	3.76	
	Ecosystem	3.97	
	Afforestation	4.41	
	Kyoto Protocol	3.85	
	Growth	3.65	
	Organic matter	3.56	
	Index	4.87	
2009–2019	Carbon sequestration	4.40	
	United States	4.38	
	Aboveground bioma	4.03	
	Climate	3.91	
	Life cycle assessment	3.63	
	China	4.04	

5 Citation Analysis

5.1 *Reference Co-citation Analysis*

Reference co-citation analysis can be used to identify core scholars and core articles in the field of environmental accounting. It uses mathematical and statistical methods to analyze citations of analysis objects, such as scientific journals, articles, and authors. Finally, it can reveal quantitative characteristics and inherent laws [14]. Based on these, use CiteSpace V to reference co-citation analysis: Time slice = 11, node type = cited reference and set the top 50 articles cited in each slice as the analysis object. The map of reference co-citation analysis is shown in Fig. 4.

According to Fig. 4, Aronsson [15] was the first scholar in the ranking of centrality. In this paper, the form of national welfare measures was linked with the operation of economic system by using Green NNP to measure the level of national welfare. Brake [12] developed a decision support system to manage about 400,000 trees planted in Canberra. It is found that the total value of energy conservation, emission reduction, pollution mitigation, and carbon sequestration from 2008 to 2012 was $20–67 million. Weitzman [16] linked the three basic concepts of sustainability, Green GDP and technological progress. It finally found that the best estimate of long-term sustainability may largely depend on the prediction of future technological progress. Gray [13] provided suggestions for sustainable development by analyzing the significance and internal contradictions of sustainable development. Masera et al. [17] used CO2FIX V.2 model to estimate total carbon balance of alternative management regimes in both even and uneven-aged forests. Caspersen et al. [18] found land use was the main factor governing the rate of carbon accumulation, and the contribution

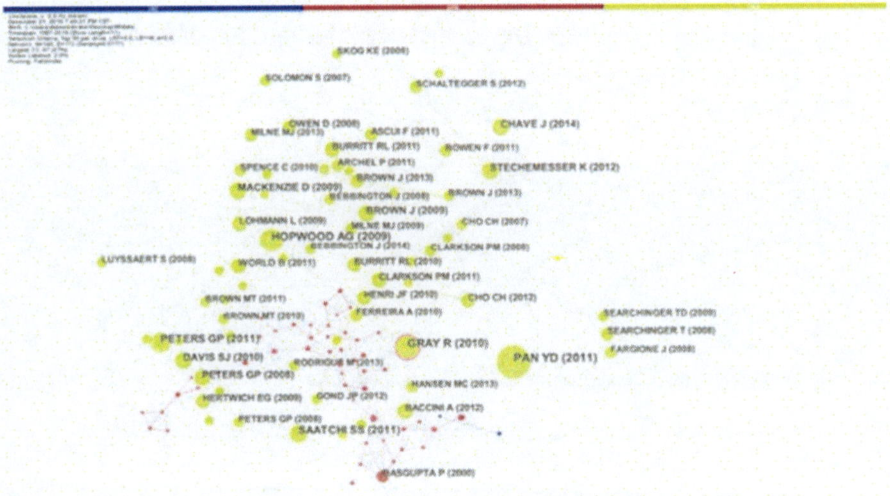

Fig. 4 Reference co-citation analysis map of environmental accounting research

of growth promotion was far less than previously reported. Paul et al. [19] used GRC3 model to simulate the change pattern of soil C after afforestation in four test cases. Costanza et al. [3] estimated the current economic value of 17 ecosystem services for 16 biomes based on published research results and some original calculations.

5.2 *Cluster Analysis*

Use CiteSpace V, citation cluster map of environment accounting research is shown in Fig. 5. The Q = 0.78 (>0.3), and the S = 0.77 (>0.5), indicating that the cluster structure is prominent and clusters are rational.

According to Fig. 5, there are 3 clusters in the environment accounting research. Cluster #0 carbon accounting, the largest cluster, contained 11 articles with a silhouette value of 0.678. Silva and Beatriz [20] added to the international research on environmental disclosure, extending the scope of the understanding of the environmental reporting practices. Cluster #1 environmental accounting contained 10 articles with a silhouette value of 0.79. Milne et al. [21] studied the tensions and contradictions between different conceptions of the meaning of carbon accounting. In the end, Cluster #2 eco-efficiency contained 6 articles with a silhouette value of 0.83. Schaltegger and Csutora [22] provided an overview of carbon accounting as a rapidly developing area of sustainability management.

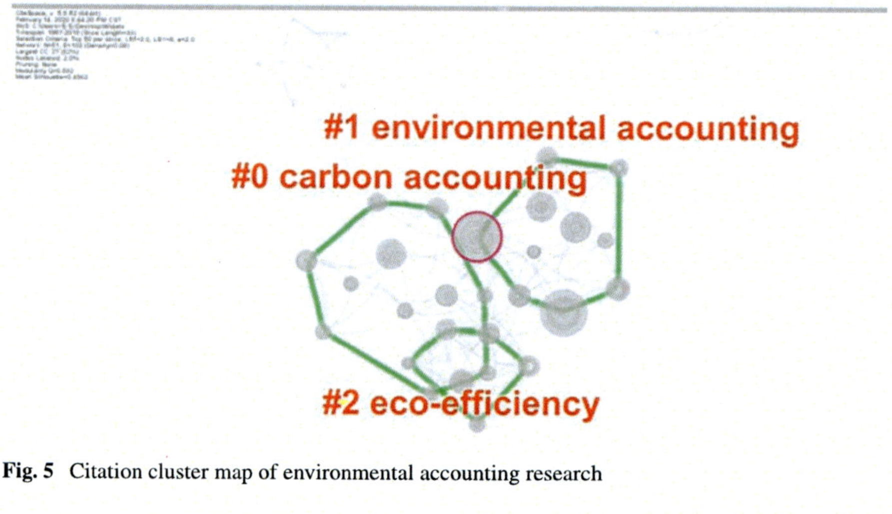

Fig. 5 Citation cluster map of environmental accounting research

5.3 Burst Analysis

In the reference burst analysis, an article has a high burst indicates that its cited frequency has a dramatically increase in a certain period. It shows that it breaks through an important proposition in the field and it is a direction with great research potential. Based on these, use CiteSpace V, the list of citation burst of environmental accounting research is shown in Table 3.

Among them, Daily et al. [1] pointed out that ecosystems were capital assets, and it can produce important goods and services. Solomon et al. [2] described human and natural drivers of climatic change based on past IPCC assessments. More specially, Costanza et al. [3] pointed out that ecosystem services and natural capital stocks were essential for the operation of the earth's life support system. Luyssaert and Schulze [4] found the net productivity of ecosystem was usually positive through consulting the literature and database of forest carbon flux estimation. Moreover, MaLer [5] found the concept of national net worth by considering the impact of environmental resources and environmental damage factors. Parker and Owen [23] reviewed the development of social and environmental accounting research. And the study found that although the current research covered a wide range, most of it focused on theoretical research and failed to effectively integrate practice. Hertwich and Peters [24] analyzed the contribution of eight categories: construction, housing, food, clothing, liquidity, manufactured goods, services and trade by quantifying greenhouse gas emissions related to final consumption of goods and services in 73 countries and 14 regions. In another respect, Wiedmann et al. [25] identified six major models that employ multi-sector, multi-region, input–output analysis in order to calculate environmental impacts embodied in international trade. Result showed that it was important to explicitly consider the production recipe, land and energy use. Odum et al. [26] put forward the comparison method of Emdollar value. By comparing the Emdollar value of six schemes of conversion of cropland to forest in Puerto Rico, the contribution of forest to the public interest and the method for selecting reforestation

Table 3 Citation burst of environmental accounting research

Author	Journal	Year	Burst
Dasgupta, P.	*Science*	2000	7.24
Solomon, S.	*Climate Change*	2007	4.67
Costanza, R.	*Nature*	1997	4.63
Luyssaert, S.	*Nature*	2008	4.15
Maler, K.G.	*Environmental & Resource Economics*	1991	4.10
Owen, D.	*Accounting Auditing & Accountability Journal*	2008	3.88
Hertwich, E.G.	*Environmental Science & Technology*	2009	3.76
Wiedmann, T.	*Ecological Economics*	2007	3.66
Odum, H.T.	*Forest Science*	2000	3.62
Richards, G.P.	*Australian Forestry*	2004	3.62

were evaluated. Richards and Brack [27] developed a hybrid method based on GIS modeling and empiricism to estimate land biomass stock and stock change in order to promote the development of Australian national carbon accounting system.

6 Conclusion and Enlightenment

6.1 Conclusion

This paper regarded 1191 articles in the WoS Core Collection database from 1987 to 2019 as the research objects. Through descriptive statistics, keyword analysis and citation analysis by using CiteSpace V, research results are as follows. (1) The research has gone through three stages, including initial stage, growth stage, and outbreak stage. And its research has focused on environmental accounting concept, environmental accounting system, and national policies. Moreover, the research frontiers mainly included carbon sequence, aboveground bioma, climate, and life cycle assessment. (2) The research paths of environmental accounting mainly included biodiverse environmental planting, Western Europe, decision support, long-term forest inventory data, green cost, using graph search algorithm, international flow, Austrias livestock system, and carbon accounting tool. (3) From the results of citation cluster analysis, there are 3 clusters in this field, including carbon accounting, environmental accounting, and eco-efficiency. (4) From the results of citation burst analysis, references with high burst included Daily et al. [1], Solomon et al. [2], Costanza et al. [3], Luyssaert and Schulze [4], MaLer [5], Parker and Owen [23].

6.2 Enlightenment

Since the 1970s, environmental accounting has completed the formulation of international laws and policies through the efforts of scholars. Moreover, it has constructed the basic theoretical framework system combining with traditional accounting, including objectives, assumptions, elements, subject setting, and information disclosure. However, the research of environmental accounting still has some limitations. Based on these, this paper puts forward the following suggestions for the development of environmental accounting from the different perspectives of micro, meso and macro:

1. At the micro level, innovative research ideas and research methods should be improved. The existing research mainly uses standardized research methods to analyze environmental accounting. As a new branch of modern accounting, environmental accounting research should focus on absorbing the achievements of other subjects, such as information technology, chemistry, economics, and ecology. That is to say, the interdisciplinary research of environmental accounting

can not only make different subjects infiltrate each other, but also comprehensively use qualitative and quantitative research. In order to enrich environmental accounting theories and improve the environmental accounting research framework, communication and cooperation between academic communities should be expanded.

2. At the meso level, we should strengthen environmental education and public supervision. It is our duty to respect nature and ecological balance. After the Third Plenary Session of the 18th CPC Central Committee, the CPC Central Committee stressed that the construction of ecological civilization should be put into practice. Moreover, it emphasized solving uncoordinated issues between economic and environmental in developing countries. It is important to promote supply-side structural reforms and establish ecological evaluation systems. Furthermore, the 19th CPC National Congress clearly put forward the proposal of “building a harmonious and beautiful China”. It put environmental protection in a prominent position. Environmental protection, as an important part of corporate social responsibility, makes higher requirements for cleaner production, reducing consumption and increasing efficiency. Therefore, we should strengthen environmental education and public supervision of corporate production emissions. On the other hand, we should actively promote the construction of unified environmental accounting system to complete environmental information disclosure guidelines and set up companies in key emission industries.
3. At the macro level, we should promote the construction of environmental accounting regulations and improve the environmental responsibility accountability system. The government plays a leading role in the country’s economic and social development. As we known, the environmental accounting legal systems of most countries have learned from the advanced experience of Japan, Britain and other countries. However, due to different ideologies, the related legal systems are not practically suitable. They still have compatibility issues. Only starting from the national conditions and setting specific legal norms can we truly achieve the development of environmental accounting. In another respect, we should establish an independent third-party environmental responsibility audit and clarify the accountability form to improve the international environmental audit system. Practice has proved that the implementation of the stringent source protection system, damage compensation system and accountability system were of great significance for improving environmental governance and ecological restoration systems.

In short, through visualization analysis, we found that environmental accounting research has gone through three stages, including initial stage, growth stage, and outbreak stage. And its research has focused on environmental accounting concept, environmental accounting system, and national policies. Besides that, the research frontiers mainly included carbon sequence, aboveground bioma, climate, and life cycle assessment. On the whole, this paper summarizes the development course, research hotspots, and the latest trends of environmental accounting through scientometric analysis. Moreover, the paper explores its evolution paths and makes some suggestions for subsequent research in this area. However, our paper still has some limitations. On the one hand, we took the English literature data of the WoS Core Collection database as the research object, but did not consider the influence of non-English articles. On the other hand, the intrinsic causes need to be solved by future research.

Acknowledgements This research was funded by the Student Science and Technology Activity Program of North China University of Technology (203051360020), Beijing Urban Governance Research Center (20XN232), the Scientific Research Starting Foundation Program of North China University of Technology (110051360002).

References

1. Daily, G. C., Soederqvist, T., Aniyar, S., Arrow, K., Dasgupta, P., Ehrlich, P. R., et al. (2000). The value of nature and the nature of value. *Science, 289,* 395–396.
2. Solomon, S., Qin, D., & Manning, M. (2007). Climate change 2007: The physical science basis. Contribution of working group I to the fourth 16 assessment report of the intergovernmental panel on climate change. In *Climate change* (Vol. 18, pp. 95–123)
3. Costanza, R., d'Arge, R., de Groot, R., Farber, S., Grasso, M., & Hannon, B. (1997). The value of the world's ecosystem services and natural capital. *World Environment, 387,* 253–260.
4. Luyssaert, S., & Schulze, E. D. (2008). Old-growth forests as global carbon sinks. *Nature, 455,* 213–215.
5. MaLer, K. G. (1991). National accounts and environmental resources. *Environmental and Resource Economics, 1,* 1–15.
6. Jie, L., & Chen, C. (2017). *CiteSpace: Text mining and visualization in scientific literature* (pp. 2–10). Beijing: Capital University of Economics and Business Press.
7. Chen, Y., & Chen, C. (2015). Methodological function of Citespace knowledge graph. *Scientific Research, 33,* 242–253.
8. Dandois, J. P., & Ellis, E. C. (2013). High spatial resolution three-dimensional mapping of vegetation spectral dynamics using computer vision. *Remote Sensing of Environment, 136,* 259–276.
9. Edens, B., & Hein, L. (2013). Towards a consistent approach for ecosystem accounting. *Ecological Economics, 90,* 41–52.
10. Maunders, K. T., & Roger, L. (1991). Accounting and ecological crisis.*Accounting, Auditing & Accountability Journal, 4.*
11. Serafy, S. E. (1997). Green accounting and economic policy. *Ecological Economics, 21,* 217–229.
12. Brack, C. L. (2002). Pollution mitigation and carbon sequestration by an urban forest. *Environmental Pollution, 116,* 195–200.
13. Gray, R. (2010). Is accounting for sustainability actually accounting for sustainability and how would we know? An exploration of narratives of organizations and the planet. *Accounting, Organizations and Society, 35,* 47–62.
14. Chen, C. (2013). Hindsight, insight, and foresight, a multi-level structural variation approach to the study of a scientific field. *Technology Analysis and Strategic Management, 25,* 619–640.
15. Aronsson, T. (1998). Green accounting in imperfect market economies. *Environmental and Resource Economics, 11,* 273–287.
16. Weitzman, M. L. (1997). Sustainability and technical progress. *The Scandinavian Journal of Economics, 99,* 1–13.

17. Masera, O. R., Garza-Caligaris, J. F., & Kanninen, M. (2003). Modeling carbon sequestration in afforestation, agroforestry and forest management projects, the CO2FIX vol 2 approach. *Ecological Modelling, 164,* 177–199.
18. Caspersen, J. P., Pacala, S. W., Jenkins, J. C., Hurtt, G. C., Moorcroft, P. R., & Birdsey, R. A. (2000). Contributions of land-use history to carbon accumulation in U.S. forests. *Science, 290,* 1148–1151.
19. Paul, K. I., Polglase, P. J., & Richards, G. P. (2003). Sensitivity analysis of predicted change in soil carbon following afforestation. *Ecological Modelling, 164,* 137–152.
20. Silva, M., & Beatriz, A. (2010). Determinants of environmental disclosure in the annual reports of large companies operating in Portugal. *Corporate Social Responsibility and Environmental Management, 17,* 185–204.
21. Milne, M. J., Ascui, F., & Lovell, H. (2011). As frames collide, making sense of carbon accounting. *Accounting, Auditing & Accountability Journal, 24,* 978–999.
22. Schaltegger, S., & Csutora, M. (2012). Carbon accounting for sustainability and management. Status quo and challenges. *Journal of Cleaner Production, 36,* 1–16.
23. Parker, L., & Owen, D. (2008). Chronicles of wasted time? *Accounting, Auditing & Accountability Journal, 21,* 240–267.
24. Hertwich, E. G., & Peters, G. P. (2009). Carbon footprint of nations: A global, trade-linked analysis. *Environmental Science & Technology, 43,* 6414–6420.
25. Wiedmann, T., Lenzen, M., & Turner, K. (2007). Examining the global environmental impact of regional consumption activities. *Ecological Economics, 61,* 15–26.
26. Odum, H. T., Doherty, S. J., & Scatena, F. N. (2000). Energy evaluation of reforestation alternatives in Puerto Rico. *Forest Science, 46,* 521–530.
27. Richards, G. P., & Brack, C. (2004). A continental biomass stock and stock change estimation approach for Australia. *Australian Forestry, 67,* 284–288.

PPP Mode and Coordinated Regional Development—Empirical Evidence from China

Bo-lu Wei, Xuemeng Guo, and Zhuo-jun Wang

Abstract Based on China's provincial panel data from 2000 to 2017, this paper investigates the role of infrastructure investment using PPP mode in promoting regional economic development. The study found that the PPP mode significantly promoted China's economic growth by breaking administrative monopolies and alleviating resource mismatches. Furthermore, after considering the influence of regional attributes, we find that the PPP mode can no longer significantly promote economic growth in eastern China. However, it can significantly boost economic growth in the central and western regions. The research provides enlightenment on how to develop the economy under the "new normal".

Keywords Infrastructure investment · PP · Economic growth · Regional attributes

1 Introduction

Since the 18th National Congress of the CPC, China's economic growth has slowed down and lacks momentum. Under the background of "three phases of superimposition", by leading the investment in infrastructure, a traditional counter-cyclical means of macroeconomic regulation, the government has failed to boost economic growth by expanding aggregate demand. Facing the current "new normal" of China's economic development, it is an inevitable choice for China to promote supply-side structural reform. Supply-side structural reform is to optimize the industrial structure in the process of increasing investment, to play the fundamental role of the market in resource allocation and to stimulate market vitality and increase economic growth

B. Wei · X. Guo (✉)
School of Economics and Management, Beijing Jiaotong University, Beijing, China
e-mail: 1031994766@qq.com

B. Wei
e-mail: blwei1991@163.com

Z. Wang
Beijing Infrastructure Investment Co., Ltd., Beijing, China
e-mail: wangzhuojun@bii.com.cn

M. Li et al. (eds.), *IEIS 2020*, https://doi.org/10.1007/978-981-33-4363-4_9

potential. The Third Plenary Session of the 18th Central Committee of the CPC proposed "allowing social capital to participate in urban infrastructure investment and operations through franchising and other means", encouraging social capital to invest in infrastructure, deepening the degree of marketization in this field, and guiding the development of Public–Private Partnership (PPP) Project. The PPP mode enables the government and social capital to reach cooperation through the principle of agency, and promotes the optimal allocation of resources. By the end of February 2019, the Ministry of Finance's PPP center management library project has accumulated a total of 8780 projects with an investment of 13.3 trillion yuan. Since the 18th National Congress of the CPC, China has entered the late stage of industrialization [1], traditionally infrastructure investment by the government as a single subject has been unable to significantly promote economic growth [2]. With the introduction of a market-oriented mechanism to deepen the degree of marketization in this field, can infrastructure investment under the PPP model that allows social capital to participate become a new driving force for economic growth? If so, given the law of diminishing marginal output, is there a suitable "degree" for this promotion of economic development?

Based on this, we use China's provincial panel data from 2000 to 2017 to examine the relationship between infrastructure investment and economic growth under the PPP mode in which social capital participates, and to study whether this relationship is different between different regions in China. The possible research contributions of this paper are: (1) Studying the relationship between infrastructure investment and economic growth under the participation of social capital, explore its mechanism; (2) Combining economic differences between regions, study the role and extent of infrastructure investment in which social capital participates in the economic growth of different regions, to provide local governments at different levels of economic development which plan to introduce the PPP mode for infrastructure investment with behavior theoretical reference, maximize the role of the government as a policy maker, supervision and creator of the macro environment.

2 Literature Review

Existing literature has made extensive and in-depth evaluations of the relationship between infrastructure investment and economic growth from both theoretical and empirical perspectives. Rostow [3] pointed out the significance of infrastructure investment for a country's economic development, especially in its infancy and takeoff stage. In the early days of reform and opening up, investment from the infrastructure sector significantly improved production efficiency, optimized resource re-allocation, and then strongly stimulated economic growth.

Since then, the World Bank has also demonstrated the relationship between infrastructure investment and economic growth from different aspects and perspectives.

In fact, in the early days of reform and opening up, investment from the infrastructure sector significantly increased production efficiency, optimized resource reallocation, and then strongly stimulated economic growth. At the same time, empirical evidence from China finds that infrastructure investment does not have a significant positive effect at all times [4]. Solow growth model illustrates that in the long run economic growth requires technological progress and infrastructure investment can hardly avoid the law of diminishing marginal returns. Canning and Pedroni [5] calculated and estimated the regional stock infrastructure and compared it with the regional economic development level, and the results show that excess investment in infrastructure exists in some regions. This finding indicates that the scale and structure of infrastructure investment need to match the level of regional economic development, otherwise the accumulation effect and scale efficiency of infrastructure investment will weaken, and the promotion effect on the latter will be reduced or even hinder economic growth. After 2013, infrastructure investment through fiscal appropriations and government debt financing has been unable to significantly promote China's economic growth [2].

The PPP mode, which originated from the UK, is essentially a hybrid organization. The public and private sectors conclude a contract through principal-agent, and form long-term stable cooperative relationships on the basis of contracts to meet the supply of public goods. Currently, as an important driver of supply-side structural reforms, it can resolve the contradiction between the demand for public goods and capital supply; by introducing diversified investment, it can reduce local government fiscal and debt pressures effectively, smooth annual fiscal expenditures and reduce long-term government debt [6]; use the sensitivity of cost–benefit form social capital to reduce the risk of government investment and force ineffective investment [7]; hedge the economic downturn, promote governance reform and innovation [8] and other advantages. Because of these, PPP mode has been valued by academia and policy makers. However, no scholars have investigated the impact of infrastructure construction on economic growth after the change of investment entities (from a single government investment entity to the cooperation between government and social capital), and what changes will occur compared to the traditional government's investment as a single subject; whether this relationship is different among different regions in China?

3 Hypothesis Development

As one of the traditional "three carriages", investment is one of the important means of macroeconomic regulation in China. The downturn of the economy is accompanied by the decline of investment. Due to the multiplier effect, investment is the most volatile variable in economic growth. At present, China is in the development stage of industrial optimization and upgrading and economic structural adjustment, infrastructure construction investment is still an irreplaceable means of stable growth [9].

3.1 The Impact of Infrastructure Investment Under PPP Mode on Economic Growth

Adam Smith, an advocate of small government and big markets, listed public goods as one of the few functions of the government, and explicitly listed "building and maintaining certain public utilities and certain public facilities" as one of the "three important businesses" that a country should have In the Wealth of Nations. Infrastructure investment, as a public and quasi-public product, is difficult to be fully supplied by the private sector due to its publicity and positive externalities. It has become the main basis for government investment, which has led to strong administrative monopoly characteristics in this field. Under the traditional mode, the government separately completes infrastructure investment construction and operation management. While the public sector invests a huge amount of construction funds, it also needs to bear the transaction costs caused by information asymmetry. At the same time, enterprises and institutions in the public sector are often divided according to the principle of regional segmentation. In the same area, there may be multiple infrastructure provision units with the same functions but belonging to different levels, which causes problems such as inefficient provision of public facilities and services, and the overall advantage cannot be exerted. As the decision-making of infrastructure investment, the fundamental goal of the government should be to improve public welfare. However, according to the viewpoint of the "grabbing hand" of local governments [10], long-term administrative monopolies make this goal deviate. Instead, it is the goal from the perspective of the economic man-self-motivated motivation, seeking to maximize the interests of the sector, and using the tendency of preferential policies to obtain financial and tax advantages and franchise rights, the so-called "cost advantage" has been used for competition, neglecting the demands of social public welfare [11], resulting in the loss of net social welfare, which in turn caused infrastructure investments to fail to effectively directly or indirectly increase total output [12], which makes it difficult to promote regional economic growth.

The focus of comprehensively deepening reforms under the new normal is economic system restructuring. The core issue is to properly handle the relationship between the government and the market. Transform government functions further, and withdraw from areas where the market should play a role or even a decisive role. As a hybrid organization, PPP has an important function of changing the role between the public sector and the market, increasing the degree of marketization in the field of infrastructure construction, by decreasing the degree of government intervention and relax administrative controls, reducing administrative monopoly caused by active and rent and has no intention of gen returns [13], thus cracking the de facto monopoly structure. At the same time, the promotion of PPP mode, on the premise of meeting the basic requirements of huge construction funding requirements and reducing government investment, through the signing of cooperation contracts between public and private sectors, has realized the sharing of information, effectively reduced the "friction" of market transactions, and improve the supply efficiency of public goods. Therefore, PPP can give full play to the advantages of social capital, use the resources of social capital, improve the supply efficiency of public goods, promote the increase of social welfare, and then promote economic growth.

In summary, we proposes the following hypothesis:

- H1: Infrastructure investment under the PPP model can promote economic growth

3.2 Impact of Regional Attributes on the Promotion of PPP Mode

The land area of China is about 9.6 million km^2. Due to the differences in the distribution of natural resources, geographical factors and development foundation among regions, there is a spatial stepwise distribution of development levels in each region. Different regions are at different stages of industrialization and the leading industries for economic development are different.

For the pre-industrial stage, the leading industry for its economic development is agriculture. For the initial stage of industrialization, labor-intensive industries such as the light industry represented by the textile industry are the leading industries for this stage of economic growth. In the middle stage of industrialization, its leading industrial system transformed into a capital-intensive heavy chemical industry. In the later stages of industrialization, the leading industrial system changes from a capital-intensive industry to a technology-intensive industry. At the same time, the productive service industry also gets accelerated development in the later industrialization period. In post-industrial economies, the leading industry of economic growth is transformed into the information economy and the knowledge economy based on the service industry.

In the middle period of industrialization, the leading industry is heavy chemical industry, whose growth is mainly through epitaxial expansion of reproduction. At this stage, investment in infrastructure directly creates demand for heavy products in the heavy chemical industry. On the other hand, the improvement of infrastructure also provides important transportation channels and market environments for the development of these industries. Therefore, in the middle of industrialization, the degree of coupling between infrastructure investment and the leading industries in this stage is well excellent. In the later stages of industrialization, the growth of leading industries mainly expanded reproduction through the connotation driven by innovation. At this stage, the development of leading industries urgently needs an innovative environment and innovative talents, and the relationship with infrastructure is irrelevant. Therefore, the ability of infrastructure investment to promote economic growth has been greatly reduced after the late industrialization.

Imbalance of regional development is a typical feature of China's development at present. The central and western regions are mostly inland provinces with a late start in economic development. Compared with the eastern provinces, which are at the forefront of reform and opening up, have an early and rapid economic development [14, 15]. In eastern region, the input structure and output structure of production factors are highly coupled, meanwhile, their regional industrial deviation is the lowest [16]. After decades of development, the eastern region has gradually entered the later stage of industrialization. However, the central and western regions are mostly in the middle stage of industrialization with high-speed industrial growth [17], and the promoting effect brought by infrastructure investment is best coupled with its current leading industries.

To sum up, this paper proposes the following two hypotheses:

- H2a: In the eastern region, infrastructure investment under the PPP mode has not significantly promoted local economic growth.
- H2b: In the central and western regions, infrastructure investment under the PPP mode has significantly promoted the region's economic growth.

4 Methodology and Data

4.1 Methodology

In order to test hypothesis H1, this paper builds a mixed panel data model that includes provincial GDP, provincial PPP project investment, and other control variables. The regression equation is defined as follows:

$$GDP_{i,t} = \alpha_0 + \alpha_1 PI_{i,t} + \sum \lambda_k Control_{k,it} + \varepsilon_{i,t} \tag{1}$$

Among them, *i* indicates the provinces, *t* indicates the year, *GDP* is the provincial GDP, *PI* is the provincial PPP project investment (hereinafter referred to as "PPP investment amount"), and *Control* indicates a series of control variables, including: *NPGR, ODR, UR, EDU, FAI, FI, EXP, CPI*. α_1, λ_k are regression coefficients; α_0 is a constant term; and ε is a residual term.

In order to test Hypothesis H2a and Hypothesis H2b, this paper adopts the 18-year historical data of 31 provinces to test. Specifically, according to the 2003 Standard of the National Bureau of Statistics, China was divided into east, central, and western regions. The above three hypotheses were verified with historical data from the eastern region, central and western regions, respectively. The regression equation is defined as follows:

$$GDP_{i,t} = \alpha_0 + \alpha_1 PI_{EAST,t} + \sum \lambda_k Control_{k,it} + \varepsilon_{i,t} \tag{2}$$

$$GDP_{i,t} = \alpha_0 + \alpha_1 PI_{M\&W,t} + \sum \lambda_k Control_{k,it} + \varepsilon_{i,t} \tag{3}$$

where PI_{EAST} is the investment amount of provincial PPP projects in the eastern region, $PI_{M\&W}$ is the investment amount of provincial PPP projects in the central and western regions. α_1, λ_k are regression coefficients; α_0 is a constant term; and ε is a residual term.

Other variables have the same meaning as Eq. (1).

4.2 Variable Definitions

1. *Explained Variable*
   $GDP_{i,t}$, This paper adopts the annual GDP of each region in China from 2000 to 2017. The data comes from the China Statistical Yearbook from 2000 to 2017.
2. *Explanatory Variables*
   Define $PI_{i,t}$ as the amount of investment in the infrastructure sector under the PPP mode of each province, autonomous region, and municipality (hereinafter referred to as the "province") from 2000 to 2017. The data comes from the World Bank PPI database.
3. *Control Variable*
   This article sets a series of control variables, including: the Regional Population Growth Rate (*NPGR*), Regional Labor Level (*ODR*), Regional Urbanization Level (*UR*), Regional Human Capital (*EDU*); Regional Fixed Asset Investment (*FAI*), Regional Financial Development Level (*FI*), Regional Economic Opening Degree (*EXP*) and Regional Consumer Price Index (*CPI*). The data comes from the China Statistical Yearbook and the China Financial Statistical Yearbook from 2000 to 2017.

The names and explanation of all variable are shown in Table 1.

4.3 Sample Selection

Considering the availability, reliability, integrity and comparability of the data, this paper selects the data of 31 provinces, autonomous regions and municipalities in mainland China from 2000 to 2017 to evaluate the impact of infrastructure investment on economic growth under the PPP mode. The sample is the balanced panel data with a total of 558 observations. Among them, the infrastructure investment data of each provincial administrative unit under the PPP mode comes from the World Bank PPI database from 1998 to 2018; the control variable data comes from the China Statistical Yearbook, the China Financial Statistical Yearbook, and the provincial and municipal statistical yearbooks.

Table 1 Explanation of variable

Variable type	Symbol	Name	Explanation
Explained variable	*GDP*	Regional annual GDP	***ln*** GDP
Explanatory variables	*PI*	Regional PPP Project Investment	***ln*** (PPP investment)
Control variable	*NPGR*	Regional population Growth rate	***Ratio*** of natural population growth
	ODR	Regional labor level	***Ratio*** of elderly dependency
	EDU	Regional human capital	***ln*** (regional education investment)
	UR	Regional urbanization rate	***Ratio*** of regional urbanization
	FAI	Regional fixed asset investment level	***ln*** (regional fixed asset investment)
	FI	Regional financial development level	***ln*** (added value of regional finance)
	EXP	Regional economic opening degree	***ln*** (regional exports by source place)
	CPI	Regional consumer price index	CPI

5 Result

5.1 Descriptive Statistics

The descriptive statistical results of each variable are shown in Table 2.

The descriptive statistical results show that the maximum value, minimum value, mean value and standard deviation of the investment in the infrastructure sector under the PPP model in China are 8.903, 0, 3.02 and 2.55, which indicate that the PPP mode varies greatly among provinces in China. Further, the eastern region from 2000 to 2017, infrastructure PPP investment has 198 observations, the maximum value of 8.903, the minimum value of 0, the standard deviation was 2.68, which indicate that there is a large difference in the strength of PPP investment in the eastern region of China between 2000 and 2017. Secondly, the average value of PPP investment in the infrastructure sector in eastern provinces is 3.31, higher than the national average, indicating that the intensity of PPP investment in the infrastructure sector of eastern region is relatively high. There were 360 observations in the central and western regions between 2000 and 2017. The maximum value is 8.352, the average is 2.46, and the median is 3.414. All of these three values are smaller than the nation's, which illustrate that the investment intensity of the infrastructure sector under PPP mode in the central and western regions is relatively low. Meanwhile their standard deviation is 2.46, reflecting the large difference in these regions, which may have a potential

Table 2 Variable descriptive statistics

Var name	Obs	Mean	SD	Median	P25	P75	MIN	MAX
GDP_{Nation}	558	8.87	1.23	9.017	8.141	9.753	4.769	11.406
PI_{Nation}	558	3.02	2.55	3.648	0.000	5.131	0.000	8.903
GDP_{EAST}	198	9.48	1.08	9.618	8.809	10.240	6.267	11.406
PI_{EAST}	198	3.31	2.68	4.198	0.000	5.400	0.000	8.903
$GDP_{M\&W}$	360	8.54	1.18	8.665	7.819	9.469	4.769	10.714
$PI_{M\&W}$	360	2.86	2.46	3.414	0.000	4.963	0.000	8.352
NPGR	558	5.59	3.04	5.630	3.280	7.660	-1.900	13.100
ODR	558	12.35	2.77	12.100	10.300	14.200	5.900	21.900
EDU	558	14.91	1.11	15.014	14.236	15.750	11.251	17.332
UR	558	49.08	15.70	47.270	38.340	56.640	19.380	91.860
FAI	558	8.30	1.34	8.427	7.338	9.354	4.160	10.919
FI	558	5.72	1.48	5.843	4.703	6.809	0.908	8.832
EXP	558	15.99	1.86	15.954	14.619	17.257	11.130	20.429
CPI	558	102.26	2.02	101.900	101.100	103.200	96.700	110.100

Data source World Bank PPI Database, China Statistical Yearbook, China Financial Statistical Yearbook

impact on the economic growth of the central and western regions, which is suitable for further analysis.

5.2 *Analysis*

This paper uses *STATA* to estimate the parameters of the constructed econometric model, established econometric model, sets the F test results according to the panel, and adopts the hybrid OLS method to method to perform a regression test on the PPP investment in the infrastructure field.

Table 3 Column 1 shows the regression results of model (1). From the empirical results, we can find GDP and infrastructure investment under the PPP mode are significant at the level of 1%, with a positive coefficient of 0.0284, which indicate that infrastructure investment under the PPP model has a significant positive impact on China's economic growth. As mentioned earlier, infrastructure investment under PPP mode has a direct impact on the original government's monopoly on the vertically integrated infrastructure construction model. It can not only improve the supply efficiency of the project itself, reduce the transaction costs, the hybrid capital structure will also be able to play the main role of market in allocating resources, alleviate resource mismatch, and then improve the total factor productivity, so as to promote economic growth. This hypothesis 1 is verified.

Table 3 Regression results of infrastructure investment under PPP mode and economic growth

Variables	National	EAST	Middle & WEST
	Model 1	Model 2	Model 3
	GDP	GDP_{EAST}	$GDP_{M\&W}$
PI	0.0284***	0.0044	0.0375***
	(4.66)	(0.87)	(4.78)
NPGR	0.0140**	0.0445***	−0.0022
	(2.43)	(7.52)	(-0.31)
UR	−0.0121***	−0.0108***	0.0030
	(−8.63)	(-6.98)	(1.40)
ODR	0.0070	−0.0011	0.0162*
	(1.09)	(−0.22)	(1.78)
EDU	−0.0814	−0.3164***	0.0478
	(−1.51)	(−5.92)	(0.63)
FAI	0.1360***	0.2760***	−0.0334
	(3.08)	(6.72)	(-0.55)
FI	0.5409***	0.6190***	0.4544***
	(21.29)	(30.18)	(12.79)
EXP	0.2465***	0.1915***	0.3468***
	(14.05)	(11.30)	(12.43)
CPI	0.0025	0.0040	−0.0012
	(0.35)	(0.60)	(−0.14)
Constant	2.0065**	4.5166***	0.1783
	(2.17)	(5.03)	(0.15)
Observations	558	198	360
R^2	0.928	0.978	0.925
R^2_A	0.927	0.977	0.923

t-statistics in parentheses
***$p < 0.01$, **$p < 0.05$, *$p < 0.1$

Column 2 and 3 of Table 3 show the regression results of model (2) and model (3). According to the measurement results, we noticed that in the eastern region, the coefficient of investment intensity in the infrastructure sector under PPP mode is positive, but not statistically significant. By contrast in central and western regions, it is significant at the level of 1%, and the coefficient is 0.0375, which is positive and higher than the national level. This illustrate that in central and western regions, the intensity of infrastructure investment under PPP mode has a significant positive impact on the regional economic development.

This is because the eastern region, which has a developed economy, is in the late stage of industrialization, and has a high level of development of the tertiary industry. After years of investment and construction, the level of its existing infrastructure construction is high. Thus, the infrastructure investment flows brought by PPP are mainly used for its existing infrastructure. In comparison, the central and western regions have higher returns on scale for direct construction of new infrastructure. The promotion effect of investment in infrastructure construction is best coupled with the leading industries in the central and western regions in the middle of industrialization, which results in a higher output efficiency, so as to promote economic growth. At the same time, due to its convenient geographical location, the eastern provinces have achieved significant results in implementing the policy of opening up to the outside world, with a high level of marketization. However, because of locating in the inland region, in central and western provinces, the degree of administrative monopoly is high, but the pace of market-oriented reforms is slow. The PPP model provides effective institutional support for improving the marketization degree of the region and stimulates market vitality, thus promoting regional economic growth. Hypothesis 2 and hypothesis 3 are verified.

6 Conclusion and Suggestion

6.1 Research Conclusion

Based on China's provincial panel data from 2000 to 2017, this paper makes a further test of the relationship between infrastructure investment under the PPP mode and economic growth by combination with regional attributes.

The research finds that infrastructure investment under the PPP model can promote economic growth. This provides inspiration for how we can develop our economy under the "new normal". At the present stage in China, investment in infrastructure construction is still an important means of "steady growth". The PPP model, as an important driver of structural reforms on the supply side, has scientifically and rationally repositioned the role of the government in the process of economic construction. Meanwhile, by introducing social capital, it actually introduces market competition into the field dominated by the government in the past. In the field, market competition is used to improve supply efficiency and reduce costs, thereby stimulating economic vitality. After adding the regional attributes, the study finds that compared with the eastern region, the infrastructure investment under PPP mode in the central and western regions has a significant promotion effect on the region's economic growth.

6.2 Suggestion

In order to further deepen the supply-side structural reform, promote the steady and healthy development of China's economy under the "new normal", and promoting the application of the PPP mode in the field of infrastructure, the following policy implications can be obtained according to the conclusions in the article:

1. *Control the Scale, Diversify the Investment*
   In the future, we should give full play to the multiplier effect of the PPP mode in stimulating economic development in infrastructure construction investment, change the old model of government unified investment, construction, and supply, rationally control the scale of infrastructure investment, and focus on the decentralization effect of public investment.
2. *Further Optimize the Input–output Structure of Central and Western Regions*
   For the central and western regions, especially the provinces, cities, and areas that urgently need to improve the level of infrastructure, appropriate investment should be made to give full play to the role of infrastructure construction as a pre-investment in economic development, create a smooth logistics channel and a good market environment, optimize the degree of coupling of their input structure and output structure to accelerate economic growth.
3. *Match Investment with Economic Aggregate of Eastern Region*
   For regions that have approached or exceeded the optimal investment scale, under the premise of ensuring the maintenance and reconstruction of old and existing infrastructure, make appropriate adjustments, pay attention to the matching of infrastructure investment with economic aggregates, and prevent infrastructure investment from becoming non-productive investment. At the same time, optimize investment efficiency, promote the development of other industries and fields, and then promote the steady growth of the overall economy.

Acknowledgements Funding for National Natural Science Foundation "Research on Performance Evaluation System of urban rail transit PPP mode based on resource "passenger-value flow"" (71973009).

Funding for Fundamental Research Funds for the Central Universities "Research on Local Government Financing Platform, Behavior Decision of Local Government's Choice of PPP Financing Mode, and Economic Growth" (2019YJS064).

References

1. Zhao, C. W., Xu, S. Y., & Zhu, H. M. (2015). The new driving force of china economic growth in the late stage of industrialization. *China Industrial Economics, 327,* 44–54.
2. Liao, M. L., Xu, S. Y., Hu, C., & Yu, C. W. (2018). Can infrastructure investment promote economic growth? Empirical test based on interprovincial panel data from 1994 to 2016. *Management World, 5,* 63–73.
3. Rostow, W. W. (1959). The stages of economic growth. *The Economic History Review, 12,* 1–16.
4. Ansar, A., Flyvbjerg, B., Budzier, A., & Lunn, D. (2016). Does infrastructure investment lead to economic growth or economic fragility? Evidence from China. *Oxford Review of Economic Policy, 32*(Autumn), 360–390.

5. Canning, D., & Pedroni, P. (2008). Infrastructure, long-run economic growth and causality tests for cointegrated panels. *Manchester School, 76,* 504–527.
6. Ye, X. S., & Xu, C. M. (2013). Review and research on PPP pattern in China. *Soft Science, 162,* 6–9.
7. Yao, D. M., & Li, J. L. (2015). The efficiency difference under the condition satisfying: Comparison of PPP model and traditional model. *Reform, 252,* 32–42.
8. Ou, C. Z., & Jia, K. (2017). Challenge and innovation of PPP in the public interest realization mechanism: From the perspective of public governance framework. *Contemporary Finance & Economics, 388,* 26–35.
9. Sun, Z., Yang, G., & Li, K. (2015). Has the investment on infrastructures promoted economic growth? Empirical evidence from the east, middle and west areas. *Economist, 8,* 71–79.
10. Stulz, R. M. (2005). The limits of financial globalization. *The Journal of Finance, 60,* 1595–1638.
11. Hu, A. G., & Guo, Y. (2002). From monopolistic to competitive markets: Profound social change. *Reform, 5,* 17–28.
12. Munnell, A. H. (1992). Policy watch: Infrastructure investment and economic growth. *The Journal of Economic Perspectives, 6*(Autumn), 189–198.
13. Guo, Y., & Hu, A. G. (2003). Administrative monopoly, rent seeking and corruption: An analysis of the corruption texture in the transition economy. *Comparative Economic and Social Systems, 106,* 61–69.
14. Coser, A., Maer-Matei, M., & Albu, C. (2019). Predictive models for loan default risk assessment. *Economic Computation and Economic Cybernetics Studies and Research, 53*(2), 149–165.
15. Liu, Y., Hu, Z., Li, H., & Zhu, H. (2019). Does preemption lead to more leveled resource usage in projects? *A Computational Study Based on Mixed-Integer Linear Programming, Economic Computation and Economic Cybernetics Studies and Research, 53*(4), 243–258.
16. Jin, T., & Chen, J. J. (2014). Transfer payments, soft constraints of state-owned enterprises and loss of efficiency—A study based on comparative perspective. *Research on Financial and Economic Issues, 365,* 89–96.
17. Huang, Q. H. (2014). 'The New Normal', the late stage of industrialization and the new power of industrial growth. *China Industrial Economics, 319,* 5–19.

Research on Key Issues of the Energy System and Mechanism in Qinghai Province

Xue Ma, Zhiqing Li, Decao Xu, and Lianghui Xie

Abstract The energy system revolution is a core component of China's new energy security strategy of "Four Revolutions and One Cooperation". The study on the key issues of the energy system and mechanism is of great significance for Qinghai Province to adapt to and promote the transformation of the energy system from a traditional energy-oriented to a clean energy-oriented one, and to support the construction of a clean energy-oriented demonstration province. This paper analyzes the principal contradictions faced by the energy system and mechanism of Qinghai Province under the new situation, analyzes the problems from four aspects, namely, the energy supply system and mechanism, the consumption system and mechanism, the scientific and technological system and mechanism, and the trans-provincial and trans-regional cooperation mechanism.

Keywords Energy system and mechanism · Energy supply · Energy consumption · Energy technology · Inter-provincial and inter-regional cooperation

1 Introduction

Qinghai Province enjoys "good light, abundant water and wind, less coal, more oil and gas, and great potential for unconventional energy". It has unique advantages in resource endowment. In recent years, made clear in the large-scale development of

X. Ma (✉) · Z. Li
Green Energy Development Research Institute (Qinghai), Xining, China
e-mail: mxue2726@qh.sgcc.com.cn

Z. Li
e-mail: lzqing2018@qh.sgcc.com.cn

D. Xu
State Grid Qinghai Electric Power Company, Xining, China
e-mail: xdcao4910@qh.sgcc.com.cn

L. Xie
School of Economics and Management, Beijing Jiaotong University, Beijing, China
e-mail: 19113031@bjtu.edu.cn

M. Li et al. (eds.), *IEIS 2020*, https://doi.org/10.1007/978-981-33-4363-4_10

new energy in Qinghai Province as the key point, to 100% of clean energy use as the goal, supported by science and technology innovation, through the construction of the smart grid security, a full chain to make clean energy construction, use and the overall plan of demonstration province [1], built to create a national clean energy demonstration province of planning policy system, create a new situation of the energy development. In August 2016, During his visit to Qinghai, General Secretary Xi Jinping made an important instruction to "make Qinghai an important new energy industry base of China", which pointed out the direction for Qinghai to establish a national clean energy demonstration province. Construction national clean energy demonstration province of Qinghai is to realize the energy advantages into economic advantages, and blaze a trail in the energy industry to realize the characteristics of the prosperous common people strong province. In this context, the traditional energy system and mechanism are no longer adapt to the energy development in Qinghai Province, must unswervingly implement change in systems and mechanisms, in the aspect of the energy management system and operational mechanism to realize the breakthrough, to support the important guarantee for the development of clean energy demonstration in Qinghai Province.

2 The Reform Process of Qinghai's Energy System and Mechanism

Since China's reform and opening-up, China's energy market main body, market structure, operation mechanism, price mechanism, investment and financing system, management system and so on a series of reform [2, 3], cancel the electric power industry, coal industry, establish two power grid companies, five power generation groups, restructuring of the three oil and gas company, introducing multivariate investment main body, promote separate government functions from enterprise management, gradually open areas such as coal prices, breaking the original enterprise unity, highly concentrated market pattern, break through the original operation mechanism under the planned economy system, greatly stimulated the market vigor, effectively promote the development of energy industry. Against the background of China's energy system and mechanism reform, Qinghai Province continues to promote energy system and mechanism reform in an orderly manner in accordance with relevant state requirements, which are mainly reflected in:

2.1 *Deepen the Reform of the Electric Power System*

The Pilot Program of Direct Trading between Power Users and power generation Enterprises in Qinghai Province was issued to carry out direct trading between power users and power generation enterprises. The "Pilot Program of Qinghai Electricity

Sales Company participating in Provincial Electricity Market Trading" has been formulated, and detailed regulations have been made on market access, trading mode, trading quantity, trading price, trading plan, and other aspects, to ensure the smooth progress of Qinghai electric power system reform. We formulated an implementation plan for the standardization of power trading institutions in Qinghai Province and took the lead in the country in launching the shareholding reform of provincial power trading institutions. We promoted the pilot business of incremental distribution networks, established the Qinghai Electric Power Market Management Committee, and promoted the formation of an open and transparent trading platform. The establishment of Qinghai Haixi Tuanyushan Electricity Sales Co., Ltd. marked the first batch of incremental power distribution business reform pilot implementation in Qinghai Province; Completed the cost monitoring and calculation of transmission and distribution prices in Qinghai Province, and determined the transmission and distribution prices in the first regulatory cycle of Qinghai power grid. After straightening out the Guoluo power management system, the assets of the Guoluo power grid have been fully taken over by The State Grid Qinghai Electric Power Company, realizing the full coverage of the state grid in Qinghai and comprehensively improving the power supply capacity in Qingnan region.

2.2 Accelerate Reform of the Oil and Gas System

Implement the Opinions of the CPC Central Committee and the State Council on Deepening The Reform of the Petroleum and Natural Gas System, study and formulate supporting policies and measures, speed up the fair and open up of oil and gas infrastructure, advance the reform of the operation mechanism of oil and gas pipeline networks, and improve the investment and operation mechanism of oil and gas reserve facilities. Introduced to carry out the national oil and gas system reform work deployment, the Qinghai Province about promoting harmonious and stable development of natural gas of the implementation of the plan, promote the reform of management for oil and gas prospecting and exploitation, establish an early warning system for the natural gas supply and demand forecast regulating mechanism, natural gas, natural gas demand-side management and development of comprehensive coordination mechanism, straighten out the natural gas price mechanism, strengthen the safe operation of the whole industry chain mechanism of natural gas.

2.3 Promote the Energy Management System

In 2013, the responsibilities of the National Energy Administration and the Electricity Regulator were integrated, and the newly established National Energy Administration inherited the regional supervision system of the agency dispatched by the regulator. At present, there are 12 provincial supervision offices and 6 regional supervision bureaus, of which Qinghai is under the jurisdiction of the Northwest Energy Regulator. In 2014, set up the energy bureau of Qinghai Province.

2.4 Explore Ways to Promote Regional Cooperation

Negotiated and reached intergovernmental framework agreements, and signed intergovernmental agreements with Jiangsu, Hubei and other provinces. In 2018, we completed the export of 1.694 billion kWh of new energy, an increase of 20.7% over the previous year, effectively guaranteeing the trans-provincial and trans-regional consumption of new energy. The efficient allocation of resources to achieve regional industry cooperation across the country, such as Qinghai and Jiangsu set up a "photovoltaic production, supply, quantity given" complementary industrial chain. Promote the reform of mixed ownership to promote inter-provincial cooperation, set up Qingyu Energy Co., LTD., strengthen cooperation between state-owned enterprises in Qinghai and Henan Provinces, and deepen the reform of mixed ownership.

3 Key Issues of Qinghai Energy System and Mechanism

In recent years, the reform of the energy system and mechanism in Qinghai Province has achieved phased results, effectively promoting the development of the energy industry. However, it should also be noted that the existing energy system and mechanism of Qinghai Province is basically a simple mapping of the national energy system and mechanism, and still takes the traditional energy system dominated by hydropower and fossil energy as the object. Although some adjustments have been made in recent years to adapt to the development of new clean energy, the original framework has not been broken on the whole. Qinghai Province to build national clean energy demonstration province for the multivariate supply of clean energy, green energy clean consumption, expand the clean energy industry, improve the clean energy output, but the construction of Qinghai Province dominated by renewable clean energy system the development needs of the energy system and the traditional energy under there is a prominent contradiction between energy systems and mechanisms, mainly reflected in the traditional energy under the dominant energy system of energy systems and mechanisms for clean energy production, circulation and consumption barriers, become the energy systems and mechanisms of Qinghai Province in urgent need of research. Therefore, the research on the key issues of Qinghai's energy system and mechanism can be divided into four parts:

3.1 Key Issues of Qinghai's Energy Supply System and Mechanism

From the perspective of energy supply, a large-scale clean energy supply system will be formed in Qinghai in the future, with clean energy electricity as the main source. Therefore, the key problems of the energy supply system and mechanism in Qinghai Province lie in the supply of clean energy power and the relationship between clean energy and traditional energy sources. First, the market system is not impeccable. It has been mainly by Qinghai electric power company the state practices monopoly for the purchase, lack of social capital into the electricity market, especially distributed constrained the development and utilization of clean energy. Second, the price formation mechanism is not sound. The price relationship has not been straightened out, and the market-oriented pricing mechanism needs to be improved. Environmental externalities are not reflected in energy prices, and the low-carbon advantages of clean energy are not fully exploited. Third, the energy regulatory mechanism needs to be improved, social supervision is relatively weak, lack of strict energy regulatory standards.

3.2 Key Issues of Qinghai's Energy Consumption System and Mechanism

From the perspective of the energy consumption, the main contradiction facing the Qinghai is energy "double control" and the contradiction between economic development needs, to solve this contradiction, must implement the profound connotation of ecological civilization thought, adjust the energy consumption structure, from dependence on the traditional oil, coal energy in clean energy, such as natural gas, green power, drive the adjustment of industrial structure on the reform of mechanism of the consumption system, guide the high pollution, high energy consumption, heavy industry towards emerging strategic industries, and giving full play to the advantages of clean energy in Qinghai, make renewable energy the driving force of economic growth, to realize the Qinghai green "overtaking corner" in the development of high quality. The key questions are: First, how to promote industrial restructuring through the mechanism of energy conservation and emission reduction? At present, the work of energy conservation and emission reduction mainly depends on the administrative mechanism. There is a widespread phenomenon of "one-size-fits-all" policies. The role of the market in energy conservation and emission reduction is not obvious. Second, how to adapt to and guide the new consumption pattern from the institutional mechanism? Currently in Qinghai has sprung up multiple new clean energy consumption patterns, clean energy, smart grid, especially in the distributed micro-grid, etc., and guided by the traditional consumption mode of the system mechanism exists prominent contradictions, needs to be set up to adapt to the new pattern and further to guide the new model to a more reasonable consumption of systems and mechanisms.

3.3 Key Issues of Qinghai's Energy Science and Technology System and Mechanism

The current lower overall level of science and technology development of Qinghai Province caused great obstacles to energy technology development. Weak energy science and technology innovation basic conditions make that there are not form research, development, production, and use of a benign circle of mutual promotion. From the perspective of system and mechanism, the main problems are as follows: first, energy science and technology management involve many departments, management functions are scattered, there is a lack of scientific, efficient, unified and coordinated decision-making and management mechanism, and the leading role of the government strategy is missing. Second, fiscal and tax policies to promote scientific and technological innovation are not sound, and industrial policies and regulations are poorly coordinated. Third, the mechanism of talent cultivation and introduction is not flexible, and there is a lack of long-term talent cultivation and introduction mechanism for scientific and technological talents in the field of energy. Fourth, there is a lack of overall management of the entire innovation chain, and there is no institutional guarantee for the cultivation and transformation of achievements, leading to the insufficient transfer of R&D to application, insufficient motivation for the application of independent technologies, and difficulty in forming the innovation value chain.

3.4 Key Issues of Qinghai's Inter-Provincial and Inter-Regional Energy Cooperation Mechanism

With the rapid development of Qinghai's new energy industry and the gradual acceleration of power supply construction, the province's power gap has changed from the annual rigid gap in the 12th Five-Year Plan to surplus. In 2018, clean energy power will be net exported. It is estimated that by 2025, 2035, and 2050, the power delivered across provinces and regions will reach 45 billion, 120 billion and 140 billion KWH, respectively. The outward power supply of Qinghai Province is mainly based on clean energy, and photovoltaic, wind power, and other new energy. The current and future quite a period, clean energy power across the province inter-district trade problems will become the key problems of energy across the province interregional cooperation in Qinghai Province, mainly displays in: one is the regional barriers between provinces, willingness to buy provincial electric power is not strong, and clean energy across the province inter-district given market mechanism has not yet been established, given between clean energy in the province. Second, the scheduling mechanism is not sound, the existing scheduling methods, market transactions, and the two detailed rules and other aspects of the lack of coordination, the operation is difficult to achieve the "landscape, water and fire storage complementary systems integration operation" requirements.

4 Direction and Key Tasks of the Energy System and Mechanism Reform in Qinghai Province

Qinghai should, in combination with its characteristics, vigorously promote the revolution of the energy system, innovate a package of systems and policies to encourage and promote efficient supply and consumption of clean energy, promote industrial upgrading and cross-regional cooperation, and blaze a new trail of clean energy development. Qinghai energy system mechanism reform has the following four aspects of ten key tasks.

4.1 Direction and Key Tasks of Qinghai's Energy Supply System and Mechanism

Based on the energy supply system and mechanism and against the backdrop of the national energy supply revolution and energy system reform, the energy supply system and mechanism reform in Qinghai Province shall be steadily and orderly promoted.

1. *Build an effective competitive energy market structure and market system*
   Improve the structure of the energy market, give full play to the leading role of the state-owned economy in the energy sector, actively develop mixed-ownership energy enterprises, and foster new types of market players. Speed up the establishment and improvement of an energy market trading mechanism, clarify the rights and responsibilities of market participants, and promote diversified competition among market participants. Promote the formulation of trading standards, standardize market-based trading behaviors, and encourage the exploration of diversified trading models such as centralized bidding in bilateral transactions. Key according to the "to hold the middle, open at both ends," [4] system architecture, strive to promote the realization of the grid from the power distribution network, distribution main body independent and trading mechanism, accelerate the shareholding system reform of Qinghai electric power trading center, to establish a scientific corporate governance structure, implementation specifications by trading independent operation, making it transparent, perfect functions, the main body of the power trading platform.

2. *Establish a market-determined energy price formation mechanism*
   Promote the reform of electric power and other energy prices, accelerate the marketization of energy prices, let go of the competitive power price links, further distribution price and sale price separately on the formation mechanism, setting the price by the "cost plus reasonable profits allowed" method [5], transmission and distribution to allow income electricity market price by the user or sell electricity through consultation, market main body and power generation enterprises bidding methods such as independence. The power transmission and distribution power price adjustment mechanism, including power supply reliability rate and service quality, should be explored.
3. *Establish a scientific market supervision system*
   Coordinate the reform of the energy management system, and move faster to put in place an energy regulatory system in which government and supervision are relatively independent, supervision is conducted at different levels, departments have a clear division of labor, and supervision powers and responsibilities are clearly defined. Improve the regulatory organization system, adjust the allocation of regulatory personnel in all areas promptly following the development and changes of the oil, gas, electricity, and coal industries, as well as the requirements of new models and new forms of business, such as multi-functional complementarity, and strengthen personnel training and exchanges in a regulatory capacity. Strengthen ongoing and post-event supervision, form a standardized, effective, open, and transparent regulatory system [6]. Strengthen the development of the social credit system and public information service platforms, and give play to the role of the public, the media, and third-party institutions in participating in supervision and self-regulation of trade organizations.

4.2 Direction and Key Tasks of Qinghai's Energy Consumption System and Mechanism

Under the guidance of energy consumption systems and mechanisms, give full play to the guiding role of the consumer side and innovate the establishment of clean energy-led consumption systems and mechanisms for demonstration.

1. *Strengthen the construction of energy conservation and emission reduction mechanisms*
   Give full play to the basic role of the market in allocating resources and the guiding and driving role of government funds, and improve policies and measures conducive to energy conservation and emissions reduction. Taking into account factors such as the level of economic development, industrial structure, the potential for energy conservation and industrial layout, set targets for total and intensity energy consumption scientifically and reasonably and strictly control them following regulation and management requirements. Further strengthen the evaluation and assessment of binding targets for energy conservation and emissions reduction, establish an assessment system for emissions reduction based on environmental quality, and incorporate "double control" of energy conservation and

emission reduction targets into the comprehensive assessment and annual assessment system for economic and social development in all regions. Strengthen incentives for energy conservation and emissions reduction, improve fiscal policies to support energy conservation, and guide financial institutions to increase credit support for energy conservation renovation projects.

2. *Strengthen institutional supply of new consumption patterns*
Improve the planning and supervision mechanism of electrical energy alternative, and according to the area with hot electricity demand, combined with cogeneration, regional efficient environmental protection boiler room, industrial waste heat utilization energy supplies a variety of ways, in the overall urban planning, developing planning fully considering power alternative energy development, guarantee instead of form a complete set of electricity grid line corridors and site land planning, where the power to develop a replacement for the construction and operation standards, establish a standardized and orderly supervision mechanism. Study and formulate suitable for distributed energy consumption, the Internet, such as microgrid pattern of specialized management regulation, simplify the management of the project examination and approval process, optimization of Internet application, improve related subject integrating resources, docking of supply and demand, the function of cooperative innovation, construction of government departments, trade organizations, enterprises and consumers and the subject of common governance consumption ecological system.

4.3 Direction and Key Tasks of Qinghai's Energy Science and Technology System and Mechanism

With the energy science and technology system and mechanism as the driving force, the development of energy science and technology shall be brought into the provincial strategic level, to make the provincial science and technology development take the lead in making breakthroughs in the energy field and cultivate new growth points driving industrial upgrading.

1. *Accelerate the improvement of the energy science and technology innovation system*
Reasonable use of the government macroeconomic regulation and control of the guide, to establish market-oriented science and technology innovation system. Promote energy technologies and a new generation of information technology, materials technology, advanced manufacturing technology, such as depth fusion, drive the solar thermal power generation, above the complementary, large-scale high-performance energy storage, wisdom, the development and application of potentially subversive technology such as comprehensive energy network. A series of innovative achievements, such as laws and regulations, industry standards, and market supervision, must be established to protect and market the environment [7], to guarantee the return of scientific research investment and form the ability of market subjects to self-cycle and sustain development.

2. *Strengthen the introduction mechanism of energy personnel training*
   Give full play to the leading role of basic research and establish a long-term and stable support mechanism for talents in key fields such as energy science. Increase support for training innovative scientific and technological personnel in key technologies such as photovoltaic and photovoltaic power generation, energy storage, and industrial energy conservation and emission reduction. Continue to plan from a global perspective and attract talented and talented people in short supply, explore and implement a flexible global talent recruitment mechanism, change "retaining persons" into "retaining talents", and build a regional highland for energy opening-up and cooperation. Encourage universities and enterprises to cooperate on new energy models, and support the early transformation of mature scientific research and technological achievements in the field of energy in Qinghai.
3. *Strengthen the international technical and financial cooperation mechanism*
   Strengthen cooperation mechanisms with the World Bank, the Asian Development Bank, and other international financial organizations regularly, and seek support for preferential financial and scientific cooperation projects. Build a more flexible, efficient, and open mechanism for international scientific and technological cooperation in energy, take the initiative to integrate into the global industrial chain, and build demonstration projects for international cooperation. Encourage provincial enterprises to actively participate in foreign cooperation and expand the export of provincial products in the fields of equipment manufacturing, new energy, energy conservation, and environmental protection. Organize the International Forum on Clean Energy Development and build a globally influential platform and exhibition brand for the exchange of ideas on the energy revolution.

4.4 Direction and Key Tasks of Qinghai's Inter-Provincial and Inter-Regional Energy Cooperation Mechanism

With the inter-provincial and inter-regional cooperation system and mechanism for energy as the starting point, efforts should be made to promote energy cooperation between provinces and regions, innovate new inter-provincial and inter-regional absorption mechanism for clean energy, and ensure that Qinghai's clean energy "can absorb output and ensure income".

1. *Optimize the trans-provincial and trans-regional operation management mechanism*
   Establish an early-warning mechanism for trans-provincial consumption of clean energy, including early-warning classification, early-warning release, early-warning disposal, early-warning cancellation, and early-warning assessment, to identify and prevent possible new energy consumption difficulties in advance. To ensure the smooth consumption of new energy generation, an assessment mechanism for the consumption of clean energy across provinces and regions shall be established. Optimize the province electric power dispatching mechanism,

clarify the provincial regulators, regional control sub-center and responsibility division at the national center for electric power dispatching communication, provincial regulators responsible for the management of new energy scheduled to run in the province, regional control sub-center is responsible for the jurisdiction, and new energy grid dispatching operation management, national electric power dispatching communication center is mainly responsible for the macroeconomic regulation and control of national power grid, the direct dispatching inter-district power grid, and the power plant, coordinate the relationship between the local power grid scheduling.

2. *Improve the cross-provincial and cross-regional macro-control mechanism*
Accelerate the promotion of the renewable energy power quota system, and the competent energy department under the State Council shall determine the minimum proportion of renewable energy power consumption in electricity consumption at the provincial level. By the targets on the proportion of non-fossil energy consumption and the development and utilization of renewable energy, a unified national renewable energy green certificate trading mechanism will be established, and the subsidy mechanism for new energy electricity will be further improved. Make overall planning for the relationship between local carbon market trials, national carbon market construction and global carbon market development, actively promote the construction of a unified national carbon emission trading market, and promote the efficient allocation of clean energy resources across provinces and regions through the establishment of a market trading mechanism that internalizes external costs.

References

1. General Office of the People's Government of Qinghai Province. Work Plan for Building A National Clean Energy Demonstration Province of Qinghai Province (2018–2020) [Z], 2018-12-29.
2. Lin, W. B., & Fang, M. (2016). Energy system revolution: Concepts and frameworks. *Study & Exploration, 3,* 71–78.
3. Guo, J. F., Gao, S. J., Tao, H., Wu, X., & Cai, S. H. (2016). The connotation and strategic goal of China's energy system revolution in 2030. *China Economic Times*, 2016-2-23 (A05)
4. CPC Central Committee, State Council. Opinions on Further Deepening the Reform of electric power System [Z], 2015-3-16.
5. National Development and Reform Commission. Pricing Method for transmission and Distribution price of provincial Power grid [Z], 2020-1-19.
6. Jing, C. M. (2016). Proposals for energy system reform during the 13th Five-Year Plan Period. *Review of Economic Research, 2016*(60), 5–10, 21.
7. Xie, X. X., Ren, D. M., & Zhao, Y. Q. (2017). Research on the institutional reform of China's energy revolution. *Energy of China, 36*(4), 16–19, 44.

Geographical Locations and Market Efficiency of Listed Companies—Analysis Based on the Chinese Market

Ying Ren, Bowen Pan, Chunyi Wang, Ruoyu Yan, and Mingyin Zhang

Abstract Several researches on behavioral finance indicate that the geographical locations of listed companies can affect investors' behaviors, return on investment or stock prices. Based on what has been achieved, this article studies whether the geographical locations of listed companies affect the weak-form market efficiency defined by Fama (J Finan 25(2):383-417, 1970 [1]). This article selects geographical and historical data of more than 1000 stocks traded on the Shanghai and Shenzhen stock exchanges between 2009 and 2018 and analyzes this issue by defining three financial centers and eight central cities and using the ADF test and variance ratio test. The test results show that the closer a listed company is to a financial center (or a central city), the more likely that its share price performance tends to conform to the weak-form efficient market hypothesis. On the contrary, the farther away a listed company is from a financial center (or a central city), the more likely that its share price performance tends to deviate from the weak-form efficient market hypothesis.

Keywords Geographical location · Efficiency market hypothesis · ADF test · Variance ratio test

Y. Ren · M. Zhang
School of Economics and Management, Beijing Jiaotong University, Beijing, China
e-mail: renying@bjtu.edu.cn

M. Zhang
e-mail: 1187595323@qq.com

B. Pan (✉)
School of Finance, Central University of Finance and Economics, Beijing, China
e-mail: 2019310269@sf.cufe.edu.cn

C. Wang · R. Yan
The Experimental High School Attached to Beijing Normal University, Beijing, China
e-mail: ryan_diepio@outlook.com

R. Yan
e-mail: wawe@263.net

M. Li et al. (eds.), *IEIS 2020*, https://doi.org/10.1007/978-981-33-4363-4_11

1 Preface

The traditional theories of finance are based on the precondition of the rational economic person. The most influential theory based on the rational economic person is the efficient market hypothesis. The efficient market hypothesis holds that information will be accurately reflected in asset prices in a timely and accurate manner. Therefore, in an effective market, investors cannot use the information available to obtain abnormal returns in the capital market. Over time, various capital market anomalies have been discovered. For example, Basu [2] discovered the value effect: those stocks with low PEs could yield positive abnormal returns. Rozeff and Kinney [3] found the January effect: the earnings of stocks in the U.S. stock market in January are significantly more than the earnings in other months. Banz [4] found the size effect: those stocks with small market caps could yield relatively high risk-adjusted earing rates. De Bondt and Thaler [5] found the long-run reverse effect: those low yielding stocks over the past 3–5 years could perform better in the next few years than those high yielding stocks in the past 3–5 years. Jegadeesh and Titman [6] found the momentum effect: those stocks with higher yields over a period of time in the past could still yield higher in the short term than those low yielding stocks in the past. These financial anomalies mean that investors can use the relevant information to obtain abnormal returns, which cannot be explained by the efficient market hypothesis and poses challenges to the efficient market hypothesis.

Since the 1980s, a new branch of research in the finance study field had gradually formed in order to explain various financial anomalies, which is behavioral finance. The behavioral finance does not assume that investors in the market are non-distinctive or rational, and it plays emphasis on individual differences and irrational behaviors. It does not believe that the market is always effective. Therefore, behavioral finance can provide reasonable and enlightening explanations to those phenomena which cannot be explained by traditional financial theories. For example, Goetzmann and Kumar [7] found that most individual stock investors had serious insufficient asset diversification problems, which, according to traditional financial theories, could only be interpreted that most individual equity investors are risk-prone investors. However, this explanation contradicted many other research findings on individual risk attitudes. From the behavioral finance perspective, the insufficient diversification may be related to a number of factors, one of which is the "local preference" of investors. "Local preference" refers to the belief of the investors that they have advantages with respect to the information of the listed companies close to where they live, though this may be an illusion about information. In addition, psychological researches have also indicated that people prefer to make investment decisions in the environments they are familiar with.

The aforementioned examples show that the geographical locations of listed companies can affect the behaviors of investors. Inspired by this point, this article will conduct researches on whether the geographical locations of listed companies in China are related to market efficiency. The market efficiency here refers to the weak-form efficiency defined by Fama [1]: the market prices of stocks have fully reflected all past historical information. In related articles, a classical method of testing whether the weak-form efficiency is valid is to test whether asset prices are consistent with the random walk model, which is also used in this article to test the market efficiency. The fundamental point of view of this article is that the geographical locations of listed companies are related to the market efficiency. The closer a listed company is to a financial center or a central city (as defined below), the easier that the information related to it can be obtained and disseminated by the market, and therefore the more the performance of its stock price is consistent with the random walk model. In contrast, the farther away a listed company is to a financial center or a central city, the more likely that its stock price deviates from the random walk model. To determine whether this view is valid, this article will use the geographical and historical data of more than 1000 stocks traded on the Shanghai and Shenzhen exchanges between 2009 and 2018 using the ADF test and the variance ratio test.

Section 2 of this article summarizes some of the existing research results related to the issues studied in this article. Section 3 will introduce the data used in the analysis and the research design of this article. Section 4 reports the results of the empirical analysis. Section 5 is the summary of this article.

2 Literature Review

Fama [1] defined three forms of market efficiency: the weak-form efficient market, the semi-strong efficient market, and the strong effective markets. In a weak-form effective market, the stock prices have fully reflected all the historical information of the stocks, including closing prices of stocks, trade volumes, and turnover rates, etc. In a semi-strong efficient market, the stock prices have fully reflected all the information about the development prospects of a company which has been made public, including the historical information of the stock and the company's management status, as well as other publicly disclosed financial information. In a strong form efficient market, the stock prices have adequately reflected all the information about a company's operations, including publicly disclosed and undisclosed internal information. Fama believed that if a market is a weak-form efficient once, the historical information of stocks cannot be used to generate regular positive abnormal returns. Therefore, the share prices should follow a random walk model. The market efficiency discussed in this article refers to the weak-form efficient. Thus, the test of market efficiency in this article is made by testing whether stock prices follow the random walk model.

This article studies whether the geographical location is related to market efficiency, in fact, there are several studies in the reference materials which indicate that the geographical location of a listed company affects the behaviors of investors, the return on investment or the stock price. The anomalies in capital markets pose challenges to the traditional hypothesis implied in the efficient market hypothesis that information spreads evenly in a specific market. Some studies have found that geographical proximity brings advantages in respect of the information. Coval and

Moskowitz [8] found that the greater the geographical distance is, the more asymmetric the information become and the more investors tend to avoid risks and uncertainties in economic transactions. The researches done by Coval and Moskowitz [9] pointed out that the stock exchanges located near the city of the mutual fund managers earn 2.67% than those stock exchanges which are farther away. A report by Malloy [10] alleged that those analysts who are near the headquarter of a U.S. public company issue more accurate stock price forecasts. Further researches have found that geographical proximity to a central area brings advantages on information. The researches done by Christoffersen and Sarkissian [11] showed that those mutual funds located in financial centers outperform other funds in terms of gross margin and risk-adjusted returns. The researches done by Anand et al. [12] showed that geographical proximity gives a capital market the ability to incorporate information into securities pricing, which is price discovery capacity. An empirical study by Cuiling [13] found that, compared with non-central cities, listed companies located in central cities can obtain higher stock returns. Geographically far away from a central area lead to disadvantages on information and information asymmetries. The researches done by Loughran [14] showed that information is spread from urban public enterprises to rural and small-scale enterprises, and that after putting the scale, industry and covering scope of analysts under control, stocks of urban companies still outperform rural and small urban stocks. Studies of "local preferences" have also confirmed that geographic location can influence investors' decisions. Jingmei et al. [15] found that local preferences have not been detected in Beijing and Shanghai, while there is a high proportion of shares of local companies in the portfolios of investors in Shenzhen, which partly confirmed the impact of geography on investors' investment activities. The studies of Xiaosong [16] found that venture capital firms show preferences to local investment in order to reduce risks of uncertainty in the face of information asymmetry.

3 Data and Research Design

3.1 Geographic Location Data of Listed Companies

This article selects the headquarter place as the geographical location of a listed company, which is the actual place of business of that company. The reason we do not select the company's registered address is that the choice of registration place may be affected by tax policies and other preferential policies of the registration place, and the actual place of business, as the contact address of the company, is the main source of company information, which has greater impacts on the company's share price than the registration place. Therefore, this article selects the actual place of business of the headquarters of listed companies for our analysis.

This article selects two ways to measure the influences of geographical locations of listed companies. One is to define three financial centers to determine the geographical location distances of listed companies from the financial centers. The other one is to define eight central cities to determine the geographical location distances of listed companies from the central cities. The financial centers are defined as follows. According to the Development Report on China Private Equity Investment Fund Industry (2017), this article selects the three cities with the largest number of private equity funds in China at the end of 2017 of the financial central cities for our studies: Beijing, Shanghai, and Shenzhen. The central cities are defined as follows. According to the National Urban System Planning (2006–2020), there are ten cities in China which play an important central and hub role in all respects of finance, management, culture and transportation, etc., which are Beijing, Shanghai, Shenzhen, Tianjin, Chongqing, Chengdu, Wuhan, Guangzhou, Zhengzhou, and Xi'an. However, because Tianjin is very close to Beijing and Guangzhou is also very close to Shenzhen, this article excludes Tianjin and Guangzhou from the list of the ten central cities. Thus, the remaining eight central cities are Beijing, Shanghai, Shenzhen, Chongqing, Chengdu, Wuhan, Zhengzhou, and Xi'an.

In this article, the calculation of the geographical distance of a listed company from a financial center or central city by is made by (a) extracting the actual business address of a listed company through the Wind database, and using XGeocoding V2 to convert the actual business address to coordinates; (b) building programs with Python and using the API of Baidu Map to calculate the ground traffic distance (self-driving) between the actual business address of the listed company and the address of the city government of each city; and (c) selecting the shortest distance between the listed company and the three financial centers as the distance from the financial center of the listed company, and selecting the shortest distance from the eight central cities as the distance from the central city of the listed company.

3.2 *Historical Data of Stocks of Listed Companies*

The historical data of listed companies originates from the Wind database. This article is based on stocks that have been traded in the market since January 1, 2009 and are still trading in the market as of December 31, 2018. On this basis, this article excludes those stocks which have been subject to ST in the ten years from 2009 to 2018. There are totally 2431 trading days during the ten years. This article excludes those stocks which have been traded for less than 60% of the total number of trading days. In addition, this article also excludes those stocks which have once been suspended from trading for more than 125 trading days consecutively over these ten years (approximately half a year's trading days). In the end, 1063 stocks remained as the samples. The empirical analysis of this article will use the daily closing price data of the 1063 stocks to test whether the market efficiency is related to the geographical locations of listed companies.

3.3 Research Design

This article studies the relationship between the geographical locations and market efficiency of listed companies from two angles. One is to consider the distances of listed companies from the three financial centers, and the other is to consider the distances of listed companies from the eight central cities. With respect to the three financial centers, three sample groups are considered: the stocks of the 100 listed companies closest to the financial centers and the 100 listed companies farthest away from the financial centers; the stocks of the 150 listed companies closest to the financial centers and the 150 listed companies farthest away from the financial centers; and the stocks of the 200 listed companies closest to the financial centers and the 200 listed companies farthest away from the financial centers. For the eight central cities, the same three sample groups are considered in this article.

This article only considers the weak-form efficiency of the market. Therefore, for any of the aforementioned sample groups, this article will test each stock to see whether the corresponding logarithmic closing price follows the random walk model. If the number of stocks of those listed company nearer which follows the random walk model is greater than the stocks of those listed companies farther away, it will indicate that the stocks of the nearest listed companies are more inclined to follow the weak-form efficient market hypothesis, which supports the fundamental point of this article. Otherwise, the point of this article cannot be supported by data.

In order to test whether the stock price follows the random walk model, this article will use two widely used statistical tests: the ADF test and the variance ratio test. The ADF test is the unit root test most widely applied, mainly for testing whether the time sequence has a unit root. Defining the logarithmic closing stock price time sequence as Pt, the random walk model tested in this article shall be

$$P_t = c + \rho P_{t-1} + \varepsilon_t \quad (1)$$

where P_t is the price of the stock on the t trading day, P_{t-1} is the price of the stock on the *t*-1 trading day, ε_t is a white noise sequence, and$c$ is the constant item. When $\rho = 1$, the time sequence P_t has a unit root, and the stock price follows the random walk model. When $\rho < 1$, P_t does not have a unit root, the stock price is a smooth sequence and does not follow the random walk model.

In the empirical study of this article, the basis for determining whether the stock price follows the random walk model is the p value of the ADF test. If the p value is greater than the given level of significance, the original assumption of the existence of unit root is accepted, and the stock price is considered to have the nature of the random walk. If the p value is less than the given level of significance, the original hypothesis of the unit root is rejected, and the stock price is considered not following the random walk.

Another test used in this article is the Variance Ratio (VR) test proposed by Lo and MacKinlay [17], which is widely used in the empirical financial papers to test whether asset prices meet the weak-form efficient market hypothesis. The VR test uses directly the return rate of assets and defines the logarithmic stock price time sequence P_t, then $R_t = P_t - P_{t-1}$, which is the return rate of the logarithmic stock price. Assuming that the stock price follows the random walk, then R_t should be a white noise sequence, and the statistic of the VR test $V(k) = \text{var}(R_t + \cdots + R_{t-k+1})/k\text{var}(R_t)$ should satisfy $V(k) = 1$, where k is a lagged variable. Otherwise, $V(k)$ will not be equal to 1, which is the basic principle that the VR test can be used to test the weak-form efficiency of stock prices.

The empirical study of this article determines whether the stock price follows the random walk by referring to uses the p value of the VR test as well. If the p value is greater than the given level of significance, the stock price is considered to follow the random walk model, and if the p value is smaller than the given level of significance, the stock price is considered not following the random walk.

4 Results of Empirical Research

4.1 Distance to Financial Centers and Market Efficiency

Table 1 is a descriptive statistic of this group of data, including the average, maximum, minimum, 25% quantile, median, and 75% quantile figures. The variables involved include the number of trading days (N-trade), the ratio of the number of trading days to the number of market opening days (R-trade), the maximum number of consecutive trading suspension days (S-max), the distance from the listed companies to Beijing (D-BJ), the distance from listed companies to Shanghai (D-SH), the distance from listed companies to Shenzhen (D_SZ), and the distance from the listed company to the financial centers (D-FC), which is the shortest distance from a listed company to Beijing, Shanghai, and Shenzhen.

As can be seen from Table 1, there are huge geographical differences between listed companies in the samples. The farthest companies are more than 3000 km away from Beijing, Shanghai, and Shenzhen, while the nearest companies are within 1 km.

Table 1 Descriptive statistics (3 financial centers)

Statistics	N_trade	R_trade	S_max	D_BJ	D_SH	D_SZ	D_FC
Average	2336	0.961	38	1231.58	1003.30	1459.33	483.49
Maximum	2427	0.998	125	3549.38	4131.23	4486.80	3549.38
Minimum	1855	0.763	1	2.55	0.51	0.54	0.51
25% quantile	2293	0.943	5	760.67	299.63	1064.76	21.83
Median	2358	0.97	24	1216.33	1055.69	1434.22	299.69
75% quantile	2404	0.989	68	1782.97	1434.58	1959.40	747.73

Note The unit of distance in the table is thousand meters

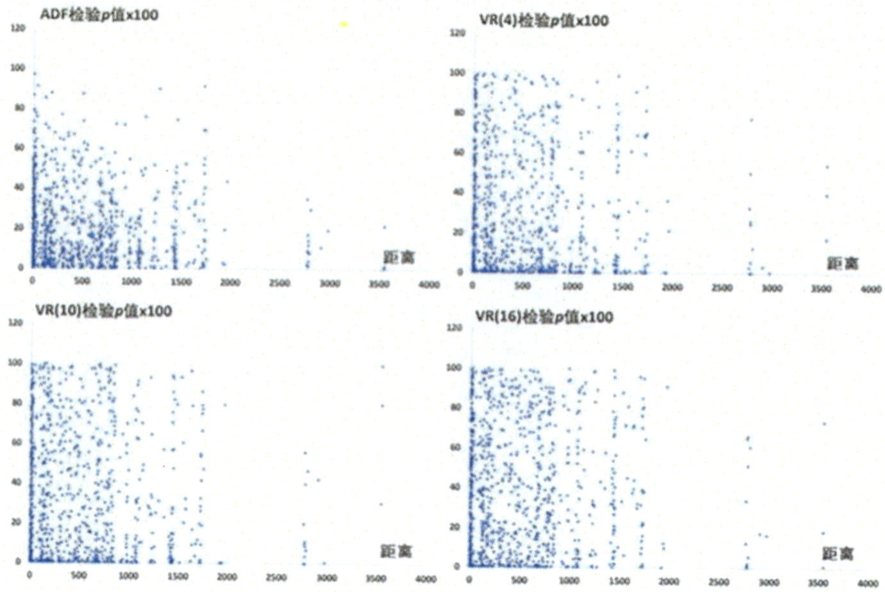

Fig. 1 Test statistics distribution (3 financial centers)

We then conduct the ADF test and VR test to the logarithmic daily closing price sequences of all listed companies. For the VR test, we select the lagged variable of $k = 4$, 10, 16, and the tests are recorded as VR (4), VR (10), and VR (16) respectively. First, the overall results of the test are shown by the four scatterplots in Fig. 1. The horizontal axis in the figure is the distance from the financial center of the listed company, the vertical axis is the 100 times enlarged *p* value of the ADF test, the VR (4) test, the VR (10) test, and the VR (16) test. Each point corresponds to the distance of a listed company from the financial center and the *p* value of the corresponding test multiplied by 100.

As can be seen from Fig. 1, when the *p* value is large, that is, the upper half of the four scatter plots, the scattered points falling on the left area are significantly more than the scattered points falling in the right area, and it can be seen that the listed companies far away from the financial centers tend to refuse their share prices to follow a random walk. However, on the whole, most of the vertical axis values of the scattered points are large, so the whole market still shows the features of weak-form efficiency. In addition, it can be seen that there is no linear relationship between geographic location and *p* value.

Based on the analysis of Fig. 1, in order to identify the impact of the geographical locations of listed companies, we will only compare the most extreme situations of geographical locations below. The data are divided into three sample groups in accordance with the distances to the financial centers following the research design made in the previous section, each of which contains *n* nearest listed companies and *n* farthest listed companies. Next, we set the significance level α at 0.05 and 0.1,

Table 2 Empirical test results (3 financial centers)

Quantity of samples	Test	$\alpha = 0.05$		$\alpha = 0.1$	
		n Nearest companies rejection rate	*n* Farthest companies rejection rate	*n* Nearest companies rejection rate	*n* Farthest companies rejection rate
$n = 100$	ADF	0.19	0.31	0.35	0.45
	VR(4)	0.31	0.47	0.40	0.49
	VR(10)	0.24	0.37	0.36	0.45
	VR(16)	**0.23**	**0.20**	**0.31**	**0.26**
$n = 150$	ADF	0.18	0.29	0.35	0.46
	VR(4)	0.31	0.40	0.41	0.45
	VR(10)	0.28	0.36	0.39	0.43
	VR(16)	**0.21**	**0.18**	**0.28**	**0.26**
$n = 200$	ADF	0.19	0.29	0.35	0.47
	VR(4)	0.36	0.40	**0.46**	**0.44**
	VR(10)	0.28	0.34	0.37	0.42
	VR(16)	**0.22**	**0.19**	**0.28**	**0.27**

and conduct the ADF test and VR test on the stock prices of listed companies in the sample group respectively. Table 2 reports the percentage of rejections by the random walk hypothesis. The higher the rejection rate is, the greater the market deviates from the weak-form efficient market hypothesis. If the rejection rate of the stock prices of the listed companies which are nearest to financial centers is lower than that of stock prices of the listed companies which are farthest to financial centers, we believe that this is a piece of evidence supporting the fundamental point of this article: the geographic locations of listed companies are related to the market efficiency, and the closer such companies are to the financial centers, the more likely their stock prices tend to follow the weak-form efficient market hypothesis.

It can be seen from the Table 2 that, when n = 100, for the significance level of 0.05 and 0.1, except for the VR test with lagged variable of 16, the stock prices of the listed companies which are closest to the financial centers follow the random walk in a greater extent than those of the listed companies which are farthest to the financial centers. This supports the fundamental point of this article. Except for the test results of VR (4) when n = 200 and $a = 0.1$, the test results when $n = 150$ and $n = 200$ are almost the same with the results when $n = 100$. Generally, among the 24 different settings in Table 2, the results of a total of 17 settings indicate that the stock prices of the listed companies closest to the financial centers are more in line with the random walk model compared with the stock prices of the listed companies which are farthest to the financial centers. As such, we believe that the results reported in Table 2 can support the fundamental point of this article.

Table 3 Descriptive statistics (8 central cities)

Statistics	D_CQ	D_ZZ	D_WH	D_CD	D_XA	D_CC
Average	1479.82	1017.91	1231.22	1646.08	1255.65	320.10
Maximum	3152.12	3225.73	3550.51	3188.41	2775.80	2744.73
Minimum	2.68	6.14	1.85	1.52	1.08	0.51
25% quantile	1303.24	693.64	772.24	1506.21	1027.48	15.59
Median	1567.76	936.46	1214.85	1779.50	1273.96	177.02
75% quantile	1719.71	1374.41	1784.11	1935.81	1499.47	510.63

Note The unit of distance in the table is thousand meters

4.2 *Distance to Central Cities and Market Efficiency*

With financial centers as the ending points in distance calculation, the analysis result of the previous group of data has exhibited the evidence supporting the fundamental point of this article. To test the steadiness of this result, we then consider the scenario where more cities are treated as ending points. Compared with the previous group, in this group of data, we add five cities, namely Chongqing, Chengdu, Wuhan, Zhengzhou, and Xi'an, which constitute the samples of the eight central cities. Table 3 is similar to Table 1, which reports the descriptive statistics for six new variables. These variables are: the distance from listed companies to Chongqing (D-CQ), the distance from listed companies to Zhengzhou (D-ZZ), the distance from listed companies to Wuhan (D-WH), the distance from listed companies to Chengdu (D-CD), the distance from listed companies to Xi'an (D-XA) and the distance from listed companies to central cities (D-CC), which is the minimum value of the distance between a listed company and the eight central cities. As can be seen from Table 3, the distance between listed companies and target cities becomes smaller after the expansion of the number of cities. A listed company which belongs to the listed companies farthest in the previous group of data may become a listed company nearest to an ending point.

Similar to Figs. 1 and 2 exhibits a graph of relations between the test results and the distance of the four tests. In the case of the eight central cities, the distance becomes smaller, so the scattered point in Fig. 2 is closer to the origin of the lateral axis. However, on the whole, the distribution pattern of scatterplots is similar to that of the three financial centers, and it can be seen that for those listed companies which are farther away from the central cities, their stock prices tend to refuse to follow the random walk.

We then use the sample listed companies which are nearest and farthest to identify the impacts of geographical locations of listed companies.

Table 4 reports the same analysis result with Table 2. It can be seen that with respect to the eight central cities, the percentages of stock prices of the listed companies following the random walk model which are nearest to the central cities are mostly higher than those of such listed companies which are farthest to the central cities. Among the 24 different settings considered, only in four situations the percentages

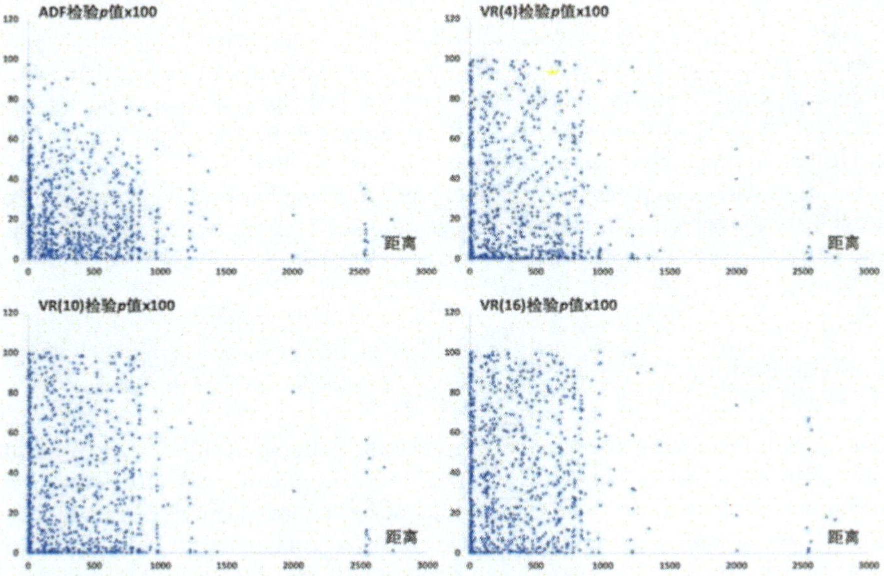

Fig. 2 Test statistics distribution (8 central cities)

Table 4 Empirical test results (8 central cities)

Quantity of samples	Test	$\alpha = 0.05$		$\alpha = 0.1$	
		n Nearest companies rejection rate	*n* Farthest companies rejection rate	*n* Nearest companies rejection rate	*n* Farthest companies rejection rate
$n = 100$	ADF	0.21	0.29	0.37	0.44
	VR(4)	0.32	0.39	**0.43**	**0.42**
	VR(10)	0.24	0.29	0.35	0.39
	VR(16)	**0.22**	**0.20**	**0.31**	**0.28**
$n = 150$	ADF	0.19	0.28	0.36	0.43
	VR(4)	0.32	0.36	**0.41**	**0.40**
	VR(10)	0.24	0.29	0.34	0.41
	VR(16)	0.19	0.21	0.25	0.31
$n = 200$	ADF	0.20	0.27	0.39	0.43
	VR(4)	0.34	0.40	0.45	0.45
	VR(10)	0.29	0.33	0.38	0.45
	VR(16)	0.20	0.22	0.27	0.29

of stock prices following the random walk model of the listed companies which are farthest are higher, which are three fewer than the results in Table 2. As such, in general, even if we use the 8 central cities as the distance ending points, the fundamental point of this article can still be supported by data. Therefore, based on all the aforementioned research results, this article holds that the geographical locations of listed companies are definitely related to the market efficiency, and the closer a listed company is to a larger city, the more likely that its stock price variations tend to follow the weak-form effective market hypothesis.

5 Conclusions

There is more and more evidence proving that the geographical locations of listed companies will affect the behaviors of investors and further affect the corresponding capital market. Based on the existing results of behavioral finance researches, this article holds that the geographical locations of listed companies are relevant to the market efficiency. The closer to financial centers and central cities the listed companies are, the more likely their stock prices tend to follow the weak-form efficient market hypothesis. On the contrary, the farther away the listed companies are, the more likely their stock prices tend to deviate from the weak-form efficient market hypothesis.

In order to verify the fundamental point of this article, this article uses the ADF test and a series of variance ratio tests to compare the efficiency of those listed companies which are near to financial centers or central cities and those listed companies which are far away from financial centers or central cities. The empirical research results are more or less consistent with the fundamental point of this article: the closer a listed company is located to a financial center or a central city, the more likely its stock price tends to follow the random walk, and the farther away a listed company is from a financial center or a central city, the more likely its stock price tends to deviate from the random walk. We believe that there are two reasons for this result: (a) since a large number of individual investors live in larger cities, which are financial centers or central cities as defined herein, there are many traders with advantages on information about those listed companies which are close to financial centers or central cities, and the transmission of information is more adequate, which cause the corresponding stock prices to be more likely to follow the random walk model. On the contrary, for those listed companies which are farther away, the corresponding stock prices are more likely to deviate from the random walk model; (b) most institutions which study the performance and prospects of listed companies and publish their study reports are located in relatively large cities. Therefore, they have distinctive advantages with respect to on site researches and information gathering about the listed companies close to them and are able to disclose information timely and accurately. As a result, although individual investors do not have too much direct access to the information of listed companies, they can obtain more accurate information in a timely manner through indirect means, which makes the corresponding stock prices able to reflect the historical information in a timely manner and to follow the random walk model.

References

1. Fama, E. F. (1970). Efficient capital markets: A review of theory and empirical work. *Journal of Finance, 25*(2), 383–417.
2. Basu, S. (1977). Investment performance of common stocks in relation to their price—Earnings ratios: A test of the efficient market hypothesis. *The Journal of Finance, 32*(3), 663–682.
3. Rozeff, M. S., & Kinney, W. (1976). Capital market seasonality: The case of stock returns. *Journal of Financial Economics, 3*(4), 379–402.
4. Banz, R. W. (1981). The relationship between return and market value of common stocks. *Journal of Financial Economics, 9*(1), 3–18.
5. De Bondt, W. F. M., & Thaler, R. H. (1985). Does the stock market over-react. *The Journal of Finance, 40*(3), 793–805.
6. Jegadeesh, N., & Titman, S. (1993). Returns to buying winners and selling losers: Implications for stock market efficiency. *Journal of Finance, 48*(1), 65–91.
7. Goetzmann, W. N., & Kumar, A. (2008). Equity portfolio diversification. *Review of Finance, 12*(3), 433–463.
8. Coval, J. D., & Moskowitz, T. J. (1999). Home bias at home: local equity preference in domestic portfolios. *Journal of Finance, 54*(6), 2045–2073.
9. Coval, J. D., & Moskowitz, T. J. (2001). The geography of investment: informed trading and asset prices. *Journal of Political Economy, 109*(4), 811–841.
10. Malloy, C. J. (2005). The geography of equity analysis. *The Journal of Finance, 60*(2), 719–755.
11. Christoffersen, S. E. K., & Sarkissian, S. (2009). City size and fund performance. *Journal of Financial Economics, 92*(2), 252–275.
12. Anand, A., Gatchev, V. A., Madureira, L., et al. (2011). Geographic proximity and price discovery: Evidence from Nasdaq. *Journal of Financial Markets, 14*(2), 193–226.
13. Cuiling, Y. (2017). Company Geography, Investor Sentiment and Stock Return. Chongqing University.
14. Loughran, T. (2007). Geographic dissemination of information. *Journal of Corporate Finance, 13*(5), 675–694.
15. Jingmei, Z., Fengyun, W., & Mei, L. (2012). Empirical research on local preferences and regional avoidance in investment decisions. Investment Research, *31*(01), 123–141.
16. Xiaosong, Y. (2018). Empirical Analysis of the Existence, Factors of Influence and Consequences of Local Preferences of Venture Capital Institutions. Shanghai Normal University.
17. Lo, A. W., & Mackinlay, A. C. (1988). Stock market prices do not follow random walks: Evidence from a simple specification test. *Review of Financial Studies, 1*(1), 41–66.

Research on the Investment Efficiency of Transport Infrastructure in Countries Along the Belt and Road

Bingyao Chen and Xuemeng Guo

Abstract Adopting Stochastic Frontier Analysis (SFA), we firstly estimate the investment efficiency of transport infrastructure in 20 countries along the Belt and Road from 2000 to 2017, and then we empirically examine the impacts of the public–private partnership (PPP) and the Belt and Road Initiative (BRI). The results show that, from 2000 to 2017, the average transport infrastructure investment efficiency of these 20 countries is 0.552, with different degrees of resource waste and efficiency loss, despite the improvement trend is sound. And the efficiency of the developed countries is higher. The results also manifest that the application of PPP, the increasing amount of and investments on PPP projects all significantly helped improve the efficiency, which implies a good effect of PPP. Moreover, the efficiency had been significantly improved since the BRI was put forward in 2013, which proves the value and effectiveness of this worldwide policy.

Keywords Stochastic frontier analysis (SFA) · Public–private partnership (PPP) · The belt and road initiative (BRI) · Transport infrastructure · Investment efficiency

1 Introduction

According to 2018 BMI's global infrastructure statistic data, the proportion of newly signed contracts of transportation ranked the highest among all industries in countries along the Belt and Road. And *The Belt and Road International Infrastructure Development Index Report* released in 2019 also claimed that the demand index for the development of transportation industry was higher than that of other industries, suggesting great investment and development opportunities regarding infrastructure connectivity for countries along the route.

B. Chen · X. Guo (✉)
School of Economics and Management, Beijing Jiaotong University, Beijing, China
e-mail: xmguo@bjtu.edu.cn

B. Chen
e-mail: 18113068@bjtu.edu.cn

M. Li et al. (eds.), *IEIS 2020*, https://doi.org/10.1007/978-981-33-4363-4_12

Projects on transport infrastructure typically anticipate rather massive investments, long recovery cycles, intensive funding and technological supports. As a consequence, the traditional investment mode with the government acting as the single agent is gradually insufficient to meet the demand for building up such infrastructure. As a worldwide popular financing mode for infrastructure, PPP has currently become mainstream in funding international cooperation of the BRI, and has been extensively applied into practice.

Can the promotion of the BRI improve the efficiency of investment? Is the application effect of PPP better than the traditional modes? Only few studies by far have concentrated on infrastructure investment efficiency by using quantitative analytical approaches. Zhongmin et al. [1] and Chenyang and Lianghai [2] measured and analyzed the transport infrastructure efficiency in countries along the Belt and Road adopting Data Envelopment Analysis (DEA). Juan et al. [3] investigated the economic growth brought by investments on road, railway and other infrastructure in countries along the and Road with a regression method. The empirical research of Dongfang [4] showed that the PPP investment was significantly positively correlated with the infrastructure efficiency. Lixin et al. [5] used the multi-period Differences-in-Differences method to assess the impact of PPP implementation on China's infrastructure output efficiency and found it causing no significant improvement. We can conclude that most scholars used DEA, a non-parameter estimation method based on linear programming to evaluate the infrastructure efficiency, so the accuracy of their results can be further enhanced. Also, few literatures have studied the relationship between PPP and the investment efficiency of infrastructure. Most of them took PPP only to be a policy shock while ignoring the impact of PPP projects' number and the scale of investments, so the comprehensiveness and robustness of their results needs to be improved.

In view of the above, we use a more precise SFA approach to estimate the investment efficiency of transport infrastructure in 20 countries along the Belt and Road from 2000 to 2017. In order to get more robust and accurate empirical results, we consider three dimensions of PPP, namely the implementation of PPP which is a dummy variable, the number of PPP projects and the investment amount. We are also interested in the policy effects of the BRI, so we add it to our regression and process it as a dummy variable. In addition, we further analyze the path and mechanism of the impact of PPP on the investment efficiency of infrastructure. On the whole, What we aim to explore is the effectiveness of the BRI policy, and our work is of practical significance to further refinement and promotion of PPP in the field of transport infrastructure.

2 SFA and Investment Efficiency Estimation

SFA and DEA are output efficiency measurement methods commonly used in economics. Unlike DEA, a non-parameter estimation method based on linear programming, SFA construct production frontiers by utilizing the production function. It requires highly precise model setting and data, and can be flexibly adjusted according to the characteristics of the research object; it is also of high pertinency and accuracy in measurement, all of which make it suitable to process large samples. What is more, SFA takes the impact of random factors on output into consideration, which avoids DEA's insufficiency in attributing the output deviation all to technical efficiency, and can make the estimation result more reliable. Regarding these, we choose to employ SFA to estimate the investment efficiency of transport infrastructure.

2.1 Model Specification

Production functions commonly used for SFA are Cobb-Douglas function and Trans-Log function. The former is concise in form and the parameters are endowed with meanings directly from economics, but only under the premise that the returns to scale remain constant; while the latter overcomes this shortcoming by considering the impact of capital and labor interaction on output, making the parameter estimation more consistent with the economic facts.

According to the basic principles of the model constructed by Battese and Coelli [6], we use logarithmic C-D function to build the analytic model:

$$\mathrm{Ln}Y_{it} = \beta_0 + \beta_1 \mathrm{Ln}K_{it} + \beta_2 \mathrm{Ln}L_{it} + v_{it} - u_{it} \tag{1}$$

Trans-Log function generalizes C-D function on the basis of formula (1):

$$\begin{aligned} \mathrm{Ln}Y_{it} = {} & \beta_0 + \beta_1 \mathrm{Ln}K_{it} + \beta_2 \mathrm{Ln}L_{it} + \beta_3 t + 1/2\beta_4 (LnK_{it})^2 \\ & + 1/2\beta_5 (LnL_{it})^2 + 1/2\beta_6 t^2 + \beta_7 \mathrm{Ln}K_{it} \mathrm{Ln}L_{it} \\ & + \beta_8 t \mathrm{Ln}K_{it} + \beta_9 t \mathrm{Ln}L_{it} + v_{it} - u_{it} \end{aligned} \tag{2}$$

$$u_{it} = e^{-\eta(t-T)} u_i \tag{3}$$

In formula (1) and (2), Y_{it} represents output, K_{it} represents capital input, L_{it} represents labor input, β_1 represents capital output elasticity, β_2 represents labor output elasticity; The error term is a composite structure, in which v_{it} obeys Independently Identically Distribution and $u_{it} \geq 0$ obeys Half-Normal Distribution to represent the impact only on individual factor.

In formula (3), η is a time-varying parameter. $\eta > 0$ indicates an efficiency increase over time, and vice versa.

The technical efficiency is defined as:

$$TE_{it} = e^{-u_{it}} \tag{4}$$

If $u_{it} = 0$, $TE_{it} = 1$, indicating that the individual production scale is on the production frontier, suggesting a state of technical efficiency. If $u_{it} > 0, 0 < TE_{it} < 1$, the individual production scale is below the production frontier, suggesting a state of technical inefficiency.

As for the selection of the function form, a model based on the Trans-Log function is generally used to perform model setting test. Based on the statistical test result, combined with the nature of the research object and whether the parameter estimation results are consistent with common sense, etc., we can make a comprehensive judgment.

2.2 *Indicator Selection*

Considering data comparability and availability, the GDP of 20 countries along the Belt and Road are selected as output, the number of employees in transportation industry as labor input, and investments on transport infrastructure (including railway, road, water and air transport) as capital input. The data derives from the World Bank database, the OECD database and *Yearbook of China Transportation and Communications.*

2.3 *Estimation Result*

The infrastructure investment system is a multi-objective, multi-variable and non-linear input-output system. That the input-output increment multiplies from the original increment by a fixed ratio does not accord with the objective conditions in reality, and does not apply to the premise that the returns to scale are constant [7]. The result of our model setting test (see Table 1) also proves that the Trans-Log production function is more suitable.

Table 1 Result of model setting test

	Coefficient	Standard-error	T-ratio
σ^2	0.277	0.202	1.372
γ	0.935***	0.047	19.737
μ	0.194	0.419	0.464
η	0.043***	0.004	11.056
Log likelihood = 158.232***			
LR statistic = 648.369***			

Note (1) ***, **, * respectively indicate statistically significance of 99%, 95%, and 90%; (2) Only the core and relevant results are listed

$\gamma = 0.935$, indicates that there is an evident composite structure in the error term. So it is necessary to employ SFA instead of simple linear analysis. $\eta = 0.043$, shows that the investment efficiency of transport infrastructure in these 20 countries has been increasing year by year.

Table 2 presents the estimation of the investment efficiency based on the Trans-Log function.

The sample countries are grouped according to their degrees of development, and the average investment efficiency of each group is listed in Table 3.

The above results show that: (1) the investment efficiency of transport infrastructure in 20 countries along the Belt and Road does not reach the optimal production efficiency and manifests a technical inefficient state, although it has been increasing with each passing year; (2) the average investment efficiency of transport infrastructure in these countries is 0.552, and that of the developed countries is 0.722, which is much higher than 0.479 of the less-developed countries. But Turkey, as a developing country, reaches an average investment efficiency as high as 0.967. Its ranking first is inseparable from Turkey's policy of massive investment in infrastructure to fuel economic growth. At present, Turkey has become one of the most potential markets in emerging economies to attract foreign investments; (3) 90% of the countries below the average investment efficiency are less-developed, and the average investment efficiency of Moldova ranks in the last with 0.160, which is rooted in its lagging economy dominated by agriculture and animal husbandry, its frequent occurrence of political conflicts and economic crises.

3 Theory Analysis and Research Hypothesis

As a project financing mode, PPP has been widely used in the field of infrastructure construction and public management. Developed countries are still the main and mature PPP markets. According to the theory of public goods, transport infrastructure belongs to the quasi-public goods, with limited non-competitiveness and non-excludability. It can generate revenue and be provided by the market. Meanwhile, as the market is not constantly efficient in the allocation of resources, it requires the government to exert powerful macro-control functions. So this provides the basis for public-private partnership in the field of transport infrastructure.

Most studies have confirmed the superiority of PPP over the traditional supply mode. Guangliang et al. [8] pointed out that PPP can effectively introduce market mechanisms into the public domain and improve the efficiency of public services. PPP is not only an innovation of financing mode, but also of public management and administrative governance [9]. PPP has inherent advantages and can improve the quality and efficiency of public service supply [10]. Stulz [11] showed that PPP, being a "semi-organized, semi-market" hybrid way of organization, on the one hand can avoid the disadvantages caused by the local government's "grabbing hands" out of long-term administrative monopolies, therefore reducing the active rent creation

Table 2 Investment efficiency of transport infrastructure in 20 countries from 2000 to 2017

Country	Year																	
	2000	2001	2002	2003	2004	2005	2006	2007	2008	2009	2010	2011	2012	2013	2014	2015	2016	2017
Turkey	0.954	0.956	0.957	0.959	0.961	0.962	0.964	0.965	0.967	0.968	0.969	0.971	0.972	0.973	0.974	0.975	0.976	0.977
Slovenija	0.918	0.921	0.924	0.927	0.930	0.933	0.936	0.938	0.941	0.943	0.945	0.947	0.950	0.952	0.954	0.955	0.957	0.959
Czech	0.720	0.730	0.740	0.749	0.758	0.767	0.775	0.784	0.792	0.799	0.807	0.814	0.821	0.828	0.834	0.841	0.847	0.853
Poland	0.712	0.722	0.732	0.742	0.751	0.760	0.769	0.777	0.785	0.793	0.801	0.808	0.815	0.822	0.829	0.836	0.842	0.848
Slovakia	0.680	0.691	0.701	0.712	0.722	0.732	0.741	0.751	0.760	0.768	0.777	0.785	0.793	0.801	0.808	0.815	0.822	0.829
Hungary	0.587	0.600	0.613	0.626	0.638	0.650	0.662	0.673	0.685	0.696	0.706	0.716	0.726	0.736	0.746	0.755	0.764	0.772
Croatia	0.546	0.560	0.573	0.587	0.600	0.613	0.626	0.638	0.650	0.662	0.673	0.684	0.695	0.706	0.716	0.726	0.736	0.745
Estonia	0.513	0.527	0.542	0.556	0.569	0.583	0.596	0.609	0.622	0.634	0.646	0.658	0.670	0.681	0.692	0.703	0.713	0.723
Russia	0.467	0.482	0.497	0.512	0.526	0.541	0.555	0.568	0.582	0.595	0.608	0.621	0.633	0.646	0.657	0.669	0.680	0.691
China	0.451	0.467	0.482	0.496	0.511	0.526	0.540	0.554	0.568	0.581	0.595	0.608	0.620	0.633	0.645	0.657	0.668	0.680
Lithuania	0.424	0.439	0.455	0.470	0.485	0.500	0.514	0.529	0.543	0.557	0.571	0.584	0.597	0.610	0.623	0.635	0.647	0.659
Latvia	0.402	0.418	0.433	0.448	0.464	0.479	0.494	0.508	0.523	0.537	0.551	0.565	0.579	0.592	0.605	0.618	0.630	0.643
Romania	0.366	0.381	0.397	0.413	0.428	0.444	0.459	0.474	0.489	0.504	0.518	0.533	0.547	0.561	0.574	0.588	0.601	0.614
India	0.286	0.301	0.316	0.332	0.347	0.363	0.379	0.394	0.410	0.425	0.441	0.456	0.471	0.486	0.501	0.516	0.530	0.544
Serbia	0.262	0.277	0.292	0.307	0.323	0.338	0.354	0.370	0.385	0.401	0.416	0.432	0.447	0.463	0.478	0.493	0.507	0.522
Bulgaria	0.233	0.247	0.262	0.277	0.292	0.308	0.323	0.339	0.354	0.370	0.386	0.401	0.417	0.432	0.448	0.463	0.478	0.493
Albania	0.210	0.224	0.238	0.253	0.268	0.283	0.298	0.314	0.329	0.345	0.361	0.376	0.392	0.407	0.423	0.438	0.454	0.469
Azerbaijan	0.112	0.123	0.134	0.146	0.158	0.171	0.184	0.197	0.211	0.225	0.240	0.254	0.269	0.284	0.300	0.315	0.331	0.346
Georgia	0.112	0.123	0.134	0.146	0.158	0.171	0.184	0.197	0.211	0.225	0.240	0.254	0.269	0.284	0.300	0.315	0.331	0.346
Moldova	0.068	0.076	0.085	0.094	0.104	0.114	0.125	0.136	0.148	0.160	0.173	0.186	0.200	0.213	0.228	0.242	0.257	0.272

(continued)

Table 2 (continued)

Country	Year																	
	2000	2001	2002	2003	2004	2005	2006	2007	2008	2009	2010	2011	2012	2013	2014	2015	2016	2017
Average	0.451	0.463	0.475	0.488	0.500	0.512	0.524	0.536	0.548	0.559	0.571	0.583	0.594	0.606	0.617	0.628	0.639	0.649

Note The investment efficiency is sorted from high to low

Table 3 Average investment efficiency of transport infrastructure of each group

Type	Country	Average investment efficiency
Developed	Slovenija; Czech; Slovakia; Hungary; Estonia; Latvia	0.722
Less-developed	Turkey; Poland; Croatia; Russia; China; Lithuania; Romania; India; Serbia; Bulgaria; Albania; Azerbaijan; Georgia; Moldova	0.479

and unintentional effects. On the other hand, it is also able to make maximum use of social resources and increase social welfare [12].

From a macro point of view, PPP can enhance the degrees of marketization in the field of transport infrastructure, effectively releasing market vitality and stimulating the quality and efficiency of economic growth. At the micro level, PPP not only relieves the government's financial pressure, but also takes full advantage of social capital's rich management experience, advanced technology and complete information to reduce transaction costs. Moreover, the incentive effect caused by private profit-seeking will help to reduce the whole life cycle cost of PPP projects, improve the operational efficiency and service level, thus achieve a win-win result between economic and social benefits. In this light, we put forward the first research hypothesis:

H1: PPP can improve the investment efficiency of transport infrastructure.

Since the BRI was first proposed in 2013, this keynote initiative from China and belongs to the world has received numerous responses and praises. A total of 65 countries along the route have joined into this initiative, making it an global cooperation and exchanges of public goods with worldwide popularity.

Infrastructure connectivity is an important part of the BRI. It demands a massive amount of investments [13]. Meanwhile, the BRI has effectively solved a series of problems especially in the shortage of domestic capital supply in relevant countries, enabling them to enjoy the financial resources from other countries along the route as well as from the international market. Wei et al. [14] found that the openness level of these countries was generally on the rise, which effectively promoted the economic growth in the area. Yujuan [15] introduced the Belt and Road dummy variable into her research and pointed out that the BRI can boost the economic along the route and optimize China's overseas investment structure. Accordingly, we put forward the second research hypothesis:

H2: The Belt and Road Initiative can improve the investment efficiency of transport infrastructure in countries along the route.

4 Empirical Test

4.1 Data, Variables and Model

1. Data. We select data of 20 member states of the BRI from 2000 to 2017 as research samples. PPP data of transport infrastructure derives from PPI database, and economic data derives from World Bank database. Individual missing data is supplemented by using linear interpolation.
2. Variables. The investment efficiency of transport infrastructure estimated by SFA is taken as the explained variable; PPP implementation, PPP projects number, PPP investment amount and the BRI are selected as explanatory variables. We also choose the levels of urbanization, financial development, the degrees of opening-up, industrial structure, as well as the economic growth of each country as control variables. The specific meaning and descriptive statistics of the above variables are presented in Table 4.
3. Model. According to the results gained from the Hausman test, we adopt the panel fixed effect model.

$$IE_{it} = \alpha_0 + \alpha_1 PPP_{nit} + \alpha_2 Control_{it} + \varepsilon_{it} \tag{5}$$

Table 4 Descriptive statistics of variables

Type	Name	Meaning	Mean	SD	Min	Max
Explained	*IE*	Transport infrastructure investment efficiency	0.552	0.234	0.068	0.977
Explanatory	PPP_1	Transport infrastructure PPP implementation or not	0.225	0.418	0	1
	PPP_2	Transport infrastructure PPP projects number	1.906	7.378	0	73.00
	PPP_3	Transport infrastructure PPP investment (logarithm)	4.547	8.526	0	24.33
	BRI	The Belt and Road Initiative	0.222	0.416	0	1
Control	*MS*	Financial development: broad money/GDP	57.93	35.41	10.38	209.50
	OP	Opening-up: (import + export)/GDP	82.20	37.13	19.31	176.20
	IND	Industrial structure: secondary industry/tertiary industry	0.53	0.156	0.216	0.947
	UR	Urbanization: urban population/total population	58.54	11.40	27.67	74.67
	EC	Economic growth: growth rate of GDP per capita	4.562	4.605	-14.27	33.00

$$IE_{it} = \beta_0 + \beta_1 BPI_{it} + \beta_2 Control_{it} + \varepsilon_{it} \quad (6)$$

In formula (5) and (6), IE_{it} is the investment efficiency of transport infrastructure. There are three measurement methods for PPP_n: PPP_1 is a dummy variable for PPP implementation, values 1 for implementation and 0 for non-implementation; PPP_2 indicates the number of PPP projects; PPP_3 indicates the investment amount on PPP. BRI_{it} is a dummy variable for the Belt and Road Initiative. In 2013, the BRI was formally proposed by China, and therefore, 2013 is taken as the node before which BRI_{it} values 0 and after which values 1. $Control_{it}$ represents a collection of control variables. ε_{it} represents the error term. α_1 and β_1 respectively measure the net effect of PPP and the BRI on the investment efficiency of transport infrastructure.

4.2 Result and Analysis

First, we use model (5) to test the impact of PPP on the investment efficiency of transport infrastructure. The regression results are shown in Table 5.

PPP implementation, PPP projects number, and PPP investment are all significantly in positive correlations with the investment efficiency of transport infrastructure, which proves that the regression results is of high validity. It is indicated that PPP can improve the investment efficiency of transport infrastructure, in line with Hypothesis 1. *MS*, *OP*, and *UR* are constantly significantly positive with 99% confidence, indicating that the financial development, the degrees of opening-up, and the level of urbanization can promote the improvement of investment efficiency in transport infrastructure. *IND*'s and *EC*'s being negative suggests that the development of the service industry has improved the investment efficiency of transport infrastructure more than the secondary industry; and that the rapid economic growth will bring more unimpeded financing channels along with a sound institutional environment, partly reducing the spillover effect of PPP. The above results are generally consistent with our expectations and conform to the general law of world economic development.

Next, we use model (6) to test the impact of the BRI on the investment efficiency of transport infrastructure. The regression results are shown in Table 6.

With or without the control variables, the *BRI* coefficient is always significantly positive, indicating that there is a significant difference in the investment efficiency of transport infrastructure among the countries before and after the promotion of the BRI, that is, the investment efficiency remarkably improved after the countries entered the initiative, confirming that this policy is of profound value and effectiveness. The significance of the control variables are consistent with the above analysis, so there would not be further discussion.

Table 5 Regression results of PPP

Explained variable: *IE*								
Variable	1	2	Variable	3	4	Variable	5	6
PPP_1	0.0455***	0.0158**	PPP_2	0.0013*	0.0008**	PPP_3	0.0025***	0.0008*
	(3.29)	(1.99)		(1.84)	(2.09)		(3.45)	(1.91)
MS		0.0035***	*MS*		0.0035***	*MS*		0.0035***
		(10.72)			(10.63)			(10.68)
OP		0.0007***	*OP*		0.0007***	*OP*		0.0007***
		(3.56)			(3.28)			(3.56)
IND		-0.0917*	*IND*		-0.1315**	*IND*		-0.0958*
		(-1.77)			(-2.53)			(-1.86)
UR		0.0038***	*UR*		0.0038***	*UR*		0.0038***
		(3.32)			(3.35)			(3.28)
EC		-0.0019***	*EC*		-0.0017***	*EC*		-0.0019***
		(-3.01)			(-2.82)			(-3.02)
Cons		0.0896	*Cons*		0.1165*	*Cons*		0.0943
		(1.30)			(1.68)			(1.37)
R^2	3.10%	71.74%	R^2	0.99%	71.79%	R^2	3.39%	71.70%

Note (1) The values in parentheses are *t* values; (2) ***, **, * respectively indicate statistically significance of 99%, 95%, and 90%; (3) the estimated results were processed by Stata14.0

Table 6 Regression results of the BRI

Explained variable: *IE*		
Variable	1	2
BRI	0.1039***	0.0459***
	(16.71)	(7.85)
MS		0.0029***
		(9.49)
OP		0.0006***
		(3.15)
IND		−0.0469
		(−1.01)
UR		0.0029***
		(2.83)
EC		−0.0017***
		(−3.06)
Cons		0.1556**
		(2.50)
R^2	45.17%	77.11%

5 Research Conclusion

1. From 2000 to 2017, the average investment efficiency of transport infrastructure in 20 countries along the Belt and Road is 0.552—far less than 1, the state of technical efficiency—which indicates that input and output of transport infrastructure is suffering from efficiency loss and resource waste, despite an overall rising trend. The average investment efficiency of the developed countries is 0.722, higher than 0.479 of the less-developed countries. Turkey, Slovenia, Czech Republic, Poland, Slovakia, and Hungary have relatively higher investment efficiencies with an average of 0.822, which is close to the optimal efficiency. 90% of the countries below the average investment efficiency are less-developed. The average investment efficiency of Moldova, ranking at the bottom with 0.160, has to do with the country's political turbulence and economical backwardness.
2. Both the implementation of PPP and the increasing number of and investments on PPP projects can enhance the investment efficiency of transport infrastructure, which proves that PPP, as a majorly market operation, enables the participation of social capital to balance the allocation of resources, to improve the resource efficiency, and to generate ample profits. It is worth noting that such a conclusion does not imply that we encourage unrestricted employment of PPP or as many investments as possible. The number of and investment on PPP projects reflect the construction volume and service scale of infrastructure just to a limited extent. Only by building more high quality PPP projects and determining reasonable

project investment scale can the infrastructure investment efficiency be improved in the long range, forming favorable scale effects and positive external effects.

3. After the BRI was put forward, the investment efficiency of transport infrastructure in countries along the route has been significantly improved. The reasons are that, on the one hand, comprehensive infrastructure development and transport interconnectivity are two priorities of the initiative. Transportation is strategically crucial as it can lead further development of all industries, and therefore resulting in a huge demand of constructions and investments of the related countries. On the other hand, the BRI connects developed and developing economies and establishes an all-round financial cooperation system to break the bottleneck of funding, which extensively promotes the exchanges of and communications on advanced experience and technology. "Financial integration", as is suggested by this initiative, has played an important role in funding and safeguarding infrastructure development as well as improving investment efficiency in the Belt and Road region.

Acknowledgements The National Natural Science Foundation of China subsidized projects "Research on the Performance Evaluation System of PPP Passenger Flow-Value" in Urban Rail Transit. (Code:71973009).

References

1. Zhongmin, Li., Deshui, X., & Yao, Yu. (2014). Analysis of transport infrastructure efficiency of the Silk Road Economic Belt—Malmquist index method based on DEA. *Seeker, 02,* 97–102.
2. Chenyang, Z., & Lianghai, L. (2018). Analysis on the investment efficiency of transport infrastructure in the countries along the Belt and Road. *Economic Research Guide, 377*(27), 75–77+81
3. Juan, Z., Hui, L., Yunfei, W., & Zuanshi, L. (2016). Comparison of transport infrastructure investment efficiency in countries along the belt and road. *Statistics and Decision, 19,* 61–63.
4. Dongfang, C. (2019). The researches on the promoting effect of PPP on infrastructure efficiency—An empirical test based on countries along the belt and road. *Technology Economics and Management Research, 07,* 1–10.
5. Lixin, W., Chuan, Z., & Cangxi, Li. (2019). Does the PPP improve the output efficiency of infrastructure? *Financial Research, 01,* 90–102.
6. Battese, G. E., & Coelli, T. J. (1992). Frontier production functions, technical efficiency and panel data: With application to paddy farmers in India. *Journal of Productivity Analysis, 03,* 153–169.
7. Xiping, R. (2017). The researches on the evaluation of urban infrastructure investment efficiency based on DEA. *Economic System Reform, 05,* 49–54.
8. Guangliang, Y., Long, C., & Dan, Y. (2018). PPP, mixed ownership enterprise and social welfare. *Social Science Front, 06*(2), 64–74 (in Chinese)
9. Shaoying, C. (2017). Public service innovation of PPP localization in China. *Jinyang Academic Journal, 04,* 121–130.
10. Ping, P. (2017). The researches on the governance value and implementation of PPP. *Theory Monthly, 10,* 124–130.
11. Stulz, R. M. (2005). The limits of financial globalization. *Journal of Finance, 60.*

12. Qian, F., & Hui, Q. (2015). The essence, generating motivation and evolutionary dynamic mechanism of PPP. *Commercial Research, 05,* 137–143.
13. Liping, Z. (2017). Infrastructure investment, financing needs and China's role under the Belt and Road initiative. *China Development Review, 19,* 18.
14. Wei, M., Xiaoshu, C., & Tao, L. (2019). Does the improvement of openness promote regional economic growth?—Empirical analysis based on panel data of countries along the Belt and Road. *Economic Survey, 5.*
15. Yujuan, Li. (2018). Analysis on the impact of the Belt and Road initiative on the efficiency of China's overseas investment. *Journal of Chongqing Jiaotong University (Social Science Edition), 18*(06), 88–93.

Does Vocational Education Benefit Manufacturing?—An Empirical Study Based on China Panel Data from 2009 to 2016

Jingcheng Li

Abstract This study uses a theoretical model for vocational education to promote manufacturing growth by promoting human capital accumulation and employment and using panel data from 30 provinces and cities in China from 2009 to 2016 to test the model empirically. The results show that the development of vocational education cannot promote manufacturing growth. The average contribution rate of vocational education to economic growth is 0.465%, which is higher than its contribution to second industry (−0.226%). The research conclusion also shows that if the development of vocational education cannot meet the needs of economic growth, it will become a factor restricting economic growth.

Keywords Manufacturing · Vocational education · Industrial hollowing

1 Introduction

China economy has experienced a fast development period with taking advantage of cheap labor, though currently China needs to face the challenge of rising cost of labor. Furthermore, the evolution of productivity requires higher labor skills, which leaves a gap in supply and demand in high-quality labor. After WWII, Germany and Japan both have successfully built their efficient vocational education system, which help their manufacturing flourish. Their experience offers a good example for China to learn from, thus On 13 February 2019, the Chinese State Council published its *Implementation plan on National Vocational Education Reform* with signaling a dramatic strengthened focus on vocational education. However, according to current studies, there are few about what role vocational education plays in Chinese manufacturing and how vocational education have influences on manufacturing. This paper proposed a model about vocational education influences manufacturing through accumulation and employment and used cross-province panel data from 2009 to 2016 in China for an empirical study.

J. Li (✉)
Beijing Laboratory of National Economic Security Early-Warning Engineering, School of Economics and Management, Beijing Jiaotong University, Beijing, China
e-mail: 19113005@bjtu.edu.cn

M. Li et al. (eds.), *IEIS 2020*, https://doi.org/10.1007/978-981-33-4363-4_13

2 Literature Review

The majority of studies regard human capital as the quality of labor. The research on the relationship between labor quality and productivity and economic growth originates from endogenous growth theory of 1980s [1]. Since the emergence of endogenous economic growth theory, a lot of results have verified the important role of labor quality (human capital) on productivity improvement and economic growth. Howitt and Aghion [2] argue that investment in education has a significant effect on long-term economic growth. A consensus has been reached that labor quality is an important factor determining the competitiveness of a country or a region.

In China, it is tricky to define if the human capital includes vocational school, due to unique Chinese education pathways which is different from Germany's. In Germany, it is students have autonomy to choose vocational colleges and universities, and they have freedom to transfer from one system to another [3]. Yet, China's vocational systems would be seen as an alternative to academic university education, students who failed or are not well-prepared in GaoKao would choose it (Fig. 1).

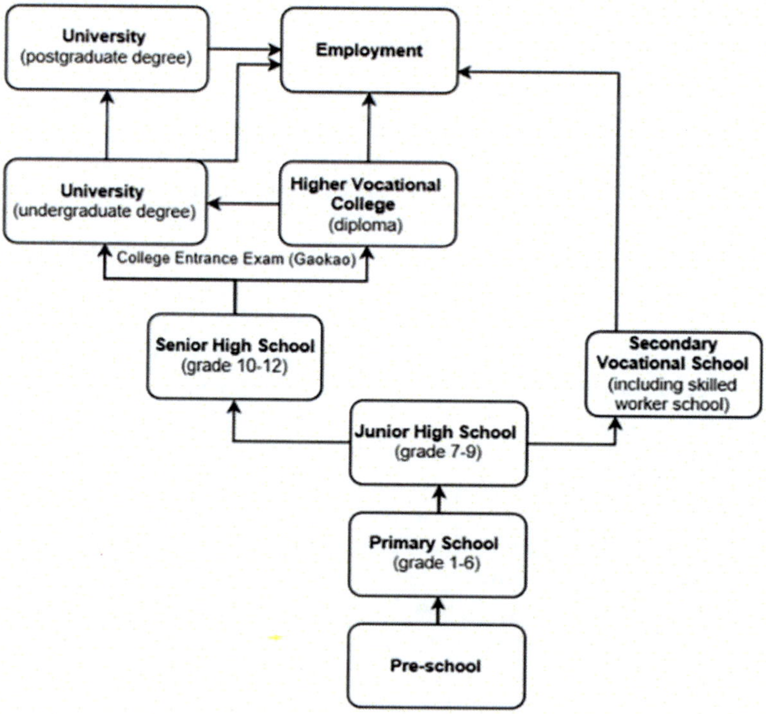

Fig. 1 Chinese education system

Existed analysis shows that there is an obvious two-way causal relationship between education investment and economic growth. From 1982 to 1993, the contribution rate of education to the average annual GDP growth rate was 7.54%. Pritchett argued that the relationship between improvement of education and economic growth is negative [4]. Yet previous study calculated the contribution of education to the economic growth rate in China from 2003 to 2014, they argued that results show that the contribution of higher education to comprehensive education level in the six provinces was all <1% within the sample period, and its contribution to economic growth was <5%, far lower than that of basic education to economic growth [5].

Due to heterogeneity of industrial structure, testing how vocational education influences manufacturing industry in different provinces is necessary to improve the pertinence and effectiveness of the reform mentioned. Wang Lei tried an empirical study from 2004 to 2007 to test the contribution of vocational education to regional economic growth, but he used middle school graduates to replace secondary vocational school, as Fig. 1 shows, those two education systems is not the same, thus the conclusion is not reliable [3].

3 Theoretical Model

According to Cobb Douglas production function, economic growth depends on the input of labor and capital. Based on human capital theory, the contribution of education to economic growth comes from the promotion of social human capital by education, with becoming one of the sources of economic growth [3]. Thus, we set the model as:

$$\mathrm{Y} = \mathrm{AK}^{\alpha}(\mathrm{LH})^{\beta} \tag{1}$$

where Y means output; A means technology level which has been assumed as a constant; K stands for capital input; L is labor input; H is the input of human capital part; α and β is capital output elastic coefficient and labor output elastic coefficient respectively, $\alpha > 0$, $\beta > 0$ and $\alpha + \beta = 1$.

Thus, we assume H is a function of education expense E, therefore $H = E^{\gamma}$. Equation (1) could be expressed as:

$$\mathrm{Y} = \mathrm{AK}^{\alpha}(\mathrm{LE}^{\gamma})^{\beta} \tag{2}$$

Taking time factor t into account, the product function would be:

$$\mathrm{Y}_{\mathrm{t}} = A\mathrm{K}_{\mathrm{t}}^{\alpha}(L_{\mathrm{t}}\mathrm{E}_{\mathrm{t}}^{\gamma})^{\beta} \tag{3}$$

Transfer (3)–(4) by taking logarithm and derivatize t:

$$\ln(\dot{Y_t}) = \ln(\dot{A_t}) + \alpha \ln(\dot{K_t}) + \beta \ln(\dot{L_t}) + \beta\gamma \ln(\dot{E_t}) \tag{4}$$

Based on (4), get the model for labor and education's contribution to economic growth.

$$C_L = \beta \frac{l}{Y} \tag{5}$$

$$C_E = \beta\gamma \frac{e}{Y} \tag{6}$$

where Y is the economic growth rate; l is the average growth of labor; C_L is the contribution rate from labor input to economic growth; e is the average growth of education scale, C_E it education's contribution rate to economic growth.

One of vocational education's purposes is to build labor reserve with qualified skills for economic development especially for manufacturing. According to the contribution of vocational education to economic growth, (3) can be decomposed into two parts. The first part is that vocational education offers trained-well labors for economic activities, which is regarded as employment part; the second part is that pre-employment education could promote the increase of human capital in the whole society, which is regarded as education part. In that case, the contribution of vocational education would be:

$$C_i = C_L + C_E \tag{7}$$

where C_L stands for the labor part of vocational education's contribution; C_E stands for the education part of vocational education's contribution. Based on (5) and (6), C_L is:

$$C_L = \beta L_v \frac{l}{Y} \tag{8}$$

L_v means the contribution part of vocational education to labor increase.

$$C_E = \beta\gamma E_v \frac{e}{Y} \tag{9}$$

E_v is the proportion of vocational education investment in education investment.

The aim of this study is to test the contribution rate of vocational education to manufacturing C_i.

4 Empirical Analysis

4.1 Data

Based on the availability of data and the purpose of this study, relevant data from provinces, municipalities, and autonomous regions in the country from 2009 to 2016 are selected as samples. The growth of second industry would stand for *Y*; Capital input would be represented by social fixed asset investment; the total employed people would stand for labour input; Education input *E* is indicated by education funding input; L_v would be indicated by graduate students from vocational education (includes both secondary vocational school and higher vocational college); the proportion of vocational education investment in education investment E_v is expressed by the proportion of education expenditure in secondary professional schools and higher vocational college to education expenditure. All data in this article are from *China Statistical Yearbook (2010–2017)* and *China Education Funds Statistical Yearbook (2010–2017)*.

4.2 Model Selection

From a statistical view, the panel-data-model not only greatly increases the number of observations and improves the degree of freedom of the sample, but also reduces the influence of the multicollinearity of the explanatory variables and reduces the estimation error. Therefore, the measurement method of panel model in this paper can explain the contribution of vocational education to economic growth more comprehensively.

After doing F test (Pro > F = 0), a fixed model should be applied. To further confirm which model to use, we did Hausman test, and the result back us to use individual random effects model.

Then, the following regression would be run on the base of:

$$\ln(\mathrm{Y_t}) = \alpha \ln(\mathrm{K_t}) + \beta \ln(\mathrm{L_t}) + \beta\gamma \ln(E_t) \qquad (10)$$

4.3 Regression Results

Based on above discussion, using Stata 16.0 to run model (10), the results are shown in Table 1.

The next step is to calculate the contribution rate of vocational education to manufacturing industry, according to (8) and (9), we could get the sum contribution of vocational education, to increase robustness, we also use graduates students with

Table 1 The impact of vocational education on manufacturing growth—analysis based on individual random effect model with variable coefficient

Dependent variable	α		β		γβ	
Region	Coef.	t	Coef.	t	Coef.	t
Bei Jing	0.4805101	2.32	−3.620145	−4.51	0.2311548	1.84
Tian Jin	0.9093611	1.71	−1.989909	−5.97	−0.0.30419	−0.73
He Bei	0.4708641	2.08	0.77596	−0.72	−0.375987	−3.24
Shan Xi	2.468364	3.05	−12.7523	−2.6	−0.581977	−0.85
Inner Mongolia	0.3776666	4.1	−0.858512	−2.98	−0.094805	−0.83
Liao Ning	0.8732100	0.1	−2.864200	−4.67	−0.460719	−8.27
Ji Lin	0.5664365	4.17	−1.605867	−4.81	−0.085648	−0.85
Hei Long Jiang	0.1528	0.26	2.141369	1.02	−1.126822	−2.46
Shang Hai	−0.7016474	−5.86	−0.2548472	−1.44	0.1082274	1.10
Jiang Su	−0.341606	−4.66	45.38806	2.72	−0.08928	−1.88
Zhe Jiang	0.1635071	0.47	−0.010926	−1.14	−0.334125	−1.10
An Hui	0.0041621	0.01	−4.367603	−1.11	0.4359031	0.94
Fu Jian	−0.004346	−0.03	0.058787	0.11	0.0252659	2.11
Jiang Xi	−0.552049	−1.49	3.671389	0.79	0.0718886	0.28
Shan Dong	−0.625496	−2.13	9.309800	1.91	−0.297411	−1.12
He Nan	0.2997883	1.35	−3.102236	−4.99	−0.0165474	−0.11
Hu Bei	−0.1851253	−0.85	5.0475330	2.71	0.1128701	0.45
Hu Nan	−0.0798628	−0.46	3.7829890	4.10	0.0397514	0.23
Guang Dong	0.1622631	0.44	0.1549240	0.34	−0.303765	−0.91
Guang Xi	0.3451369	6.64	−0.2844363	−1.02	−0.224835	−5.68
Hai Nan	0.2718745	2.93	−0.793140	−2.40	−0.2091794	−1.77
Chong Qing	−0.1640000	−2.41	0.902067	2.71	0.024336	0.34
Si Chuan	0.7340625	2.92	−19.16641	−2.66	−0.2211794	−0.72
Gui Zhou	0.5630796	4.04	−0.206238	−0.86	−0.571685	−3.61
Yun Nan	−0.179287	−0.52	3.377952	1.08	−0.2324095	−0.85
Shanan XI	0.3932808	1.60	−0.347169	−1.70	−0.444848	−2.14
Gan Su	0.5988967	1.96	−7.195458	−3.38	−0.6971215	−1.97
Qing Hai	0.0123325	0.16	−5.297276	−3.71	0.2229541	0.16
Ning Xia	−0.089135	−1.74	−0.4297829	−0.80	0.1288378	2.32
Xin Jiang	0.1364355	0.91	−2.104364	−3.96	0.4443300	1.57
$R^2 = 0.936$						

certificates of competency to stand for L_v. Furthermore, we try to get a more comprehensive examination of vocational education's impact with taking GDP into account, Table 2; Figs. 2, 3 and 4 shows the results of contribution rate in different context:

- From the perspective of country level, the contribution of vocational education to manufacture industry is −0.226% on average, this result is nearly to the result

Table 2 The influence of vocational education

Region	Contribution rate		
	Manufacture	GDP	Q graduate
Bei Jing	0.489543	−0.192460	0.604399
Tian Jin	−0.398191	0.617638	−0.223411
He Bei	−0.502549	0.014328	−0.524614
Shan Xi	−0.558634	0.535527	−1.735990
Inner Mongolia	−0.123706	0.746019	−0.312925
Liao Ning	−1.316150	1.557320	−0.815939
Ji Lin	−0.120238	0.360396	0.047306
Hei Long Jiang	0.183984	1.405420	−0.306686
Shang Hai	−0.138768	0.424711	0.102722
Jiang Su	−0.384741	0.338585	−0.819373
Zhe Jiang	−0.436045	0.427749	−0.411111
An Hui	0.584154	0.005040	0.439115
Fu Jian	−0.726348	0.134515	−0.720947
Jiang Xi	−0.008638	−0.061340	0.069327
Shan Dong	−0.664927	1.041190	−0.626939
He Nan	0.151932	−0.101010	−0.119736
Hu Bei	0.225966	0.057011	0.294430
Hu Nan	0.037568	0.143743	0.031876
Guang Dong	−0.469067	0.579699	−0.474788
Guang Xi	−0.331478	1.238160	−0.409804
Hai Nan	−0.622987	1.204200	−0.788574
Chong Qing	0.049965	0.061866	−0.001338
Si Chuan	−0.565503	0.045992	−0.809737
Gui Zhou	−0.884102	0.128721	−0.891038
Yun Nan	−0.030219	0.446087	0.332060
Shanan XI	−0.325529	0.674096	−0.425671
Gan Su	−0.886242	−0.36954	−1.292920
Qing Hai	0.243189	0.246441	0.216654
Ning Xia	0.213886	1.30403	0.152174
Xin Jiang	0.517024	0.950009	0.182101

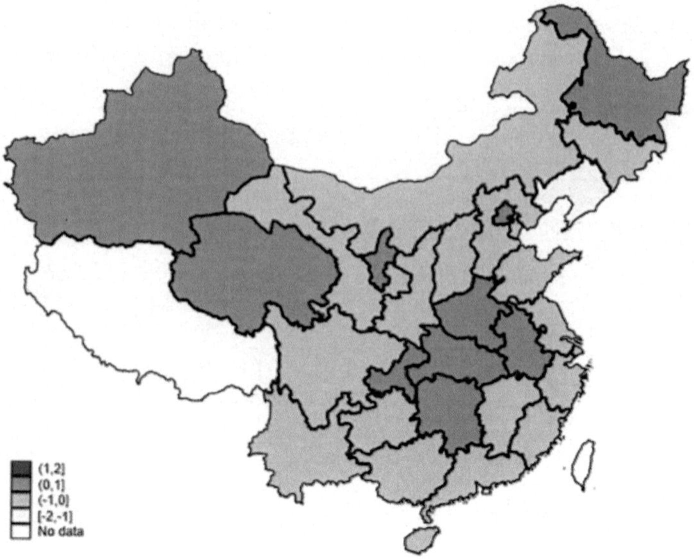

Fig. 2 Contribution of vocational education to manufacturing industry China, 2009–2016

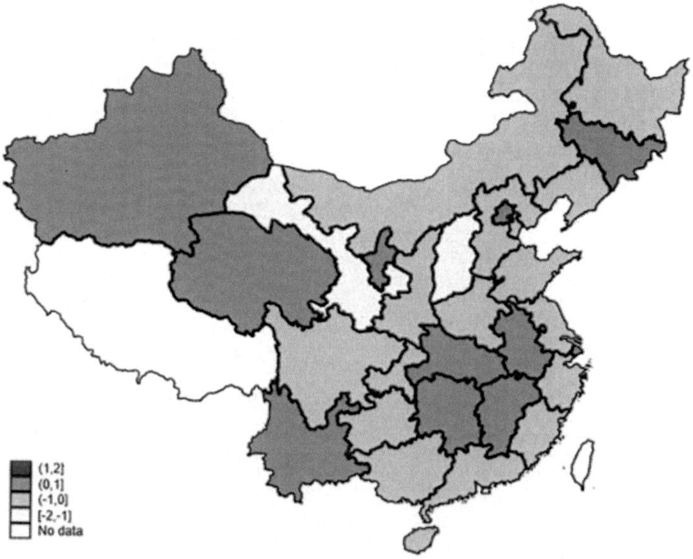

Fig. 3 Contribution of vocational education to manufacturing industry (using qualified graduates for robustness test) China, 2009–2016

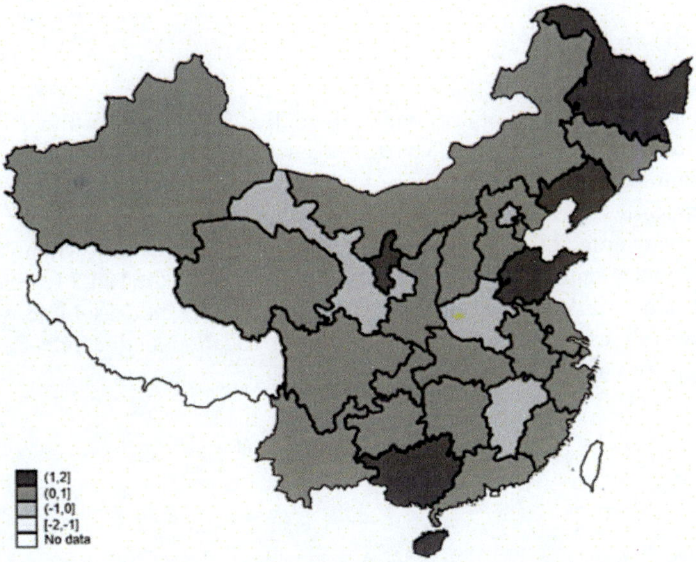

Fig. 4 Contribution of vocational education to GDP China, 2009–2016

of previous researches. It is worth noting that the contribution rate is negative, which means the current vocational education system could not fit the manufacture industry structure of China. This could explain why the government launch the reform plan of vocational education.

- The influences of vocational education has strong regional heterogeneity: for manufacture part, on one hand, vocational education plays a positive role in central China provinces such as Hu Bei, An Hui; on the other hand, vocational education could not offer positive help to east provinces and west provinces. An possible explanation for this phenomenon is that central regions undertakes industrial transfer from coastal areas, meanwhile the current vocational training system could not offer suitable second-industry labor to coastal areas which has finished industrial upgrade to some extent.
- When it comes to overall GDP, vocational education has 0.45% contribution rate to economic growth, which means vocational education could support the development of service industry. Combining with Fig. 4, it shows that vocational education supports development in the eastern region better. This phenomenon is consent with the trend of vocational education after 2008, the number of students who majored in teaching, nursing and accounting has increased dramatically.

5 Conclusion

China is changing he mode of economic growth that relies on investment and labor-intensive industries. Meanwhile, vocational education is changing its focus from training manufacturing labor to train service labor. The change of vocational education is a concern since vocational education seems to lose its navigation: it cannot contribute to manufacturing industry except in central regions. But when vocational schools shift its attention to third-industry-related training, the lack of sophisticated workers would be an emerging factor to deteriorate China industrial hollowing.

This research also indicates that the reform of vocational education should take full account of regional differences.

References

1. Mankiw, N., Romer, D., & Weil, D. (1992). A contribution to the empirics of economic growth. *The Quarterly Journal of Economics, 107*(2), 407–437.
2. Howitt, P., Aghion, P. (1998). *Journal of Economic Growth, 3*(2), 111–130.
3. Wang, L. (2011). The contribution of vocational education to regional economic growth. *Journal of Central University of Finance & Economics, 11,* 80–85.
4. Pritchett, L. (2001). Where has all the education gone? *The World Bank Economic Review, 15*(3), 367–391.
5. Zhu, T., Peng, H., & Zhang, Y. (2017). The influence of higher education development on economic growth: Evidence from central China. *Higher Education Policy, 31*(2), 139–157.

Research on the Factors Affecting the Income of Internet Monetary Funds in China

Chaoxiang Jia and Xinyi Mei

Abstract In this paper, the seven indicators: the Shanghai interbank offered rate, the Shanghai composite index, the government bond index, the establishment time of the monetary fund, the size of the monetary fund, the proportion of the underlying bond, and the proportion of cash invested in the monetary fund, were selected as variables. Based on the impact of the previous period of monetary fund returns to the current period, a dynamic panel regression model was established to compare and analyze the returns of 28 Internet monetary funds in China. The results showed that the last term yield of Internet monetary funds, Shanghai interbank offered rate, Shanghai stock index, and government bond index has a significant positive impact on the yield of monetary funds. In contrast, the establishment time of monetary fund, fund size, and the proportion of underlying bonds and cash harm the yield of monetary funds. Finally, this paper put forward reasonable suggestions from the perspectives of fund regulatory authorities, marketing agencies, and investors.

Keywords Internet finance · Monetary funds · Earnings

1 Introduction

As a new financial business model of financing, payment, investment, and information intermediary services, Internet finance plays an irreplaceable role in accelerating the formation of China's new financial ecosystem [1]. Among them, Internet financial monetary funds play a crucial role, which is based on Internet platform sales. Different from other traditional monetary funds, the Internet monetary fund realizes direct marketing through Internet channels, which reduces the custody fees of intermediary organizations such as banks and wealth management companies and

C. Jia (✉)
National Academy of Economic Security, Beijing Jiaotong University, Beijing, China
e-mail: jcxdante@163.com

X. Mei
Beijing Laboratory of National Economic Security Early-Warning Engineering, Beijing Jiaotong University, Beijing, China
e-mail: meixinyi1234@126.com

M. Li et al. (eds.), *IEIS 2020*, https://doi.org/10.1007/978-981-33-4363-4_14

increases the yield rate of Internet monetary funds. With the further promotion of the "Internet plus" strategy in China, Internet monetary funds are likely to lead the development direction of future monetary funds. China's Internet monetary fund emerged in 2013. Although some Internet monetary funds appeared in the early stage, due to the immature financial management awareness of the users and the backward Internet technology, the market did not pay much attention to the monetary funds. After the launch of Yu'E Bao, monetary funds became the primary way for the public to deposit their spare money. Yuyan Wang considers that monetary funds may replace the role of traditional bank savings, and according to the needs of different products, the current and time deposit models of Banks may be realized [2]. However, the returns of monetary funds have generally been declining in recent years. Taking Yu'E Bao as an example, the returns have dropped from 6% to less than 3%. While bearing the fluctuation of earnings, the comparative study of the performance of Internet monetary funds can help investors better choose suitable monetary funds for their money management.

The essence of the Internet financial monetary fund is the collection of Internet sales channel and monetary fund. Furthermore, the Internet financial monetary fund has more convenient operation and high liquidity characters. That mainly stems from the fact that the Internet financial monetary fund implements the T + 0 trading mode, which allows investors to redeem online or pay for online shopping on the day of purchase. Consequently, there is also liquidity risk for Internet monetary funds, "rigid payment" such as sizeable short-term redemption or improper fund management will lead to a liquidity crisis and then affect the yield rate. Some scholars have studied the possible risks of Internet monetary funds. Hongmei Zhang found that individual investors dominate the client structure of Chinese Internet monetary funds. When major redemption windows such as holidays or financial markets are unstable, the possibility of sizeable short-term redemption of monetary funds is very high [3]. Danping Qiu believes that the current liquidity risk of Internet money market funds is vast, and there is no guarantee of the deposit reserve system and deposit insurance system corresponding to the traditional financial industry. There are no effective response solutions to short-term liabilities and capital outflows [4]. By building the GARCH model and calculating the VaR value of the comparison sample funds, Lingyan Fu found that the return volatility risk of Internet monetary fund products was less than that of traditional currency products and monetary fund products [5]. Wanli Yang built the GARCH model to calculate the VaR value and used the RAROC index to evaluate the performance of the sample funds. Then, he concluded that the risk-adjusted return rate of Internet monetary funds is much higher than that of traditional monetary funds [6]. Fuyun Zhu uses the EGARCH-GED model and finds that the return risk of Internet financial products is very similar to the return risk of corresponding monetary funds [7]. Yuting Wang found that the third-party payment system of Internet monetary funds has the highest returns, moderate risks, and more substantial investment potential [8].

Some scholars have analyzed the income of Internet financial monetary funds and various influencing factors. Zhipeng Li found that pure internet-based money market funds had the lowest VaR value, the smallest average return volatility, and the smallest relative return risk [9]. Through the research of the GARCH-CoVaR model, Ke Chen found that both traditional money market funds and Internet monetary market fund products are susceptible to the impact of macroeconomic and financial markets and have the same cycle characteristics [10]. Jingyun Lu studied the trend of Yu'E Bao yield rate and interbank lending rate and then found that the interbank lending rate had a positive influence on Yu'E Bao yield [11]. According to the time series analysis, An Liu found that the yield of financial products is negatively correlated with the money supply, and positively correlated with the interest rate of time deposit and new credit in the current month [12]. Qianhong Wang's research found that the scale of banks issuing financial products and whether the financial products issued a promise to protect the principal have an impact on the income of financial products [13]. Qing Xiao concluded that the overall performance of the Internet commodity base after considering risk factors is not as good as that of traditional internet-based monetary funds, so risk control should be strengthened [14]. Hong Fan mentioned that the Internet monetary funds could improve the stability of the existing banking system, while the proportion of Internet monetary fund savings transfer is not high [15]. Xiaofeng Tu claimed that the size of money market funds and the system importance of positive correlation. Funds with higher returns in the previous year will indeed continue to enjoy this advantage in the current year, and vice versa [16].

Most of the current research results on China's monetary funds are the analysis of a single factor. Due to the complexity and diversity of the financial market, the analysis of a single factor can easily deviate from the essence of things. Therefore, this paper collects the yield data of Internet financial monetary funds and selects several independent variables such as Shanghai composite index, government bond index, Shanghai interbank offered rate, and monetary fund size for empirical analysis. Finally, this paper makes suggestions to the fund supervision departments, marketing agencies, investors from the macro and micro point of view.

2 Variable Selection and Empirical Analysis

2.1 Variable Selection

In this paper, the rate of return of 28 Internet monetary funds is selected as the dependent variable. All the selected monetary funds have the corresponding sales platform or Internet monetary fund products (see Table 1 for details). To ensure that the research conclusion has a specific value and universality, different monetary funds were selected from the three dimensions of the establishment time, size, and Internet sales platform. Furthermore, include the iconic products of Internet financial products, Alibaba Yu'E Bao and Tencent Financier.

Through the official website of each fund, Everyday fund net and wind database can collect seven days' annualized rate of return of Internet monetary funds and other relevant data. The period of the samples is selected from January 1, 2017, to December

Table 1 Samples of internet financial monetary funds

Abbreviation of fund	Corresponding product or sales platform	Fund code	Fund size (100 million)	Establishment time
Zhongjia money	Tiantian bao	000331	12.60	2013/10/21
Icbc currency	Cash fast line	482002	1038.27	2006/3/20
Nuoan currency	Noam cash Bao	320002	1.31	2004/12/6
Yin hua currency	Zhonglu cash Bao	180008	420.75	2005/1/31
Jiaoyin currency	Jianyin cash Bao	519588	2.40	2006/1/20
Jianxin currency	Jianxin value-added Bao	530002	32.50	2006/4/25
Xinhua Yinuo Bao	Yinuo Bao	000434	4.59	2013/12/3
Yinhe Yinfu currency	Beili Bao	150005	182.78	2004/12/20
Haifutong currency	Hexun current surplus	519505	4.48	2005/1/4
Debangdeli currency	Debang wage Bao	000300	0.79	2013/9/16
Hui Tianfu full treasure	WeChat Financier	000397	909.52	2013/12/13
Huatai Borui currency	Huatai Borui cash Bao	460006	42.49	2009/5/6
Fuguo Tianshi currency	Daily wealth management Bao	100025	3.23	2006/6/5
Pingan day profit currency	Ping An Ying	000379	1083.06	2013/12/3
Hua'an daily Xin currency	Weiqian Bao	040038	1616.70	2012/11/26
Tianhong Yu'E Bao currency	Yu'E Bao	000198	10935.99	2013/5/29
Guangfa every day red currency	Change Bao	000389	28.08	2013/10/22
Jiashi current treasure currency	Baidu 100 profit roll profit	000464	154.11	2013/12/18
China fortune treasure currency	WeChat Financier	000343	714.99	2013/10/25
Rongtong easy to pay currency	Rongtong cash Bao	161608	311.18	2006/1/19
Hui tianfu cash treasure currency	Micro-wealth piggy bank	000330	428.59	2013/9/12
Efonda easy financial management	WeChat Financier	000359	1554.74	2013/10/24

(continued)

Table 1 (continued)

Abbreviation of fund	Corresponding product or sales platform	Fund code	Fund size (100 million)	Establishment time
Zhaoshang cash appreciation currency	Zhaohang fortune bao	217004	93.35	2004/1/14
South cash to increase interest currency	Ping An Ying	202301	94.84	2004/3/5
Huaxia cash to increase interest currency	Huaxia Circulation	003003	291.33	2004/4/7
Dacheng cash to increase interest currency	Xingye Bao	090022	310.99	2012/11/20
Minsheng plus silver cash treasure currency	Minsheng ruyi bao	000371	197.71	2013/10/18
Shenwan Lingxin revenue treasure currency	Cash bag	310338	1.74	2006/7/7

Data source https://fund.eastmoney.com/

31, 2019, and the data includes all trading days. Because this paper adopts a panel measurement method to analyze the rate of return of Internet monetary funds, it does not need the high-frequency data of the one-day level in terms of time. However, it chooses the quarter as the unit of time. Therefore, the 7-day annualized rate of return of the one-day level is converted into the quarterly rate of return (R). The conversion method is to calculate the average value, that is:

$$\text{Quarter level yield} = \Sigma\left(\text{The quarterly yield per trading day}_i\right)/\text{the number of trading days} \quad (1)$$

In essence, the Internet monetary fund is an investment tool. Through horizontal comparison, it is evident that its yield will be affected by other investment products to some extent. In China, the current investment tools are stocks, bonds, bank time deposits. In general, the Shanghai composite index can reflect the stock market changes. The treasury bond index can reflect the changes in the bond market.

Meanwhile, the Shibor can explain interbank liquidity. On the other hand, the development of the Internet monetary fund itself will also affect the income of the monetary funds. Generally speaking, the size of the monetary fund is proportional to its market bargaining power. The length of time the monetary fund is issued also has a positive impact on the income. The monetary fund with a long issuance time may have a relatively significant social effect and lead to income changes, such as

Alibaba Yu'E Bao. From the perspective of the proportion of investment, China's monetary funds mainly invest in Treasury bonds, central bank bills, corporate bonds, and other investment objects with strong liquidity. These investment objects can be mainly divided into bonds and currencies, so the invested proportion of bonds and currencies will also have an impact on the income of Internet monetary funds.

To sum up, in this paper, the independent variables are Shanghai interbank offered rate (Shibor), Shanghai composite index (SCI), Government bond index (GBI), Establishment time of monetary fund (Etime), Bonds rate in monetary funds (Brate), Cash rate in the monetary fund (Crate), and Net monetary fund size (Na). The data of SCI and GBI are from Netease financial website, and the Shibor is from the website of the Shanghai interbank offered rate. The source data of SCI, GBI, and Shibor is the single-day level, so the quarterly data of SCI, GBI, and Shibor can be calculated by the average of adding up each quarterly data. The data of Etime, Brate, Crate, Na is obtained through everyday fund net and wind database inquiry. The selected period of the above data is from January 1, 2017, to December 31, 2019 (excluding non-trading days), with a total of 12 quarters after data processing. To reduce the possibility of heteroscedasticity, this paper takes logarithms of SCI, GBI, and Na, namely:

$$\text{LSCI} = \text{In}\,(\text{SCI}) \tag{2}$$

$$\text{LGBI} = \text{In}\,(\text{GBI}) \tag{3}$$

$$\text{LNa} = \text{In}\,(\text{Na}) \tag{4}$$

This paper selects more data individuals, but the period is short, is a "short panel" data format. Since the historical rate of return of monetary funds may have an impact on the current rate of return. This paper uses the dynamic panel data model to analyze the factors affecting the return of China's Internet financial monetary funds.

2.2 *Empirical Analysis*

Panel unit root check. In order to make the data more stable and avoid false regression, the unit root test is conducted on the panel data first. The conventional methods of panel unit root test include LLC, IPS, ADF-fisher, and PP-fisher. The LLC method allows the panel data to have different time trends and intercepts and is suitable for the panel unit root test of medium dimensions. However, the LLC test method has certain limitations, which requires that the autoregressive coefficients of each individual are equal. It fails to take into account the possible heterogeneity of data of different individuals. The test method of the IPS panel unit root can solve the heterogeneity situation. Therefore, this paper adopts IPS, ADF-fisher, and PP-fisher

Table 2 Unit root test results of panel data

Variable	IPS test	ADF-fisher test	PP-fisher test
R	−4.27 (0.00***)	196.89 (0.00***)	328.15 (0.00***)
Brate	−4.35 (0.00***)	167.03 (0.00***)	262.71 (0.00***)
Crate	−2.6739 (0.00***)	151.46 (0.00)	215.13 (0.00***)
LNa	−5.72 (0.00***)	148.87 (0.00***)	117.92 (0.00***)

Remark *$P<$ 0.05, **$P<$ 0.01, ***$P<$ 0.01; same as below

methods to conduct unit root test on the data of Internet monetary fund and related independent variables and makes a comparative analysis.

In this paper, Shibor, SCI, GBI, the three indicators are only 12 different data when converted into quarterly data, which every individual at the same time corresponding these data are the same, so there is no need to conduct unit root test. Correspondingly, more samples in the Etime have the same year of establishment, so the unit root test is not required. Therefore, the unit root test is only needed for four variables: Internet financial monetary fund yield (R), Brate, Crate, and fund size (LNa). The test results are shown in Table 2. The results show that R, Brate, Crate, and LNa are all stable series, which can carry out panel regression.

Dynamic panel model. According to the above unit root test, there is no unit root phenomenon in the data between the income of Internet monetary funds and related variables in China so that the panel data model can be established. However, in real life, the income of financial products is often affected by the previous income, which is called the phenomenon of wave accumulation. This paper attempts to build a dynamic panel data model combined with the factor. The general form of the dynamic panel is:

$$y_{it} = \alpha + \rho_1 y_{i,t-1} + \rho_2 y_{i,t-2} + \cdots + \rho_p y_{i,t-p} + x_{it}^{'}\beta + z_i^{'}\delta + u_i + \varepsilon_{it} \tag{5}$$

The difference Generalized Method of Moments (GMM) is used to estimate the returns of Internet monetary funds, and the estimation results are shown in Table 3.

From the estimation results, it can be concluded that the dynamic panel model of Internet monetary funds is:

$$\begin{aligned} R_{it} = &-63.0716 + 0.2845R_{it-1} + 0.5694Shibor_{it} \\ &- 0.3163Etime_{it} - 0.8891Brate_{it} - 0.3887\text{Crate}_{it} \\ &+ 1.1339LSCI_{it} + 12.8093LGBI_{it} - 0.1495LNa_{it} + \varepsilon_{it} \end{aligned} \tag{6}$$

Table 3 Estimation results of the dynamic panel data

Variable	DIF-GMM		
	Coefficient	Standard deviation	*P*-value
L.R	0.2845	0.0507	0.00***
Shibor	0.5694	0.0479	0.00***
Etime	−0.3163	0.0898	0.00***
Brate	−0.8891	0.1912	0.00***
Crate	−0.3887	0.1831	0.03**
LSCI	1.1339	0.1021	0.00***
LGBI	12.8093	7.2240	0.08*
LNa	−0.1495	0.0413	0.00***
Constant term	−63.0716	33.8876	0.06*

Remark $*P< 0.05$, $**P< 0.01$, $***P< 0.01$

Empirical result analysis. According to the calculation results of the above dynamic panel data model, the rate of return of China's Internet monetary funds has a lag effect, and the rate of return of the last quarter has an impact on the rate of return of the next quarter. It can be seen from the empirical results that the return coefficient of the last period is positive and significant at the level of 1%. It can be proved that the rate of return of China's Internet monetary funds is affected by its historical rate of return. Shibor has a positive impact on the yield of Internet monetary funds. A rise in Shibor means an increase in the central bank's reserve requirement, usually accompanied by deflation. In this case, if the monetary funds want to attract more investors to invest, it must raise its yield. Etime of the monetary funds harm the return of the monetary funds. The earlier a monetary fund is set up, the larger its size (Na) will generally be, and thus the public will pay more attention to it. When investors further recognize the investment efficiency of the monetary fund, its yield will fall. Take Ali Yu'E Bao as an example, which was established in 2013. It has two advantages: the demand deposit and higher bank interest rate. While rapidly attracting a large number of investors, its yield is also falling.

The regression coefficients of Brate and Crate in monetary funds were negatively significant. It shows that the yield of Internet monetary funds decreases as the ratio of bonds to cash deposits increases. The main reason for the phenomenon is that bonds and cash deposit interest rates are relatively low. Once the monetary funds increase the proportion of the two, the yield will naturally fall. Nevertheless, also, because bonds and cash have a high degree of stability, so they are favored by the prudent investment group. In the regression equation (6), both LSCI and LGBI are positively significant. In the process of social and economic development, the rise of SCI and GBI usually means the looser monetary policy. At this time, there is a large amount of currency in circulation on the market, the investment varieties and investment objectives of monetary funds will be diversified, the investment yield rate will also increase so that the yield rate of Internet monetary funds rise.

3 Conclusions and Suggestions

3.1 Conclusion

Through the empirical analysis of the dynamic panel above, the following conclusions can be drawn: the income of Internet monetary fund has a positive impact of lagging one period. The persistence performance characteristic of the income is that the previous period of the monetary fund yield is high. The increased possibility of the next period yield will also greatly enhance, and vice versa. The rise of interbank offered rate, SCI, and GBI will raise the yield of monetary funds. Loose monetary policy can give monetary funds more investment varieties and projects, thereby increasing the possibility of yield. However, the time of fund establishment, fund size, the invested proportion of bonds and cash deposits in the monetary fund all have adverse effects on the yield. Among them, there is a strong linkage between the fund establishment time and the fund size, so the monetary fund with the earlier establishment is more likely to win the favor of the public, and thus gain a large amount of investment to expand the fund size. However, the investment field of monetary funds is relatively fixed, such as treasury bonds, central-bank bills, certificates of deposit, corporate bonds (with high credit ratings), and so on, that will limit the rise of the yields.

3.2 Suggestions

For the supervision department of the monetary funds, the establishment of funds flow risk prevention mechanism, and the formulation of standardized monetary fund investment and operation policies can effectively avoid the decline of the yields of the monetary funds. Specifically, the measures include: strengthening liquidity control of monetary funds, preventing the risk of massive redemption of monetary funds, and standardizing the types and portfolios of investment projects of monetary funds. In terms of regulating liquidity, institutional investors are sensitive to emergencies, and their funds are rather concentrated, while the retail investors are relatively slow, and their funds are somewhat dispersed. Therefore, institutional investors often play the primary role of redemption in fund flow in the tight market and should strengthen their capital flow control.

It should be noted that Internet property is not equal to high risk. The monetary funds only use the Internet as the sales channel, which did not cause significant fluctuations in returns due to the ease of Internet operation. In the prevention of enormous redemption risk, a study points out that the holiday "rigid consumption," "rigid payment," has a cluster effect. Therefore, every great holiday, the Internet monetary funds need to prepare in advance the risk of the vast redemption early warning work, set the redemption limit, or extend to account time. Absorb the advantages of the traditional monetary funds' redemption system, combined with the nature of the

Internet, and introduce other redemption rules. In the aspect of regulating monetary fund investment projects, it is necessary to guard against the potential risks of portfolio investment. The investment target of the monetary fund was initially limited to the various range of high safety coefficient and stable income. However, in reality, there is the situation of investing in unstable products, which also makes the return rate of monetary funds fluctuate considerably.

For monetary fund marketers, it is necessary to keep abreast of the macroeconomy trend and strengthen the management of Internet traditional monetary funds. In terms of the macroeconomy, it is particularly important to pay attention to policy adjustment and market dynamics, especially the inter-bank lending rate, stock index, and bond index, which can directly affect the return of monetary funds. When the market money supply has an increasing trend, actively adjust the investment strategy, to a greater extent, to protect the earnings of enterprises and investors. The investment subjects of the monetary funds also need to be timely adjusted themselves. When the systemic market risk is high, increase the proportion of treasury bonds, central bank bills, and other safer investment subjects. Secondly, the marketing institutions should set the rate of return of monetary funds around the market interest rate to adequately and reasonably allocate funds to deal with various possible uncertainties. For instance, the interest rate cut down by the central bank will affect the yield of monetary fund deposits, and then affect the yield of monetary fund products themselves. Besides, it is worth noting that more and more traditional monetary funds are expanding Internet sales channels, and they are transforming to Internet financial monetary funds. However, after connecting with the Internet, their earnings are continually declining. How to expand sales channels and still maintain the stability of earnings is a significant problem faced by monetary fund marketing institutions. On the other hand, the fund marketing institutions of the fund management fees, sales service fees, and other operational and input indicators need to be optimized and strive to use a small cost to obtain higher returns.

For the majority of ordinary investors, the Internet financial monetary fund has become a collection of savings, investment, consumption in a new way of financial management. In order to choose a safer, more stable, higher income Internet financial monetary fund, it can be analyzed from the following aspects. Firstly, observe the establishment time and size of the monetary funds. When the two indicators are relatively high, the safety and stability coefficient of the monetary fund is also relatively high, such as Ali Yu'E Bao. For moderates, there is no need to worry too much about the volatility of Internet financial monetary funds, which are even less risky than the traditional monetary funds. For investors in pursuit of higher returns, they can choose the monetary fund established for a short time, which can generally provide relatively high returns. When deciding to invest, we need to consider the historical return rate of the product and Shibor lag data, and timely adjust the investment strategy. Investments should be reduced when historical yields and Shibor lag data decline and vice versa. The pursuit of high returns cannot ignore the risk of the product itself at the same time, so the investors should not blindly believe the short-term index, such as the Internet company propaganda of a 7-day annualized rate of return. Investors should analyze the risk before investment.

Acknowledgements This paper is supported by China Postdoctoral Science Foundation (No. 2019M650476).

References

1. Ma, H. (2016). The connotation, characteristics and risks of Internet finance in the context of China. *Business Economics Research, 21,* 165–167.
2. Wang, Y. (2019). Analysis of monetary fund risk and innovative development model. *China Collective Economy, 31,* 104–105.
3. Zhang, H., & Ding, Z. (2016). Discussion on the development and supervision of China's Internet monetary fund—a case study of Yu'E Bao. *Academia, 3,* 65–75.
4. Qiu, D., Xue, D., & Liu, S. (2017). Return analysis of Internet money market fund based on Var. *Management Observation, 25,* 142–143.
5. Fu, L. (2018). Empirical analysis on the comparison between fund products and non-treasure products based on VaR. *Science and Technology, 4,* 112–115.
6. Yang, W. (2018). Risk analysis of internet money fund based on GARCH model. *Times Finance, 14,* 208–209.
7. Zhu, F. (2018). Empirical analysis on risk and performance of internet finance products——Based on VaR method with EGARCH-GED model. *Journal of Shanghai Lixin University of Accounting and Finance, 6,* 60–69.
8. Wang, Y. (2019). Research on the risk spillover effect of Internet money fund on commercial Banks. *Financial Theory and Practice, 03,* 7–14.
9. Li, Z., & Yao, X. (2015). A comparison of return volatility risk of China's Internet currency fund. *Monthly Journal of Finance and Accounting, 26,* 120–124.
10. Chen, K., & Zhang, J. (2017). Research on risk spillover effect of internet money market fund and financial market—based on the GARCH-CoVar model. *Financial Theory and Practice, 09,* 41–46.
11. Lu, J. (2016). Empirical research on the influence of interbank lending rate on the yield rate of Yu'E Bao. *Credit Investigation, 11,* 71–74.
12. Liu, A. (2017). Analysis on the factors influencing the expected rate of return of financial products. *Journal of Inner Mongolia University of Finance and Economics, 4,* 52–56.
13. Wang, Q. (2016). Research on the yield rate of personal finance products of commercial Banks and its influencing factors. *Price Theory and Practice, 1,* 129–132.
14. Xiao, Q., & Wang, H. (2019). Empirical research on the performance of China's Internet monetary fund based on DEA model. *Journal of Hubei University of Economics (Humanities and Social Sciences), 12,* 50–52.
15. Fan, H. (2018). Modeling and simulation of bank network system with Internet monetary fund. *Journal of System Simulation, 4,* 1237–1244.
16. Xiaofeng, T. (2019). Research on the systemic risk of China's money market fund. *Finance and Economics, 6,* 29–36.

Analysis of the Social Capital Financial Characteristics of Chinese PPP Projects

Yufei Qin and Xuemeng Guo

Abstract Based on the data of Chinese listed companies, this paper selects listed companies that have participated in or are participating in PPP projects, which are called PPP concept stocks, as the research object, and uses the propensity score matching method to find matching companies for them. From the micro level, it quantitatively analyzes which financial characteristics of the social capital side affect the degree of participation in the PPP project. The following conclusions: The four characteristics of an enterprise's solvency, profitability, asset management capability, and nature of equity have a significant negative impact on its participation in PPP projects, of which the impact of solvency, is particularly important. The statistical significance of the profitability indicator will become more significant when the company focuses on its ability to repay its current liabilities.

Keywords Financial characteristics · PPP concept stocks · Propensity score matching · Solvency

1 Introduction

Now that China's economy has switched to a medium-to-high-speed growth model, the government has begun to gradually explore a new supply model, the PPP model, in order to optimize the supply structure of public goods and services. According to statistics from the China PPP Comprehensive Information Management Database, as of December 31, 2019, the total investment in the management library project has reached 14.392 trillion yuan. However, the new financing model encountered a huge challenge in the promotion process of low social capital participation [1]. In fact, a large part of the project has been in the project identification stage for a long time,

Y. Qin · X. Guo (✉)
School of Economics and Management, Beijing Jiaotong University, Beijing, China
e-mail: 14275018@bjtu.edu.cn

Y. Qin
e-mail: 19113064@bjtu.edu.cn

M. Li et al. (eds.), *IEIS 2020*, https://doi.org/10.1007/978-981-33-4363-4_15

and it has not been successfully implemented and a substantial investment has been formed.

For a PPP project, its investment payback period is relatively long, often as long as ten or even decades. Therefore, the project usually has to wait for the transfer, and the stakeholders will get their respective expected benefits in the whole life cycle of the project before they can be considered truly successful. Social capital bears the project risks together with the government during the long project period and continuously provides resources to maximize the overall benefits of the project. Therefore, choosing an optimal partner is very important for the successful implementation of the PPP project. Financial characteristics are important indicators for judging the financial strength of social capital. If the management ability of the use of funds is weak, the funds cannot be reasonably financed and planned, thereby affecting project construction [2]. Therefore, this paper intends to make statistics on the listed companies in the social capital of China participating in PPP projects, analyze the financial characteristics of the enterprises and their reasons, with a view to providing a financial perspective for the government's selection and evaluation of social capital.

2 Literature Review

In recent years, due to the financial burden caused by the debt crisis, the infrastructure construction of each country has been greatly affected. Due to its unique advantages, the PPP model has been widely favored by governments of various countries. Scholars have also gradually shifted from studying the concept of PPP to focusing on PPP practical issues, such as how to make PPP projects smoothly implemented. Some scholars have analyzed the literature on the key factors for the success of PPP and found that the second most important factor affecting the implementation of PPP projects is the powerful private sector [3]. The International Department of the Chinese Ministry of Finance [4] also pointed out that according to IFC's global experience, the long-term financing capacity of social capital is one of the key factors for the successful implementation of PPP projects. This shows that most scholars affirm that the strength of the social capital side will have an important impact on the success of the PPP project.

Financial characteristics are important indicators for judging the strength of social capital. If the management ability of the use of funds is weak, the funds cannot be properly financed and planned, which will affect the project construction [2]. Existing researches on the financial characteristics of social capital of PPP projects mostly adopt questionnaire surveys and expert interview methods [5], use AHP model [6] or principal component analysis method [7] to construct a preliminary index system for social capital, and then study finance The degree of influence of factors such as status on the success of a PPP project, and the conclusion that financial status significantly affects initial trust. However, there is not much research on quantitative analysis of the impact of financial characteristics of social capital on PPP project participation from a micro level.

The so-called financial characteristics refer to the characteristics and symptoms reflected in corporate financial statements. Relative data can more easily reveal the current movements and future trends of corporate finance, so they have been widely used in academia [8]. Theoretically, the fundamental reason why an enterprise can win a PPP project is its reasonable capital structure, high asset management capability, outstanding main business effects, and strong market competitiveness [9]. Therefore, the financial characteristics of successful PPP projects may be reflected in three aspects: solvency, profitability and asset management capability. This paper establishes a measurement model to quantitatively analyze which financial characteristics of social capital affect the successful bid of a PPP project from a micro perspective, so as to provide reasonable suggestions for enterprises to better participate in PPP project construction, and also provide a valuable reference for the government to choose a partner.

3 Theoretical Hypothesis

3.1 Solvency

PPP projects generally generate huge capital requirements, so ensuring low financial risks is a basic condition for ensuring smooth project execution. Whether the interests of stakeholders can be protected is a key factor affecting the successful implementation of PPP projects. Some studies have pointed out that the basic financing structure of PPP projects is 90% debt financing and 10% equity financing [10]. For enterprises, the lower the debt to asset ratio, the higher the degree of protection of the company's stakeholders, the lower the financial risk, indicating that the stronger the company's solvency, the more likely it is to successfully intervene in PPP projects for business and investment activities.

H1: The higher the company's solvency index, the more likely it is that the company will successfully participate in a PPP project.

3.2 Profitability

Because social capital is profit-seeking, capital tends to flow to projects with higher returns. According to the "2019 China Top 500 Private Enterprises Analysis Report", it is known that the average return on equity of the top 500 private enterprises in 2018 was 13.02%, which is much higher than the current expected maximum return of 9% for PPP projects in China. Therefore, for companies with better profitability status, compared with other projects with higher return on investment, the possibility of enterprises choosing to enter PPP projects will be reduced.

H2: The higher the company's profitability index, the less likely it is that a company will successfully participate in a PPP project.

3.3 Asset Management Capability

The higher the enterprise's asset management capability, the better it is able to coordinate and manage assets and achieve a good allocation of resources. However, because the current assessment of business managers is biased towards operating performance indicators, compared with those PPP projects with longer investment recovery periods, larger funding amounts, higher financing risks, and lower operating efficiency [11], for good company managers, they are more likely to be willing to participate in projects that do not affect their operating performance indicators. At this time, the opportunities for enterprises to participate in PPP project construction are weakened, and the possibility of enterprises choosing and entering PPP projects will be reduced.

H3: The higher the company's asset management capability index, the less likely it is that a company will successfully participate in a PPP project.

3.4 Nature of State-Owned Equity

Because state-owned enterprises are subject to multiple regulations and their autonomy is poor, the update of various management systems needs to be reported at various levels for approval. Decisions are slow and inefficient, and they cannot well meet the requirements of enterprise development under the new economic situation [12]. Therefore, the social capital of state-owned enterprises in the process of cooperation with government departments has a significant disadvantage compared with nonstate-owned enterprises, that is, the government will be more inclined to choose nonstate-owned enterprises.

H4: State-owned enterprises are less likely to successfully participate in PPP projects.

4 Model and Sample

4.1 Regression Model

In order to study which financial characteristics of an enterprise will affect its likelihood of successfully participating in a PPP project, this article uses a binary logit model for regression analysis:

Table 1 Variable explanation

Value Code		Value Name	Defination
Solv	LEV	Debt to asset ratio	LEV = Total liabilities / Total assets
	CR	Current ratio	CR = Current assets / Current liabilities
	QR	Quick ratio	QR = (Current assets Inventory) / Current liabilities
Prof	ROA	Return on assets	ROA = Net profit / Total asset balance
Turn	TAT	Total assets turnover	TAT = Operating income / (End balance of total assets + End balance of total assets of the previous year) / 2
Pc	GOV	State-owned enterprises	If the enterprise is a wholly state-owned enterprise or a state-owned enterprise, the value is 1; otherwise, it is 0

$$\text{PPP} = \alpha_0 + \alpha_1 Solve_i + \alpha_2 Prof_i + \alpha_3 Turn_i + \alpha_4 Pc_i + \varepsilon \tag{1}$$

Among them, α_0 is a constant term and ε is an error term. See Table 1 for variable descriptions.

(1) *Explained variable*: The explained variable (PPP) is a binary variable used to indicate whether an enterprise participates in a PPP project. The value of participation (that is, a PPP concept stock enterprise) is set to 1, otherwise it is 0.

(2) *Explanatory variables*:

(a) *Solvency (Solv)*: Solvency reflects the risk level of a company's liability and its ability to continue operations, and is an indispensable indicator for measuring business risks [13]. This article draws on existing research in the academic world [14], and selects the debt to asset ratio (LEV), current ratio (CR), and quick ratio (QR) as proxy variables to measure the company's solvency.

(b) *Profitability (Prof)*: This article selects the return on assets (ROA) as the proxy variable for the indicator of corporate profitability, which is used to indicate how much profit each unit of assets brings. At the same time, the robustness test was performed on the regression results using the earnings per share (EPS) variable.

(c) *Asset Management Capability (Turn)*: Asset management capability refers to the capital operation efficiency of an enterprise, that is, how a manager can obtain the maximum return with the smallest cash flow. This article chooses the total assets turnover (TAT) to measure asset management capability indicators in accordance with common practices in academia.

(d) *Nature of state-owned equity (Pc)*: This article focuses on the impact of the government's choice of social capital to participate in PPP projects when companies may have slow decision-making and low efficiency. Therefore, this article chooses the nature of the company's equity as a proxy variable. If the enterprise is a wholly state-owned enterprise or a state-owned enterprise, the value of the variable state-owned equity is 1; otherwise, it is 0.

4.2 Sample Selection

Listed companies with financial strength can often be more easily selected than ordinary nonlisted companies, and thus win the bid for a PPP project. Therefore, this article selects listed company which has participated or is participating in the PPP project, that is, companies whose concept is "PPP concept stock" as the research object. According to straight flush statistics, as of December 31, 2019, China has 156 PPP concept stocks. It means that 156 listed companies have participated or are participating in PPP projects.

Governments often choose partners based on factors such as business size and efficiency. Obviously, this process is not random. Therefore, in order to eliminate the selection bias, we used the propensity score matching method to pair the "PPP concept stocks". The initial matching ratio is 1:2, and the matching basis is four variables: the industry to which the company belongs, the size of employees, total assets and operating income.

This article excludes ST and *ST companies, financial companies, companies with negative net profit, newly listed companies in 2019, and companies with missing data before pairing. The financial characteristic indicator data comes from CSMAR, and some missing data are supplemented by manual collection. Except for the nature of state-owned equity used in the 2018 annual data, the rest of the data are from the company's third quarter 2019 financial statements. Duplicate samples were deleted after pairing, and finally 375 sample companies were obtained, including 235 companies which have not participated in the PPP project (nonPPP concept stocks), and 140 companies which have participated in or are participating in the PPP project (PPP concept stocks).

5 Empirical Analysis

5.1 Descriptive Statistical Analysis

The sample of 156 listed companies, which have participated or are participating in PPP projects, includes 104 private enterprises, accounting for 66.67% of the total sample, and the rest are state-owned holding companies. Obviously for the listed companies that won the PPP project, the number of nonstate-owned enterprises accounted for a larger proportion. From this we infer that listed companies are to some extent equivalent to real social capital and realize the cooperative relationship between the government and society. This article finally selected 375 listed companies as a sample. The descriptive statistical results are shown in Tables 2, 3 and 4.

As can be seen from the above table, the average debt to asset ratio of PPP concept stock companies is 0.563 for the corporate solvency index, while the average value of nonPPP concept stock companies is only 0.450. For current ratio and quick ratio, the average value of concept stock companies is significantly lower. For nonPPP concept stock companies. From this, it can be preliminarily judged that listed companies participating in PPP projects often have lower solvency.

The average profitability index of PPP concept stock companies is 0.025, which is lower than 0.037 of nonPPP concept stock companies. For the asset management

Table 2 Basic characteristics of all sample data

Code		Name	Mean	Std	Min	Max
Solv	LEV	Debt to asset ratio	0.492	0.199	0.048	0.936
	CR	Current ratio	1.906	1.475	0.396	16.469
	QR	Quick ratio	1.428	1.387	0.099	15.337
Prof	ROA	Return on assets	0.033	0.032	0.000	0.304
Turn	TAT	Total assets turnover	0.486	0.378	0.020	4.054
Pc	GOV	State-owned enterprises	0.411	0.493	0	1

Table 3 Sample data characteristics of listed companies in PPP concept stocks

Code		Count	Mean	Std	Min	Max
Solv	LEV	140	0.563	0.179	0.086	0.909
	CR	140	1.593	1.070	0.518	7.904
	QR	140	1.169	1.020	0.217	7.891
Prof	ROA	140	0.025	0.023	0.000	0.155
Turn	TAT	140	0.409	0.237	0.077	1.489
Pc	GOV	140	0.350	0.479	0	1

Table 4 Sample data characteristics of listed companies with nonPPP concept stocks

Code		Count	Mean	Std	Min	Max
Solv	LEV	235	0.450	0.198	0.048	0.936
	CR	235	2.092	1.644	0.396	16.469
	QR	235	1.583	1.547	0.099	15.337
Prof	ROA	235	0.037	0.035	0.001	0.304
Turn	TAT	235	0.531	0.436	0.020	4.054
Pc	GOV	235	0.447	0.498	0	1

capability index, the average value of the former is 0.409, which is significantly lower than the average of the latter by 0.531. The descriptive statistical results have initially confirmed the original hypothesis. Due to the limited expected return rate and turnover efficiency of PPP projects, it has limited appeal to companies with high returns on assets or high total assets turnover rates.

The average value of state-owned equity of 0.350 for PPP concept stocks companies is lower than 0.447 for nonPPP concept stocks companies, that is, the government tends to prefer nonstate-owned companies to avoid slow and inefficient social capital decisions.

This shows that the financial characteristics of companies participating in PPP projects and those not participating in PPP projects have significant differences. From the descriptive statistical results alone, the mean differences of most variables are consistent with expectations, but there are also weak conclusions, so we further test hypotheses through model regression.

5.2 *Analysis of Regression Results*

This paper uses Logit and Probit regression methods to conduct an empirical analysis of the binary explanatory variable PPP. The regression results are shown in Table 5.

As can be seen from Table 5, the logit model has a correct prediction percentage of 67.47%, which is higher than the 66.40% of the probit model. Therefore, the logit model is correctly set and fits well. Obviously, the sign of the variable coefficients

Table 5 Basic regression results

Value	Forecast Symbol	Logit Model Regression		Probit Model Regression	
		Coefficient	Odds ratio	Coefficient	Average marginal effect
LEV	–	3.052^{***} (0.690)	21.158	1.864^{***} (0.410)	0.628
ROA	–	−6.142 (4.994)	0.002	−3.518 (2.905)	−1.185
TAT	–	-1.012^{*} (0.404)	0.363	-0.620^{*} (0.239)	−0.209
GOV	–	-0.792^{***} (0.246)	0.453	-0.483^{***} (0.147)	−0.163
Cons_		-0.301^{*} (0.544)	0.740	-0.193^{*} (0.328)	–
Count		375	375	375	375
Percent correctly predicted		67.47%	– d	66.40%	–

Note Standard errors are in parentheses, ***represents $p < 0.01$, **represents $p < 0.05$, and *represents $p < 0.1$

after the regression of the two models are completely the same, and the p-value difference is small, which does not affect the significance level, so the regression results are basically the same. To ensure that the regression results are intuitive, this paper calculates the odds ratio and average marginal effect of the logit model.

The symbols and the coefficient of the profitability index, the asset management capability index and the nature of state-owned equity are the same as expected, and only the return on assets variable is not statistically significant. It can be seen that the higher nature of state-owned equity or asset management capability of an enterprise, the less likely it is that the enterprise will participate in a PPP project, confirming the hypothesis H3 and H4.

The sign of the profitability index coefficient is negative, to some extent, the higher the profitability of the enterprise, the less likely it is to participate in PPP projects. But because it is not significant, it shows that the return on assets is not a key factor affecting the possibility of enterprises participating in PPP projects.

The sign of the coefficient of the debt to asset ratio is positive, that is, the higher the solvency, the less likely a company will participate in a PPP project, which is inconsistent with the hypothesis H1. This may be because the higher the debt to asset ratio reflects the higher the company's ability to borrow, it can raise more funds for PPP projects, and ensure the stability of the project's funding chain.

The odds ratio of the solvency indicator variable is 21.158, which shows that solvency is a key factor influencing the possibility of enterprises participating in PPP projects. Therefore, for enterprises that want to participate in PPP projects, they need to match enough sources of debt financing to meet the huge funding requirements of the project.

We further replaced the solvency index with the current ratio and quick ratio for regression (Table 6). The coefficient signs of both are negative and statistically significant. At this time, the profitability index (return on assets) is statistically significant. The higher the current ratio or quick ratio, the higher solvency of current debt and accordingly the weaker ability to borrow current liabilities, and the less able to ensure the stability of the capital chain. Therefore, companies with higher solvency are often more difficult to obtain opportunities to participate in the construction of PPP projects.

5.3 *Robustness Test*

We changed the matching ratio to 1:3 to test the robustness of the regression results. In order to avoid missing the variables, here we add the corporate growth indicator (total asset growth rate, TAG) variable [15] for robustness test. At the same time, since the return on assets in the profitability indicator is not statistically significant, here we choose the EPS variable per share instead of the return on assets to perform a robustness test. The test results are shown in Table 7.

It can be known from Table 7 that after changing the matching ratio, the signs of the solvency index current ratio and quick ratio are still negative, and the coefficient

Table 6 Solvency indicator regression results

Value	Forecast Symbol	Logit Model Regression		
LEV	–	3.052*** (0.690)		
CR	+		−0.327*** (0.121)	
QR	+			−0.219* (0.120)
ROA	–	−6.142 (4.994)	−11.033** (5.037)	−11.863** (5.137)
TAT	–	−1.012* (0.404)	−1.08*** (0.403)	−0.944** (0.389)
GOV	–	−0.792*** (0.246)	−0.672*** (0.235)	−0.657*** (0.234)
Cons_		-0.301* (0.544)	1.844*** (0.477)	1.495*** (0.441)
Count		375	375	375
Percent correctly predicted		67.47%	66.93%	67.47%

Note Standard errors are in parentheses, ***represents $p < 0.01$, **represents $p < 0.05$, and *represents $p < 0.1$

coefficients and significance levels of each variable have not changed. After adding the corporate growth index variable to participate in the regression, it is found that the sign of the total asset growth rate coefficient is positive, but it is not statistically significant, and the signs and significance levels of the remaining variables have not changed, indicating that there is no problem of missing variables. After the profitability index is revised, the index coefficient is significant, and the signs are consistent with the previous ones, and the signs and significance levels of the other variables have not changed. In summary, the regression results in this article are robust.

6 Conclusion

The characteristics of the nature of state-owned equity and asset management capability of listed companies have a negative impact on their participation in PPP projects. We initially believe that because state-owned enterprises are subject to multiple regulations, they are slower and less efficient than private enterprises and more difficult to successfully participate in PPP projects. And the current assessment of business managers is biased towards operating performance indicators, so it is more difficult for managers of companies with good operating performance to participate in long investment recovery periods and low operating efficiency PPP project. Among the financial characteristics indicators, solvency has a particularly

Table 7 Robustness test regression results

Value		Forecast Symbol	1:3 Match			
Solv	LEV	−			3.066^{***} (0.669)	3.367^{***} (0.610)
	CR	+	-0.283^{**} (0.111)			
	QR	+		-0.214^{*} (0.116)		
Prof	ROA	−	-10.523^{**} (4.752)	-11.105^{**} (4.858)	-5.137 (4.756)	
	EPS[a]	−				-0.560^{*} (0.326)
Turn	TAT	−	-1.094^{***} (0.390)	-0.981^{**} (0.379)	-1.052^{***} (0.390)	-1.091^{***} (0.383)
Pc	GOV	−	-0.579^{***} (0.223)	-0.568^{**} (0.223)	-0.714^{***} (0.235)	-0.708^{***} (0.235)
Grow	TAG[b]	+			0.097 (0.448)	0.199 (0.448)
Cons_			1.278^{***} (0.441)	1.011^{**} (0.413)	-0.779^{*} (0.514)	-0.904^{**} (0.455)
Count			461	461	461	461
Percent correctly predicted			70.50%	69.41%	70.72%	70.28%

[a]EPS = net current value / values of paid-up capital end of the period
[b]TAG = (end value of the current period of the total assets—the beginning value of the current period of the total assets) / (the beginning value of the current period of the total assets)
Note Standard errors are in parentheses, ***represents $p < 0.01$, ** represents $p < 0.05$, and * represents $p < 0.1$

important negative impact on its participation in PPP projects. We believe that the higher solvency, the weaker raise-debt ability its have, which does not have sufficient capacity to ensure the stability of the project funding chain. So it is more difficult to choose to enter a PPP project. Therefore, for a listed company that intends to participate in a PPP project, the company should simplify the decision-making process as much as possible and vigorously improve its ability to raise debt. Although the symbol of the ROA is negative and not statistically significant, when more attention is paid to the solvency of the company's current liabilities, the profitability indicator is statistically significant, so the profitability indicator is still one of the influencing factors in the participation of PPP projects.

When selecting social capital, the government should focus on the solvency of the enterprise and ensure that the company has sufficient debt-bearing capacity. At the same time, the government should also improve the profitability and operating efficiency of PPP projects, increase the enthusiasm for social capital participation, and achieve optimal sharing of benefits.

References

1. Liu, Q., & Ren, J. (2017). A study of the characteristics of social capital participating in public private partnerships——Evidence from Chinese listed firms. *Review of Economy and Management, 33*(6), 38–46.
2. Zhang, H., Zhang, Y., & Ouyang, H. (2016). Selection of infrastructure projects under PPP mode partners. *Journal of Engineering Management, 30*(4), 85–89.
3. Osei-Kyei, R., & Chan, A. P. C. (2015). Review of studies on the critical success factors for public-private partnership (PPP) projects from 1990 to 2013. *International Journal of Project Management, 33,* 1335–1346.
4. International Department of the Ministry of Finance. (2014). Key factors for successful implementation of ppp projects from international experience. *China State Finance, 15,* 44–45.
5. Han, L., Song, W., & Xu, Y. (2017). Research on the index system of social capital selection of PPP projects. *Engineering Economy, 27*(12), 64–67.
6. Chen, X., Zhang, G., & He, C. (2017). Study on the trusting evaluation index of social capital selection in PPP project. *Management & Technology of SME, 5,* 25–26.
7. Du, Y., & Yan, P. (2014). An empirical study about formation mechanism of initial trust in PPP projects. *China Civil Engineering Journal, 47*(4), 115–124.
8. Yang, B., Huang, Z., & Wei, J. (2019). The financial characteristics and corporate value of all roads and companies. *Communication of Finance and Accounting, 33,* 56–61.
9. Yuan, J., Ji, C., Li, Q., & Skibniewski, M. J. (2010). KPI Evaluation standard setting research and case analysis of infrastructure PPP projects. *Modern Management Science, 12,* 24–27+49.
10. Wall, A., & Connolly, C. (2009). The private finance initiative. *Public Management Review, 11*(5), 707–724.
11. Ji, F. (2020). Research on PPP model of tourism promotion in ethnic Areas in China. *Journal of Southwest Minzu University (Humanities and Social Science), 41*(1), 16–21.
12. Wang, P. (2020). Thinking about strengthening the asset management of state-owned enterprises. *China Journal of Commerce, 4,* 182–183.
13. Lopes, A. I., & Caetano, T. T. (2015). Firm-level conditions to engage in public-private partnerships: What can we learn? *Journal of Economics and Business, 79,* 82–99.
14. Zhang, Z., & Song, J. (2010). Comprehensive evaluation and analysis of financial indexes of listed companies—Data of communication and cultural companies. *Communication of Finance and Accounting, 24,* 15–17.
15. Laux, V. (2008). Board independence and CEO turnover. *Journal of Accounting Research, 46*(1), 137–171.

The Role of Industrial Policy: The Case of Tendering Intermediary Services in China

Weiwei Hao, Zongqing Liu, and Hongyan Gao

Abstract This paper examines the impact of industrial policy on the development of tendering intermediary services in China. Firstly, this paper attempts to establish a conceptual model of the impact from industrial policy on the development ability of tendering intermediary services, and puts forward 16 research hypotheses. In order to verify the theoretical hypotheses, an empirical analysis is conducted on the China's tendering intermediary services with questionnaire data by making use of structural equation model. The results indicate that there are some positive relationships between industrial policy and the development ability of tendering intermediary services. Industrial structure policy, industrial organization policy, industrial technology policy and guarantee implementation policy all have some significant positive impacts on the development ability of tendering intermediary services. The conclusion is that industrial policy can promote the development ability of tendering intermediary services. Therefore, it is significant to make full use of industrial policy to promote the development of tendering agency service industry in China.

Keywords Tendering intermediary services · Industrial policy · Development ability

1 Introduction

In recent years, with the rapid development of tendering intermediary services, it has become an important means of allocating economic resources. International experience indicates that public policy is an important factor affecting the development of

W. Hao (✉) · Z. Liu · H. Gao
School of Economics and Management, Beijing Jiaotong University, Beijing, China
e-mail: vivihao1014@163.com

Z. Liu
e-mail: 12113103@bjtu.edu.cn

H. Gao
e-mail: hygao@bjtu.edu.cn

M. Li et al. (eds.), *IEIS 2020*, https://doi.org/10.1007/978-981-33-4363-4_16

modern service industry. Especially in the current period of China's economic transformation, there may be a crucial impact from institutional and economic policy changes on the development of tendering intermediary services in the future. Therefore, to explore the impact of policy factor on the development of tendering intermediary services, is of great significance towards promoting the development ability of tendering intermediary services in China, with the aim to realize professional, healthy and sustainable development of the tendering intermediary services.

The rest of this paper is organized as follows. Section 2 provides the theory analysis and proposes research hypotheses. Section 3 is empirical analysis, including data description and empirical model. Section 4 contains the conclusion and some policy implications.

2 Theory Analysis

2.1 *The Connotation of Industrial Policy*

The industrial policy in this study are divided into three categories. The first category is target selection policy, including industrial structure policy, industrial organization policy, industrial technology policy and industrial distribution policy for a particular industry. The second category is guarantee implementation policy, including fiscal and tax policy, industrial financial policy, price policy and wage policy for a particular industry. The last category contains other industry regulation policies.

2.2 *The Development Ability of Tendering Intermediary Services*

In China, the tendering proxy agency is the work unit that by obtaining the professional certificates of relevant business in accordance with laws and regulations, can effectively carry out the supervision, design, procurement and other bidding type activities on the projects after being entrusted by the tenderee.

Based on the industrial characteristics of tendering intermediary services, the development ability is defined as the core competitiveness of China's tendering intermediary services, by comprehensively making use of the production factors of political, economic, social and environmental systems and creating more development power through the innovation, the improvement of specialization level and extension of services scope. As the development ability of tendering intermediary services is promoting, it is more conductive to gaining the multi-level growth efficiency of market entry ability, talent support ability, sustainable innovation ability and policy support ability, which can realize the rationalization of industrial structure and the sustainable development of tendering intermediary services.

Thus, it can be seen that market entry ability, talent support ability, sustainable innovation ability and policy support ability all constitute the main contents of the development ability of tendering intermediary services.

2.3 *Theoretical Hypotheses*

This paper attempts to establish a conceptual model of the impact from industrial policy on the development ability of tendering intermediary services, which provides the basis for the design of questionnaire and analytical frame for the structural equation model later. First of all, a conceptual model between industrial policy and the development ability of tendering intermediary services is built and shown in Fig. 1.

Based on this, a structural equation model of industrial policy affecting the development ability of tendering intermediary services can be built (Fig. 2).

According to the structural equation model, combined with the characteristics of tendering intermediary services, the following research hypotheses are proposed.

2.3.1 The Hypotheses of the Relationship Between Industrial Structure Policy and the Development Ability of Tendering Intermediary Services

H1: There is a positive relationship between industrial structure policy and market entry ability of tendering intermediary services. The more direct or indirect industrial structure policy issued by the government are effective, the more conducive to improving the market entry ability of tendering intermediary services; otherwise, it is not.

H2: There is a positive relationship between industrial structure policy and talent support ability of tendering intermediary services.

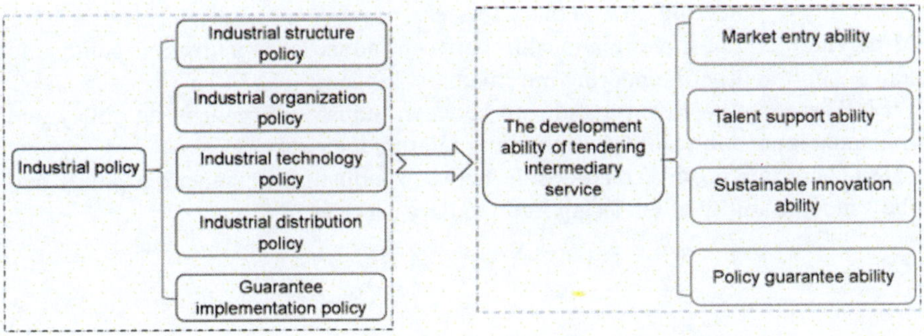

Fig. 1 Conceptual model of the effect from industrial policy on development ability of tendering intermediary services

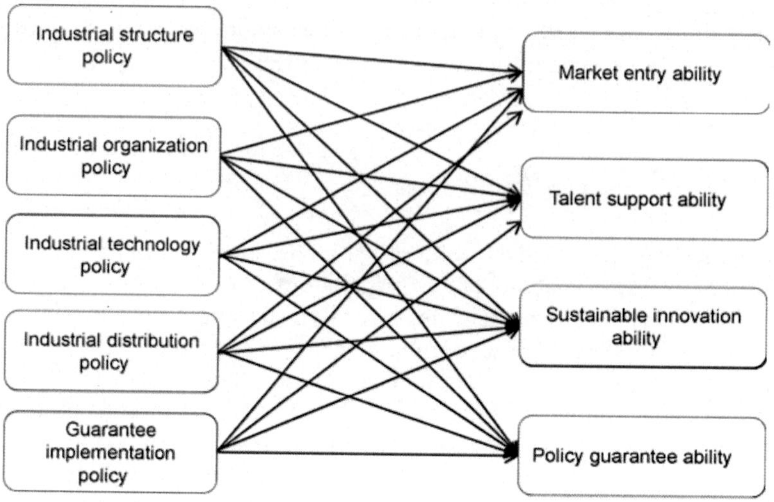

Fig. 2 Structural equation model of the effect from industrial policy on development ability of tendering intermediary services

H3: There is a positive relationship between industrial structure policy and sustainable innovation ability of tendering intermediary services.

H4: There is a positive relationship between industrial structure policy and policy support ability of tendering intermediary services.

2.3.2 The Hypotheses of the Relationship Between Industrial Organization Policy and the Development Ability of Tendering Intermediary Services

H5: There is a positive relationship between industrial organization policy and market entry ability of tendering intermediary services.

H6: There is a positive relationship between industrial organization policy and talent support ability of tendering intermediary services.

H7: There is a positive relationship between industrial organization policy and sustainable innovation ability of tendering intermediary services.

H8: There is a positive relationship between industrial organization policy and policy support ability of tendering intermediary services.

2.3.3 The Hypotheses of the Relationship Between Industrial Technology Policy and the Development Ability of Tendering Intermediary Services

H9: There is a positive relationship between industrial technology policy and market entry ability of tendering intermediary services.

H10: There is a positive relationship between industrial technology policy and talent support ability of tendering intermediary services.

H11: There is a positive relationship between industrial technology policy and sustainable innovation ability of tendering intermediary services.

H12: There is a positive relationship between industrial technology policy and policy guarantee ability of tendering intermediary services.

2.3.4 The Hypotheses of the Relationship Between the Guarantee Implementation Policy and the Development Ability of Tendering Intermediary Services

H13: There is a positive relationship between guarantee implementation policy and market entry ability of tendering intermediary services.

H14: There is a positive relationship between guarantee implementation policy and talent support ability of tendering intermediary services.

H15: There is a positive relationship between guarantee implementation policy and sustainable innovation ability of tendering intermediary services.

H16: There is a positive relationship between guarantee implementation policy and policy guarantee ability of tendering intermediary services.

3 Empirical Analysis

3.1 Variables Identification

3.1.1 The Measurement of Industrial Policy

To explore the direct impact of industrial policy on tendering intermediary services, the policy variable can be determined as an explanatory variable, so the measurement indicators are shown in Table 1.

Table 1 Relevant indicators of industrial policy measurement

Classification	Index	Corresponding problem
Industrial structure policy (ISP)	The industry structure policy of tendering intermediary services (X1)	The government encourages tendering intermediary enterprises to conduct specialized operation
		The government encourages tendering intermediary enterprises to conduct diversified operation
		The compulsory qualification requirements from the government have a great influence on the business of tendering intermediary enterprises
		The compulsory professional talents requirements of government have a great influence on the business of tendering intermediary enterprises
		The government encourages tendering intermediary enterprises to carry out the electronic bidding business
	The industrial structure policy of related industry(X2)	The policies on the post and telecommunications industry have a great impact on the demand for bidding agency
		The policy of encouraging the development of new energy industry has a great influence on the demand of bidding agency The policy on water conservancy industry has a great influence on the demand of bidding agency
		The policy of encouraging the development of water conservancy industry has a great influence on the demand of bidding agency
		The policies on the real estate industry have a great impact on the demand for bidding agency

(continued)

Table 1 (continued)

Classification	Index	Corresponding problem
		The policy on municipal facilities industry has a great influence on the demand of bidding agency
		The policy on transportation industry has a great influence on the demand of bidding agency
Industrial organization policy (IOP)	Competition promotion policy (X3)	Collusion in bidding has a great impact on the tendering intermediary services
		The policy of regulating collusion in bidding is good for the tendering intermediary services
		The regulation policy of illegal divulging can regulate the bidding agency market
		The regulation policy of malicious low-priced bidding, such as encircling bidding, is good for the tendering intermediary services
	Industrial rationalization policy (X4)	The policy of qualification identification on bidding agency is conducive to fair competition in bidding agency market
		The existing evaluation mechanism of tendering intermediary services is scientific and reasonable
		The regulation policy of illegal behavior is effective
Industrial technology policy (ITP)	The industry technology policy of tendering intermediary services (X5)	Standardized bidding is beneficial to the development of tendering intermediary services
		The government encourages the use of electronic bidding system, which is beneficial to the tendering intermediary services

(continued)

Table 1 (continued)

Classification	Index	Corresponding problem
		The use of public resource trading platform is good for the tendering intermediary services
	The industrial technology policy of related industry (X6)	National technical innovation support policy is good for tendering intermediary services
		Energy innovation policy is good for tendering intermediary services
		The policy of cutting-overcapacity in traditional industries has a great influence on the tendering intermediary services
		The policy orientation of "Internet plus" has a great influence on the tendering intermediary services
Guarantee implementation policy (GIP)	The guarantee policy of related industries (X7)	The government's subsidy policy to the new energy industry is good for the tendering intermediary services
		The government's direct investment policy for related industries is good for the tendering intermediary services
		The government's preferential tax policy for related industries is good for the tendering intermediary services
		Value added tax policy is good for tendering intermediary services
	The industry guarantee policy of tendering intermediary services (X8)	The policy of "bidding agent" is beneficial to the development of tendering intermediary services
		Industry standards are conducive to the development of tendering intermediary services

Table 2 The measurement indexes of the development ability of tendering intermediary services

Target	Index	
The development ability of tendering intermediary services	Market entry ability C1	The scale of bidding market C11
		The prospect of subdivided related industries C12
	Talent support ability C2	Talent input C21
		Professional level C22
		Qualification level C23
	Sustainable innovation ability C3	Technological innovation ability C31
		Management innovation ability C32
	Policy guarantee ability C4	Quality control system C41
		Policy supply ability C42

3.1.2 The Measurement on the Development Ability of Tendering Intermediary Services

By analyzing the internal and external factors that affect the development ability of tendering intermediary services in China, this paper constructs the evaluation index system of the development ability of tendering intermediary services, and evaluates it's development ability in China from 2005 to 2017 by using the entropy method. Whether it is the direct or indirect impact of policy on the tendering intermediary services, the development ability will be taken as the explained variable to participate in the construction of the model.

Learn from the relevant achievements of performance research, the index to measure the development ability of the tendering intermediary services is subdivided into market entry ability (MEA), talent support ability (TSA), sustainable innovation ability (SIA) and policy guarantee ability (PGA) as in Table 2.

3.2 Data Sources

3.2.1 Fundamental Data

The data about the development ability of tendering intermediary services from 2005 to 2017 are all from China Statistical Yearbook and Statistics Bulletin from Ministry of Housing and Urban-Rural Development of China.

3.2.2 Questionnaire Data

In order to carry out structural equation model analysis, we designed questionnaires, conducted field surveys and obtained some survey data. A total of 300 questionnaires were distributed in this study. The objects are mainly managers, employees, technical personnel and relevant industry experts from the tendering intermediary services. More direct, timely and effective survey data can be obtained through field interviews. A total of three rounds of questionnaires were distributed, and the way of combining concentration and decentralization was adopted to conduct interviews to ensure the questionnaire recovery rate and interview quality.

After analyzing the collected questionnaires, we found that 18 of them had problems and could not be used for data analysis. Compared with 300 questionnaires, the effective rate was 94%. Further professional analysis of the questionnaire, as far as possible to exclude simple random sampling and other errors, the comprehensive analysis found that 21 questionnaires were useless for model analysis. Finally, 261 questionnaires were available, which met the requirements of model analysis.

3.3 Hypothesis Testing

First of all, through the reliability analysis, exploratory factor analysis and confirmatory analysis on the variables of industrial policy and tendering intermediary services' development ability, it shows that the data has good reliability and validity, which is suitable for factor analysis.

3.3.1 Original Model and Its Results

In view of the characteristics of industrial policy, AMOS software is used to get the model results as shown in Fig. 3.

As Fig. 3 and Table 3 illustrate, some relationships and paths are non-significant between industrial structure policy and policy guarantee ability, between industrial organization policy and sustainable innovation ability, between industrial technology policy and market entry ability, between industrial technology policy and talent support ability, between industrial technology policy and policy guarantee ability, between implementation guarantee policy and market entry ability, between implementation guarantee policy and talent support ability, between implementation guarantee policy and sustainable innovation ability. Thus, it is necessary to modify the model and eliminate the non-significant paths.

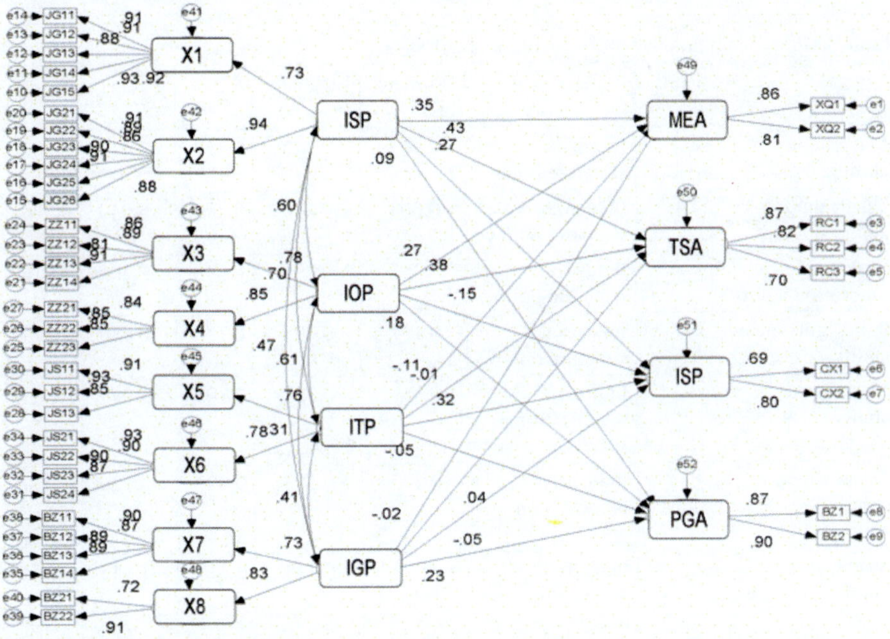

Fig. 3 Original model's results

3.3.2 Modified Model and Its Results

Combined with the original model, AMOS software is used to get the model results as shown in Fig. 4.

The results indicate that CMIN/DF is 1.875, GFI is 0.865, AGFI is 0.846, CFI is 0.957, RMR is 0.036, EMSEA is 0.046, IFI is 0.957, and NFI is 0.912. All the indexes of adaptation degree fix the standard well (Table 4).

As the path table shows that the coefficients of paths are significant, they meet the needs of model analysis. From the path analysis of the modified model, the following conclusions can be drawn:

(a) *The relationships between industrial structure policy and the development ability of tendering intermediary services after the amendment.*

The standardized path coefficient between industrial structure policy and market entry ability is 0.278, P is 0.000, which is significant. Therefore, it indicates that industrial structure policy has a significant positive impact on market entry ability. H1 is proved.

The standardized path coefficient between industrial structure policy and talent support ability is 0.442, P is 0.000, which is also significant. Thus, it shows that industrial structure policy has a significant positive impact on talent support ability. H2 is proved.

Table 3 Empirical parameters of structural model

	Paths		Estimate (ns)	Estimate (s)	S.E	C.R	P
Market entry ability	<--	Industrial structure policy	0.467	0.346	0.137	3.419	0.000
Talent support ability	<--	Industrial structure policy	0.604	0.430	0.120	5.040	0.000
Sustainable innovation ability	<--	Industrial structure policy	0.351	0.268	0.146	2.409	0.016
Policy guarantee ability	<--	Industrial structure policy	0.137	0.085	0.155	0.880	0.379
Market entry ability	<--	Industrial organizational policy	0.319	0.268	0.105	3.049	0.002
Talent support ability	<--	Industrial organization policy	0.475	0.383	0.093	5.126	0.000
Sustainable innovation ability	<--	Industrial organization policy	−0.171	−0.148	0.111	−1.543	0.123
Policy guarantee ability	<--	Industrial organization policy	0.260	0.184	0.122	2.126	0.034
Market entry ability	<--	Industrial technology policy	−0.132	−0.108	0.131	−1.007	0.314
Talent support ability	<--	Industrial technology policy	−0.007	−0.005	0.112	−0.061	0.951
Sustainable innovation ability	<--	Industrial technology policy	0.382	0.321	0.148	2.586	0.010
Policy guarantee ability	<--	Industrial technology policy	−0.075	−0.052	0.153	−0.491	0.624
Market entry ability	<--	Guarantee implementation policy	−0.020	−0.016	0.092	−0.218	0.827
Talent support ability	<--	Guarantee implementation policy	0.053	0.039	0.079	0.674	0.500
Sustainable innovation ability	<--	Guarantee implementation policy	−0.067	−0.053	0.098	−0.680	0.496

(continued)

Table 3 (continued)

	Paths		Estimate (ns)	Estimate (s)	S.E	C.R	P
Policy guarantee ability	<--	Guarantee implementation policy	0.359	0.235	0.115	3.136	0.002

Note Estimate (ns) represents non standardized path coefficient, Estimate(s) represents standardized path coefficient, and S.E. represents standard error

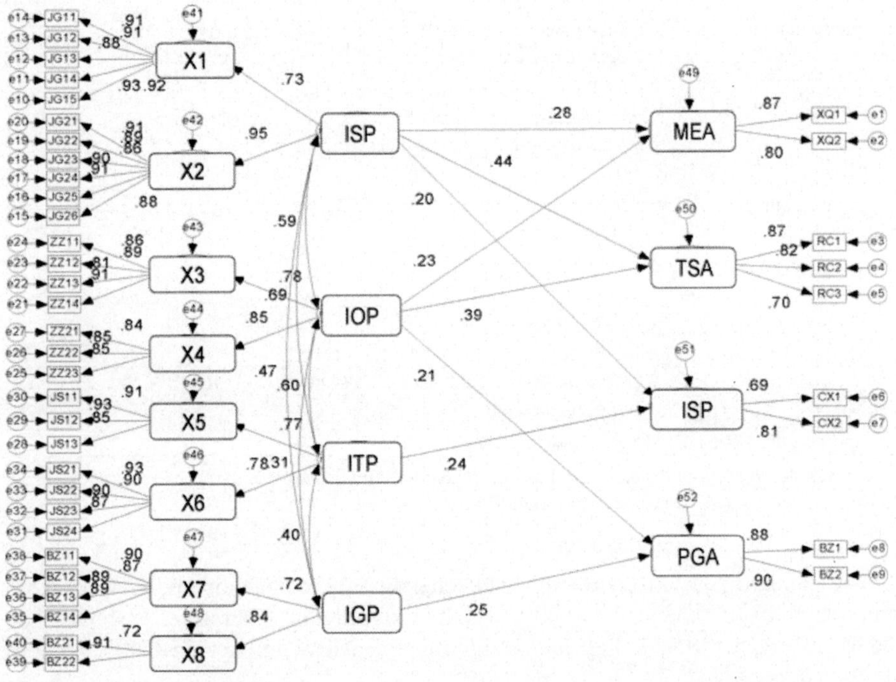

Fig. 4 Modified model's results

The standardized path coefficient between industrial structure policy and sustainable innovation ability is 0.201, P is 0.044, which is also significant. Hence, there is a significant positive impact from industrial structure policy on sustainable innovation ability. H3 is proved.

(b) *The relationships between industrial organization policy and the development ability of tendering intermediary services after the amendment.*

The standardized path coefficient between the industrial organization policy and the market entry ability is 0.231, P is 0.003, which is significant. Therefore, it shows that the industrial organization policy has a significant positive impact on the market entry ability. H5 is proved.

Table 4 The empirical parameters of the modified structural model

	Paths		Estimate (ns)	Estimate (s)	S.E	C.R	P
Market entry ability	<--	Industrial structure policy	0.382	0.278	0.103	3.697	0.000
Talent support ability	<--	Industrial structure policy	0.624	0.442	0.093	6.710	0.000
Sustainable innovation ability	<--	Industrial structure policy	0.262	0.201	0.130	2.018	0.044
Market entry ability	<--	Industrial organizational policy	0.277	0.231	0.095	2.924	0.003
Talent support ability	<--	Industrial organizational policy	0.476	0.387	0.085	5.586	0.000
Policy guarantee ability	<--	Industrial organizational policy	0.295	0.210	0.089	3.314	0.000
Sustainable innovation ability	<--	Industrial technology policy	0.284	0.242	0.127	2.238	0.025
Policy guarantee ability	<--	Guarantee implementation policy	0.390	0.251	0.105	3.699	0.000

Note Estimate (ns) represents non standardized path coefficient, Estimate(s) represents standardized path coefficient, and S.E. represents standard error

The standardized path coefficient between industrial organization policy and talent support ability is 0.387, P is 0.000, which is significant. Therefore, it shows that industrial organization policy has a significant positive impact on talent support ability. H6 is proved.

The standardized path coefficient between the industrial organization policy and the policy support ability is 0.210, P is 0.000, which is significant. Therefore, it shows that the industrial organization policy has a significant positive impact on the policy support ability. H8 is proved.

(c) *The relationships between industrial technology policy and the development ability of tendering intermediary services after the amendment.*

The standardized path coefficient between industrial technology policy and sustainable innovation capability is 0.242, P is 0.025, which is significant. Therefore, it shows that industrial technology policy has a significant positive impact on sustainable innovation capability. H11 is proved.

(d) *The relationships between guarantee implementation policy and the development ability of tendering intermediary services after the amendment.*

The standardized path coefficient between the guarantee implementation policy and policy support ability is 0.251, P is 0.000, which is significant. Therefore, it shows that the guarantee implementation policy has a significant positive impact on policy support ability. H16 is also proved.

4 Conclusion

Using the structural equation model, we have assessed the influence from industrial policy on the development ability of tendering intermediary services. Overall, the results point to a positive, significant effect of industrial policy on the development ability of tendering intermediary services. In detail, industrial structure policy has a significant positive impact on the market entry ability, talent support ability and sustainable innovation ability; industrial organization policy has a significant positive impact on the market entry ability, talent support ability and policy support ability; industrial technology policy has a significant positive impact on the sustainable innovation ability; guarantee implementation policy has a significant positive impact on policy support ability. Indeed, our empirical results suggest that industrial policy can promote the development ability of tendering intermediary services.

In summary, we should give full play to the positive role of industrial policy in the development of tendering intermediary services by formulating, executing and making good use of industrial structure policy, industrial organization policy, industrial technology policy and guarantee implementation policy, so as to promote the healthy and sustainable development of tendering intermediary services.

References

1. Zhang, L., Zhu, G. S., Li, S. G., & Li, X. Y. (2019). Market environment, key industrial policies and enterprise productivity difference.*Management World, 3*, 114–126
2. Cheah, S. L., & Ho, Y. P. (2020). Effective industrial policy implementation for open innovation: The role of government resources and capabilities. *Technological Forecasting & Social Change, 151*(2), 119845.
3. Yang, J. D., & Luo, L. B. (2018). Industrial policy, regional competition and distortion of resources spatial allocation. *China Industrial Economy, 12*, 5–22.
4. Meng, X. L. (2019). Research on development capability of modern service industry from the perspective of sustainable development. *Journal of Changchun Normal University, 38*(10), 131–135.
5. Li, F. M., Liu, X. L., Gao, Y. C., & Zhu, L. (2017). The influence of industrial policy on the innovative performance and economic performance of Chinese photovoltaic enterprises. *Science of Science and Management of S.&.T, 38*(11), 47–60.
6. Jiang, C. Y. (2019). Connotation definition and promotion strategy of high-quality development of service industry. *Reform, 304*(6), 41–52.
7. Yan, L. P., Gu, L. X., & Chen, Y. (2016). Evaluation of the development ability of high-tech industry under the innovation driven strategy. *Enterprise Economy, 430*(6), 54–58.

8. Xu, K., & Geng, C. X. (2018). Research on institutional environment and development of innovation capability of strategic emerging industries. *Journal of Technical Economics & Management, 267*(10), 106–111.
9. Liu, J. B., Wang, Q., & Ma, H. Y. (2015). Impact of the institutional environment for emerging industries of strategic-innovation capability. *Science & Technology Progress and Policy, 32*(23), 54–61.
10. Tang, R., Gu, N. H., & Tan, Z. L. (2019). Industrial policy, market structure and positioning of the enterprise value chain. *Industrial Economics Research, 98*(1), 12–26.
11. Lin, Y. F., Xiang, W., & Yu, M. J. (2018). Placed-based industrial policy and firm productivity. *China Economic Quarterly, 17*(2), 781–800.

Research on the Effectiveness of the American Financial Crisis Rescue Policy

Xuan Lv, Xinyi Mei, Menggang Li, and Haotian Wu

Abstract This paper studies the effectiveness of the monetary policy of the United States, before the financial crisis (2002.01–2007.08), the financial crisis period (2007.09–2014.10), and after the financial crisis (2014.11 2019.12), by using the vector autoregressive model (VAR), with the monthly federal funds rate of 2002–2019, M2, industrial gross domestic product, the consumer price index, total imports and exports, the Standard & Poor's 500 index, and the NYSE volume data as samples. Thus, to reflect the American financial crisis adopted a series of measures to the macroeconomic, foreign trade, and stock market rescue effect. The empirical results show that before the financial crisis, monetary policy was basically effective, but it also failed to restrain the overheating of the economy. During the financial crisis, monetary policy is ineffective, unable to stimulate economic development effectively. In the post-financial crisis period, the monetary policy of the United States can restore its effectiveness and promote the stable development of the economy, foreign trade, and stock market. From this point of view, during the financial crisis, the formulation and implementation of a series of policies in the United States effectively contained the worsening of the financial crisis and effectively rescued the American economy.

Keywords VAR · Financial crisis rescue · Monetary policy · Impulse response

X. Lv (✉)
School of Economics and Management, Beijing Laboratory of National Economic Security Early-Warning Engineering, Beijing Jiaotong University, Beijing, China
e-mail: 18113013@bjtu.edu.cn

X. Mei
National Academy of Economic Security, Beijing Jiaotong University, Beijing, China
e-mail: meixinyi1234@126.com

M. Li
Beijing Laboratory of National Economic Security Early-Warning Engineering, National Academy of Economic Security, Beijing Jiaotong University, Beijing, China
e-mail: mgli1@bjtu.edu.cn

H. Wu
Cornell University, Ithaca, USA
e-mail: hw675@cornell.edu

M. Li et al. (eds.), *IEIS 2020*, https://doi.org/10.1007/978-981-33-4363-4_17

1 Introduction

Although more than a decade has passed since the 2008 financial crisis and global financial stability has been gradually restored, new financial risks have also begun to gather. According to the IMF's January 2020 <World Economic Outlook>, global economic growth is expected to rise from an estimated 2.9% in 2019 to 3.3% in 2020 and 3.4% in 2021, with an average annual decrease of 0.1 percentage points in 2019 and 2020 and a decrease of 0.2 percentage points in 2021, compared with the October 2019 forecast. The current global economic situation is still not optimistic, so it is necessary to revisit the 2008 financial crisis and review the American financial crisis rescue policies. At present, domestic and foreign scholars have done numerous researches on the effects of financial crisis policies.

Mishkin believes that the monetary policy is more effective during financial crises than in standard times. It is because the active monetary policy is less likely to generate a negative feedback loop [1]. Chard believed that, in the short term, the central bank's rescue measures to inject liquidity into the market were effective. In the case of severe liquidity shortage of financial institutions, the rescue measures could prevent the assets from being sold at ultra-low prices, thus effectively preventing the vicious spread of the crisis in the financial system [2]. Coibion used the standard VAR method to analyze the shock effect of the monetary policy and believed that the shock of the monetary policy might play a more significant role in the actual economic fluctuations [3].

Ma and Sun found that the operation of traditional and non-traditional monetary policy tools effectively improved the currency market in the crisis. However, the rescue of individual financial institutions and the large-scale asset purchase plan did not improve the currency market, which resulted in the widening spread of LIB-OIS [4]. Wang et al. summarized various assistance measures in the crisis process. They believed that the rescue policy adopted by the federal reserve had an ideal effect in the short term [5]. Wu and Yang found that the monetary policy can play a role through the two-way channels of finance and trade, and the effect of price-based monetary policy tools is noticeable. At the same time, fiscal policy is determined by the previous trade situation and can only play the role of relief through the trade channels [6]. Wu and Wang believed that the monetary policy of quantitative easing in the United States performed well in both domestic market rescue and international spillover effects. Neither conventional monetary policy nor fiscal policy is effective when used alone. However, the coordinated application of the two will help economic recovery [7].

In general, there are different opinions on the measures and effects of the financial crisis rescue by the federal reserve [8]. This paper studies the effectiveness of American monetary policy from three aspects: macroeconomics, foreign trade, and the financial market. From the crisis rescue policy impact on the effectiveness of monetary policy perspective, before the financial crisis, monetary policy can effectively increase the macroeconomy, foreign trade, and the stability of financial markets. However, during the financial crisis, monetary policy fundamental failure appears even an adverse effect, but after the rescue, the effectiveness of the monetary policy of market regulation was restored. In summary, the rescue policies implemented during the financial crisis effectively freed the American economy from the predicament and continued to achieve healthy and stable development.

2 The Financial Crisis Relief Policy Review

Before 2007, as the American economy lose the monetary policy and relax the regulation of the subprime mortgage market, it led to the credit rapidly expanding, and cause the housing market bubble burst. The housing slump, rates tend to rise, subprime loan default rates have risen sharply, banks non-performing assets increase rapidly; each big financial institution causes the massive loss, then formed the subprime crisis. Since then, a series of vicious events including the near-bankruptcy of New Century Financial led to the subprime mortgage crisis, becoming a severe global Financial crisis. Because of this situation, the Federal Reserve took steps to pump liquidity into the financial system and rescue the stock market. This stage was the formation stage of the subprime mortgage crisis, and its destructive power and influence were limited. The US authorities hoped that the market mechanism would adjust itself, and their policies failed to prevent the crisis from spreading further.

On September 15, 2008, Lehman Brothers, the fourth-largest investment bank in the United States, filed for bankruptcy protection, and the subprime crisis turned into a financial crisis. On the same day, the bank of America bought Merrill Lynch, America's third-largest investment bank. Then the government took over the American international group and provided $85 billion in short-term emergency loans. On September 20th, the Bush administration authorized the Treasury to buy up to $700 billion of toxic mortgage assets. With the evolution of the crisis, the government began to implement a large number of active rescue policies. The government's rescue package raised the original total from $700 billion to $850 billion, extended tax cuts and raised the ceiling on bank deposit insurance to $250,000. It cut interest rates by 50 basis points to 1.5% and cut the benchmark rate to 0–0.25%. The American authorities coordinated the fiscal policy and the monetary policy to stimulate the economy on a large scale. The innovated monetary policy tools and the unconventional monetary policy were the main contents of the rescue policies at this stage. Nevertheless, the rescue results are not optimistic; the stock index is still falling, credit crunch, consumption downturn, the market is still filled with pessimism.

With the continuous spread of the crisis, the financial crisis gradually spread from the virtual economy to the real economy. It caused considerable damage to the real economy of the United States. Although the American government took aggressive rescue efforts, the unemployment rate still raised to 9.4%, a 26-year high point, in May 2009. The global economy is weakening due to the financial crisis and the recession in the real economy. At this time, the domestic assistance of the United States could not meet the governance of the global economic recession, and the goal of this stage was to promote economic recovery through international cooperation jointly.

Besides, the American launched four rounds of quantitative easing monetary policies on November 25, 2008, November 4, 2010, September 14, 2012, and December 13, 2012, respectively. The quantitative easing policy of the federal reserve has played an essential role in promoting the economic recovery of the United States while increasing the issuance of base money, collecting seigniorage, and supporting the government to expand fiscal spending. It was not until late 2014 that the federal reserve officially announced the end of quantitative easing.

Overall, the evolution of the U.S. financial crisis relief policy has experienced a gradual process of change. From the U.S. government not tend to rescue the subprime crisis, a massive relief to the financial crisis broke out, and then to the development of global economic recession, international collaboration, and mutual governance crisis fully embodies. The content of aid policy is closely related to the nature and development degree of the crisis.

3 Empirical Research

3.1 Data Sources

The data of this article is from the wind database. We selected the monthly data of the three representative indicators of the macroeconomy, foreign trade, and stock market from 2002 to 2019 as samples. We selected January 2002 to August 2007 as a non-crisis period, September 2007 to October 2014 as a financial crisis period, November 2014 to December 2019 as a post-financial crisis period to examine the effect of the American financial crisis rescue policy.

For the measurement of the monetary policy, we selected the federal funds rate (FFR) and the money supply (M2).In terms of macroeconomic research, we selected the gross industrial product (GI) and the consumer price index (CPI). For foreign trade, we selected import (IM) and export (EX) totals. For the stock market, we selected the closing price of the S&P 500 index (SPX) and the daily average volume of the New York stock exchange (NYSE).

3.2 Model Setting

VAR model, its principle is to express the current variable as a regression form of the following changes, in order to find the dynamic relationship between variables.

The VAR model expression containing the lag k period of N variables is shown in Eq. (3.1):

$$Y_t = c + \Phi_1 Y_{t-1} + \Phi_2 Y_{t-2} + \cdots + \Phi_k Y_{t-k} + u_t,\ u_t \sim iid(0, \Omega) \quad (3.1)$$

VAR model with single variable first-order lag, VAR(1), which is expressed as follows:

$$\begin{bmatrix} y_{1t} \\ y_{2t} \end{bmatrix} = \begin{bmatrix} \Phi_{11} & \Phi_{12} \\ \Phi_{21} & \Phi_{22} \end{bmatrix} \begin{bmatrix} y_{1t-1} \\ y_{2t-1} \end{bmatrix} + \begin{bmatrix} u_{1t} \\ u_{2t} \end{bmatrix};\ \mathrm{var}(u_t) = \mathrm{var}\left(u_t^{'}\right) = \begin{bmatrix} \sigma_1^2 & \sigma_{12} \\ \sigma_{21} & \sigma_2^2 \end{bmatrix} \quad (3.2)$$

The condition to ensure the stability of the VAR model is that the characteristic roots of the following matrices fall outside the unit circle.

$$\begin{bmatrix} Y_t \\ Y_{t-1} \\ Y_{t-2} \\ \vdots \\ Y_{t-k+1} \end{bmatrix}_{Nk\times1} = \begin{bmatrix} c \\ 0 \\ 0 \\ \vdots \\ 0 \end{bmatrix}_{Nk\times1} + \begin{bmatrix} \Phi_1 & \Phi_1 & \cdots & \Phi_1 & \Phi_1 \\ I & 0 & \cdots & 0 & 0 \\ 0 & I & \cdots & 0 & 0 \\ \cdots & \cdots & \ddots & \cdots & \cdots \\ 0 & 0 & \cdots & I & 0 \end{bmatrix}_{Nk\times Nk} \begin{bmatrix} Y_{t-1} \\ Y_{t-2} \\ Y_{t-3} \\ \vdots \\ Y_{t-k} \end{bmatrix}_{Nk\times1} + \begin{bmatrix} u_t \\ 0 \\ 0 \\ \vdots \\ 0 \end{bmatrix}_{Nk\times1} \quad (3.3)$$

This paper establishes the following VAR model:

$$Y_t^1 = (GI_t, CPI_t, M_{2t}, FFR_t)'$$

$$Y_t^1 = (EX_t, IM_t, M_{2t}, FFR_t)'$$

$$Y_t^1 = (SPX_t, NYSE_t, M_{2t}, FFR_t)'$$

3.3 Empirical Results

3.3.1 Model Establishment and Test

Before establishing the VAR model, the ADF test method was used to test the stationarity of the data. The test results showed that the original data had unit root, and the data was not stable, and the data after taking the logarithm also had unit root, and the data was still unstable. After further first-order difference, the ADF test results of the data show that the test statistics are all smaller than those corresponding to the significance level of 1%, indicating that the data are stable. Further with DFFR, DM2, DGI, DCPI, DEX, DIM, DSPX, and DNYSE VAR models are established. Including DFFR means the federal funds rate, DM2 means money supply rate, DGI means industrial production rate, gross DCPI means consumer price level change rate, DEX means total exports rate, DIM means rate, total imports DSPX means standard & poor's 500 rates, closing DNYSE means NYSE volume change rate.

According to AIC and SC information criteria, the lag order of the VAR model was determined. VAR model was established by checking the data stationarity and determining the order of the model.

The following equation expresses the VAR model before the financial crisis from January 2002 to August 2007:

$$\begin{aligned} DGI_t = {} & -0.514559DGI_{(t-1)} - 0.0108411DGI_{(t-2)} + 0.00256DCPI_{(t-1)} \\ & - 0.0342081DCPI_{(t-2)} - 0.1982478DM2_{(t-1)} + 0.0769991DM2_{(t-2)} \\ & + 0.0282651DFFR_{(t-1)} - 0.0102905DFFR_{(t-2)} + 0.3547877 \end{aligned}$$

$$\begin{aligned} DCPI_t = {} & 0.0973267DGI_{(t-1)} + 0.0179462DGI_{(t-2)} + 0.630968DCPI_{(t-1)} \\ & - 0.4641878DCPI_{(t-2)} + 0.0597573DM2_{(t-1)} - 0.0255684DM2_{(t-2)} \\ & + 0.0128238DFFR_{(t-1)} - 0.0120266DFFR_{(t-2)} + 0.1530193 \end{aligned}$$

$$\begin{aligned} DIM_t = {} & -0.3999821DIM_{(t-1)} + 0.1707218DEX_{(t-1)} \\ & + 0.0768918DM2_{(t-1)} + 0.0104066DFFR_{(t-1)} + 1.039755 \end{aligned}$$

$$\begin{aligned} DEX_t = {} & -0.157879DIM_{(t-1)} - 0.1690807DEX_{(t-1)} \\ & - 0.4258365DM2_{(t-1)} + 0.045312DFFR_{(t-1)} + 1.271821 \end{aligned}$$

$$\begin{aligned} DNYSE_t = {} & -0.3749448DNYSE_{(t-1)} - 0.5267365DNYSE_{(t-2)} \\ & + 0.140385DSPX_{(t-1)} + 0.1096326DSPX_{(t-2)} \\ & - 0.3793417DM2_{(t-1)} - 4.842462DM2_{(t-2)} \\ & + 0.287564DFFR_{(t-1)} - 0.056486DFFR_{(t-2)} + 4.476745 \end{aligned}$$

$$\begin{aligned} DSPX_t = {} & 0.0820072DNYSE_{(t-1)} - 0.0333853DNYSE_{(t-2)} \\ & + 0.0979686DSPX_{(t-1)} + 0.1006547DSPX_{(t-2)} \\ & + 0.9501926DM2_{(t-1)} + 0.5846333DM2_{(t-2)} \\ & + 0.0531977DFFR_{(t-1)} - 0.0678931DFFR_{(t-2)} - 0.6597295 \end{aligned}$$

It can be found from the regression results that the rate of change of DGI status in the past two periods will affect the rate of change of current DGI in the opposite direction, and DGI is largely affected by itself.

It can be found from the regression results that the rate of change of DCPI conditions in the past two periods will affect the rate of change of current DCPI in the opposite direction, and DCPI is affected mainly by itself.

According to the regression results, the coefficient of DIM hysteresis order 1 is negative, and DIM is affected mainly by itself.

It can be found from the regression results that the rate of change of DEX status in the prior one period will have a negative influence on the current rate of change of DEX. DEX is affected mainly by itself.

It can be found from the regression results the change rate of DNYSE status in the past two periods will negatively affect the change rate of current DNYSE, and DNYSE is primarily affected by itself.

It can be found from the regression results that the rate of change of DSPX in the past two periods will have a positive impact on the rate of change of DSPX, and DSPX is largely affected by itself.

The following equation expresses the VAR model during the financial crisis from September 2007 to October 2014:

$$\begin{aligned}DGI_t = &-0.0399664DGI_{(t-1)} - 0.8259817DGI_{(t-2)} + 1.270091DGI_{(t-3)}\\ &+ 0.0193982DCPI_{(t-1)} + 0.0117414DCPI_{(t-2)} - 0.0039554DCPI_{(t-3)}\\ &+ 1.571833DM2_{(t-1)} - 3.846426DM2_{(t-2)} + 4.104107DM2_{(t-3)}\\ &+ 0.0095766DFFR_{(t-1)} - 0.2059479DFFR_{(t-2)}\\ &+ 0.3303342DFFR_{(t-3)} - 3.644723\end{aligned}$$

$$\begin{aligned}DCPI_t = &-0.2590101DGI_{(t-1)} - 3.100385DGI_{(t-2)} + 3.307584DGI_{(t-3)}\\ &+ 0.0235155DCPI_{(t-1)} + 0.1656876DCPI_{(t-2)}\\ &+ 0.090001DCPI_{(t-3)} + 4.492359DM2_{(t-1)}\\ &- 12.32998DM2_{(t-2)} + 12.34511DM2_{(t-3)} - 0.0158395DFFR_{(t-1)}\\ &- 0.6588619DFFR_{(t-2)} + 0.9604374DFFR_{(t-3)} - 11.55147\end{aligned}$$

$$\begin{aligned}DIM_t = &-0.2566656DIM_{(t-1)} - 0.0989599DIM_{(t-2)} + 0.4433973DEX_{(t-1)}\\ &+ 0.6334949DEX_{(t-2)} - 0.3916535DM2_{(t-1)} - 0.329151DM2_{(t-2)}\\ &+ 0.0424814DFFR_{(t-1)} - 0.0074499DFFR_{(t-2)} + 0.42617\end{aligned}$$

$$\begin{aligned}DEX_t = &\,0.2194897DIM_{(t-1)} + 0.2045057DIM_{(t-2)} - 0.1095548DEX_{(t-1)}\\ &+ 0.0078806DEX_{(t-2)} - 0.2168082DM2_{(t-1)} + 0.0205577DM2_{(t-2)}\\ &+ 0.0386291DFFR_{(t-1)} - 0.0139849DFFR_{(t-2)} + 0.57462\end{aligned}$$

$$
\begin{aligned}
DNYSE_t = {} & 0.3781337DNYSE_{(t-1)} + 0.1776892DNYSE_{(t-2)} \\
& + 0.4356518DNYSE_{(t-3)} - 0.4495138DNYSE_{(t-4)} \\
& + 0.505546DSPX_{(t-1)} + 1.154699DSPX_{(t-2)} \\
& - 0.9797201DSPX_{(t-3)} - 0.3732758DSPX_{(t-4)} \\
& + 4.749311DM2_{(t-1)} - 8.63049DM2_{(t-2)} + 25.52358DM2_{(t-3)} \\
& + 43.36769DM2_{(t-4)} + 0.2059368DFFR_{(t-1)} - 1.499421DFFR_{(t-2)} \\
& + 2.902606DFFR_{(t-3)} - 0.5114352DFFR_{(t-4)} - 51.63412
\end{aligned}
$$

$$
\begin{aligned}
DSPX_t = {} & -0.4101623DNYSE_{(t-1)} - 0.1362032DNYSE_{(t-2)} \\
& - 0.3654586DNYSE_{(t-3)} + 0.0836512DNYSE_{(t-4)} \\
& + 0.6641995DSPX_{(t-1)} - 0.7516985DSPX_{(t-2)} \\
& + 0.417957DSPX_{(t-3)} + 0.0220601DSPX_{(t-4)} \\
& + 0.4444344DM2_{(t-1)} + 16.40054DM2_{(t-2)} \\
& - 13.03228DM2_{(t-3)} - 29.88789DM2_{(t-4)} \\
& - 0.0038918DFFR_{(t-1)} + 0.9328308DFFR_{(t-2)} \\
& - 1.690363DFFR_{(t-3)} + 0.1743013DFFR_{(t-4)} + 29.21211
\end{aligned}
$$

According to the regression results, the rate of change of DGI status in the past three periods will affect the current rate of change of DGI in the opposite direction, and DGI is mostly affected by itself.

It can be found from the regression results that the rate of change of DCPI status in the past three periods will affect the rate of change of current DCPI in the opposite direction, and DCPI is primarily affected by itself.

It can be found from the regression results that the change rate of DIM conditions in the past two phases will negatively affect the change rate of current DIM, and DIM is affected mainly by itself.

It can be found from the regression results that the rate of change of DEX state in the past two periods will have a positive influence on the current rate of change of DEX. DEX is primarily affected by itself.

It can be found from the regression results that the change rate of DNYSE status in the past four periods will affect the change rate of current DNYSE in the opposite direction, and DNYSE is primarily influenced by itself.

According to the regression results, the rate of change of DSPX in the prior four periods will affect the rate of change of DSPX in different directions, and DSPX is influenced mainly by itself.

The following equation expresses the VAR model during the post-financial crisis period from November 2014 to December 2019:

$$
\begin{aligned}
DGI_t = {} & -0.4030982DGI_{(t-1)} - 0.228418DGI_{(t-2)} - 0.09837DCPI_{(t-1)} \\
& + 0.7479664DCPI_{(t-2)} - 0.0172942DM2_{(t-1)} - 0.2664587DM2_{(t-2)}
\end{aligned}
$$

$$+ 0.0177323 DFFR_{(t-1)} - 0.0054159 DFFR_{(t-2)} + 0.0489642$$

$$\begin{aligned} DCPI_t = & -0.0353118 DGI_{(t-1)} - 0.0057522 DGI_{(t-2)} + 0.4838493 DCPI_{(t-1)} \\ & - 0.1925194 DCPI_{(t-2)} + 0.0769047 DM2_{(t-1)} + 0.0726516 DM2_{(t-2)} \\ & + 0.0021543 DFFR_{(t-1)} + 0.0018884 DFFR_{(t-2)} + 0.0303598 \end{aligned}$$

$$\begin{aligned} DIM_t = & -0.4150873 DIM_{(t-1)} - 0.2080339 DIM_{(t-2)} + 0.1334506 DEX_{(t-1)} \\ & + 0.198003 DEX_{(t-2)} - 0.3859815 DM2_{(t-1)} - 0.0103384 DM2_{(t-2)} \\ & + 0.0079073 DFFR_{(t-1)} - 0.0181683 DFFR_{(t-2)} + 0.4377775 \end{aligned}$$

$$\begin{aligned} DEX_t = & 0.0849364 DIM_{(t-1)} + 0.0396853 DIM_{(t-2)} - 0.0909257 DEX_{(t-1)} \\ & + 0.0157657 DEX_{(t-2)} - 0.1078799 DM2_{(t-1)} + 0.3431117 DM2_{(t-2)} \\ & - 0.0164097 DFFR_{(t-1)} - 0.0022883 DFFR_{(t-2)} + 0.0967228 \end{aligned}$$

$$\begin{aligned} DNYSE_t = & -0.2328759 DNYSE_{(t-1)} - 0.104598 DNYSE_{(t-2)} \\ & - 0.1471402 DSPX_{(t-1)} + 0.2007423 DSPX_{(t-2)} \\ & - 5.840767 DM2_{(t-1)} + 3.218552 DM2_{(t-2)} \\ & + 0.1921439 DFFR_{(t-1)} - 0.1324482 DFFR_{(t-2)} + 0.7634143 \end{aligned}$$

$$\begin{aligned} DSPX_t = & -0.0924827 DNYSE_{(t-1)} - 0.0806688 DNYSE_{(t-2)} \\ & - 0.3933321 DSPX_{(t-1)} - 0.1723077 DSPX_{(t-2)} \\ & + 1.216839 DM2_{(t-1)} - 0.4598403 DM2_{(t-2)} \\ & - 0.0793992 DFFR_{(t-1)} - 0.0454354 DFFR_{(t-2)} + 0.986608 \end{aligned}$$

According to the regression results, the rate of change of DGI status in the past two periods will negatively affect the rate of change of current DGI, and DGI is mostly affected by itself.

Through the regression results, it can be found that the rate of change of DCPI conditions in the past two periods will affect the rate of change of current DCPI in the opposite direction, and DCPI is primarily affected by itself.

It can be found from the regression results that the coefficients of order 1 and order 2 of DIM lag are negative, and the change rate of DIM conditions in the past two phases will negatively affect the change rate of current DIM, and DIM is primarily affected by itself.

According to the regression results, it can be found that the rate of change of DEX conditions in the past two periods will affect the rate of change of DEX in different directions, and DEX is affected mainly by itself.

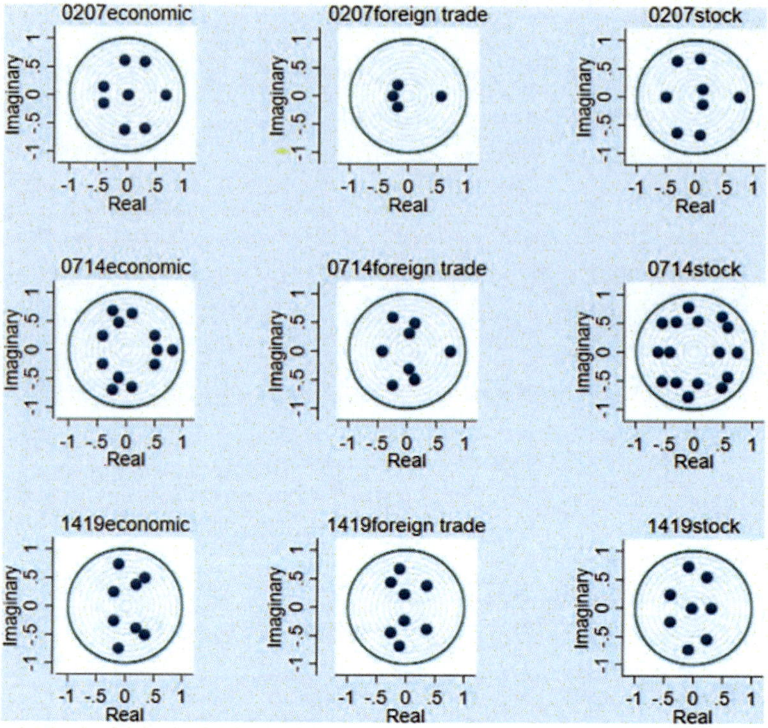

Fig. 1 Stability test diagram of VAR model

It can be found from the regression results that the change rate of DNYSE status in the past two periods will negatively affect the change rate of current DNYSE, and DNYSE is largely affected by itself.

It can be found from the regression results that the rate of change of DSPX in the past two periods will have a negative impact on the rate of change of DSPX, and DSPX is largely affected by itself.

The stability basis of the VAR model requires that all unit roots in the model should be placed in the unit circle with radius 1. As shown in Fig. 1, the results show that the unit-roots of the models are all in the unit circle, which indicates that the models have passed the stability test.

3.3.2 Impulse Response Analysis

Impulse response refers to the dominant influence on the current and future values of endogenous variables when a random disturbance term is impacted by one standard deviation. By observing the impulse response results and discovering its reaction situation and trend direction, the following conclusions are drawn:

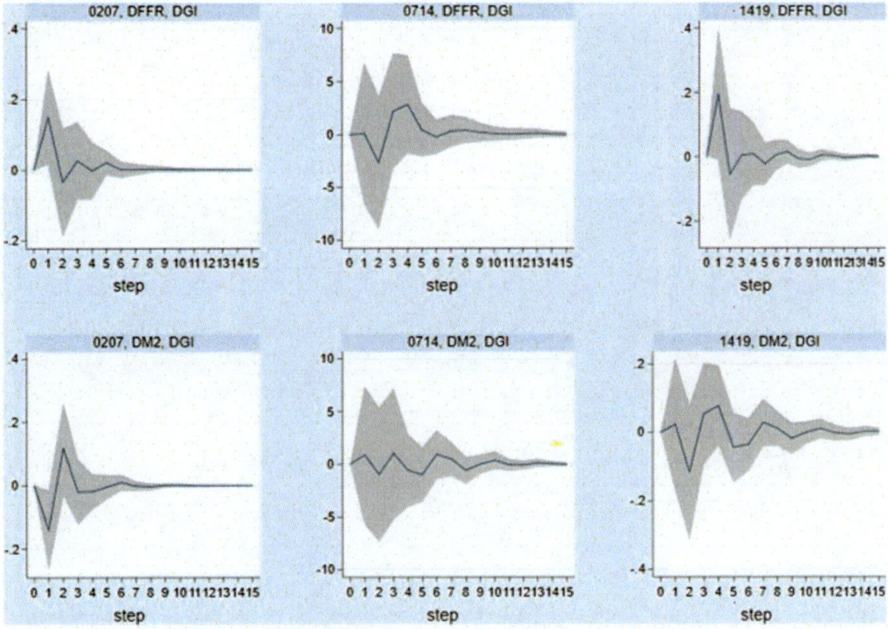

Fig. 2 Response of DGI to DFFR and DM2

First, in terms of the macroeconomy, the monetary policy plays a prominent role in output during a non-financial crisis. Before and after the financial crisis, the monetary policy had similar effects, while during the financial crisis, the monetary policy basically failed or even had a reaction. In the period of the non-financial crisis, economic development can provide timely and practical feedback to the monetary policy. However, during the financial crisis, the change of monetary policy causes great fluctuation of economic development and price level (Figs. 2 and 3).

Second, from the perspective of foreign trade, to raise the federal funds rate. Enterprises directly increase the cost of financing, financing difficulties, is not conducive to enterprises to expand production, further influence the import and export, on the other hand, in the open economy system, raising interest rates will increase people's propensity to save, and reduce consumption, is unfavorable to the import and export. The increase of money supply makes it easier for enterprises to obtain funds and reduce the cost of using funds, which is conducive to the expansion of production and thus creates conditions for the expansion of import and export trade. Also, with the support of the loose monetary policy, enterprises can invest and build factories abroad through residential loans, which has a positive impact on import and export trade. However, before the financial crisis, the domestic market in the United States was hot, and more money supply would only be invested in the domestic market. During the financial crisis, money was mainly used for domestic market rescue, so the monetary policy could not effectively affect the import and export trade before and during the financial crisis. In the post-financial crisis period, with the economy

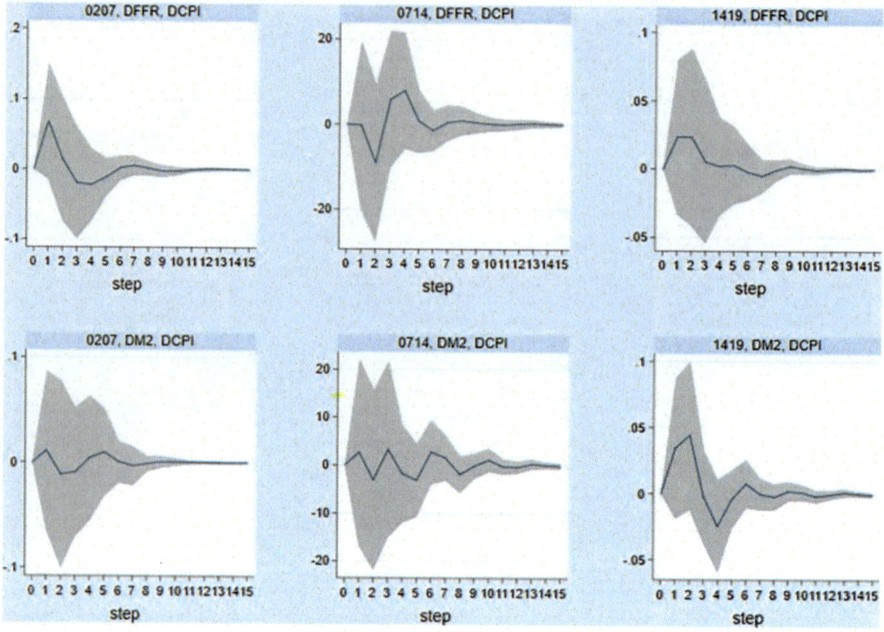

Fig. 3 Response of DCPI to DFFR and DM2

entering the stage of stable development, the monetary policy can effectively adjust the import and export trade (Figs. 4 and 5).

Thirdly, from the perspective of the stock market, after the interest rate is raised, the financing cost of enterprises increases, the production scale is cut, and the profit is also reduced, which will lead to the decline of the stock price. Besides, the discount rate of the stock value will also rise with the interest rate so that the stock value will fall, and the stock price will fall accordingly. Moreover, higher interest rates divert some funds from the stock market to bank deposits and bonds, thus reducing the demand for stocks and causing stock prices to fall. The increase of money supply leads to the change of the number of loanable funds, further reduces the operating costs of enterprises, promotes the growth of corporate performance, and more robust profitability, and then stimulates the rise of stock prices, further promoting the rise of stock prices (Figs. 6 and 7).

3.3.3 Variance Decomposition

Whether before, during, or after the financial crisis, the shocks of DGI, DCPI, DEX, DIM, DNYSE, and DSPX themselves are the main sources of variance. Before the financial crisis, DM2 and DFFR impact of DGI, DCPI, DEX, DIM, DNYSE and DSPX change in 13%, 5%, 7% and 2%, respectively, 8% and 7%, and during the financial crisis, the impact is respectively 2%, 3%, 13%, 14%, 3% and 3%, after

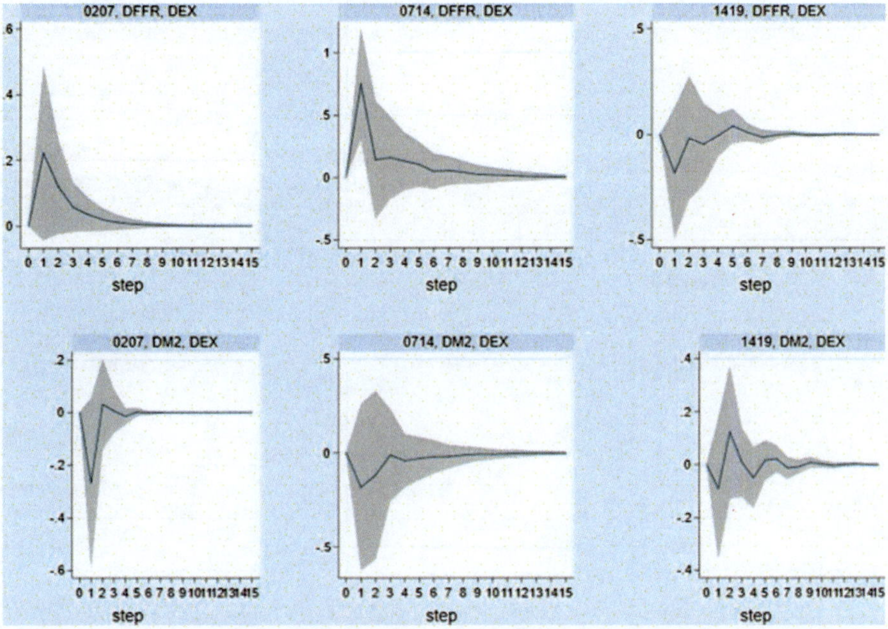

Fig. 4 Response of DEX to DFFR and DM2

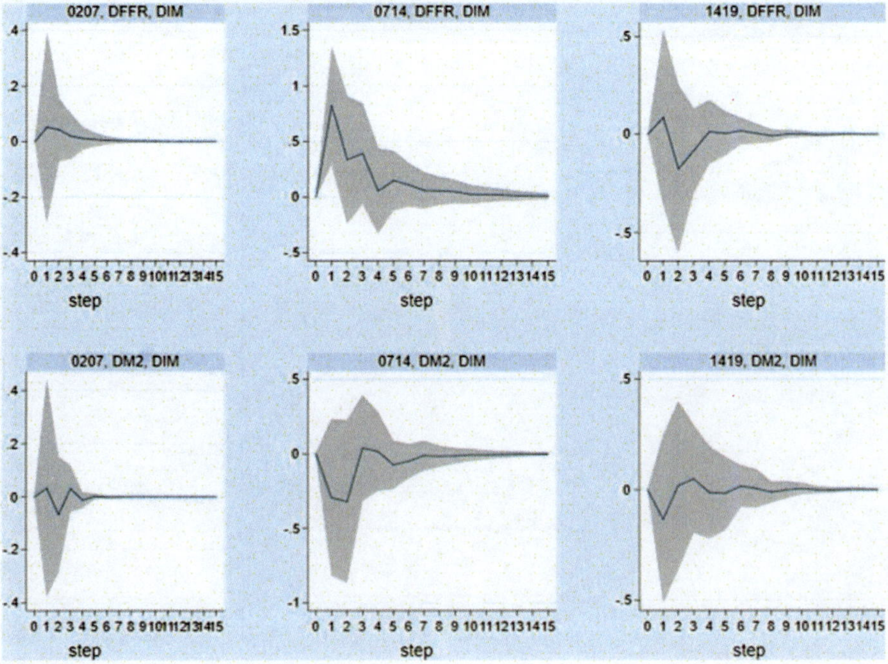

Fig. 5 Response of DIM to DFFR and DM2

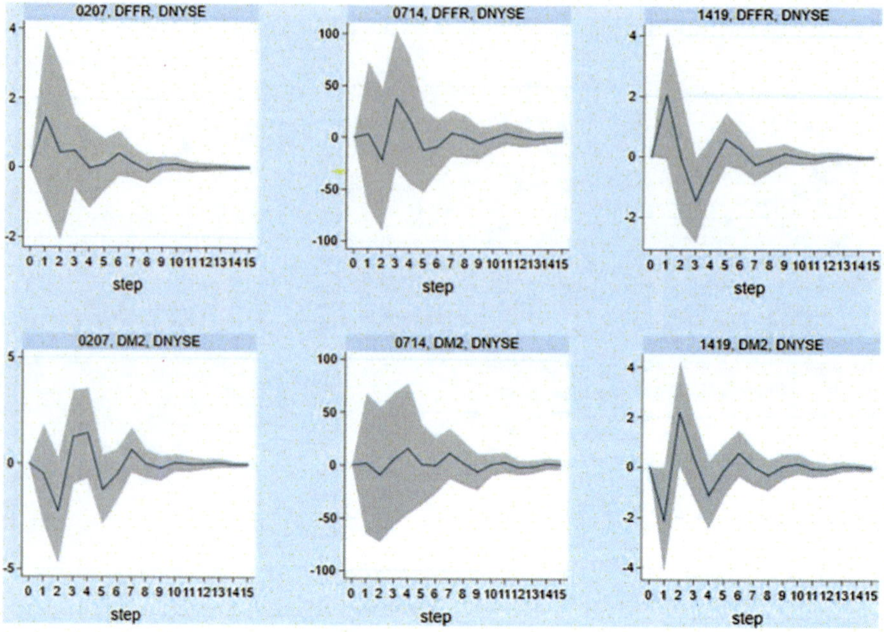

Fig. 6 Response of DNYSE to DFFR and DM2

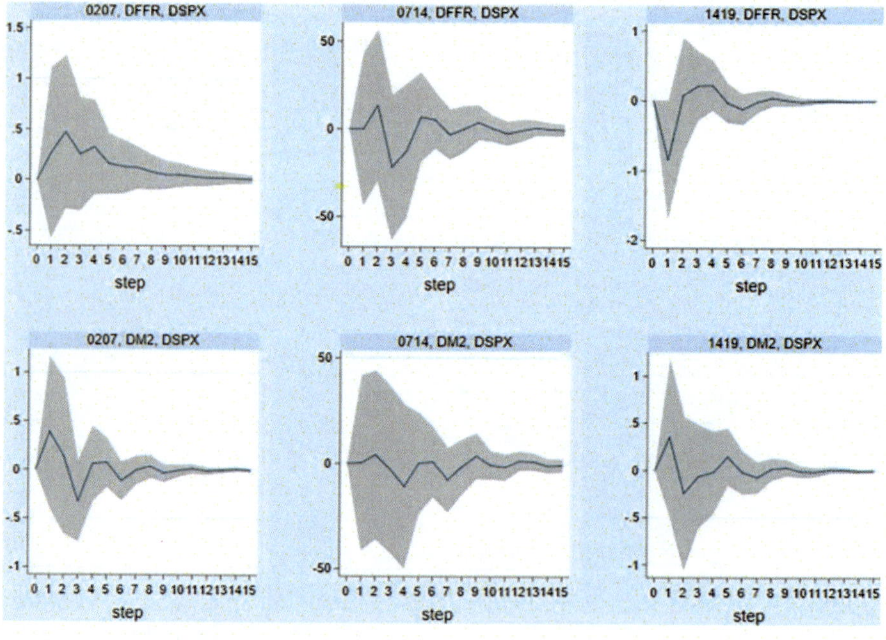

Fig. 7 Response of DSPX to DFFR and DM2

the financial crisis, the impact is respectively 7%, 7%, 4%, 2%, 18% and 7%. In a non-financial crisis period, the monetary policy impact has a significant effect on industrial output GI, consumer price index CPI, export EX, NYSE trading volume NYSE and standard & poor's 500 indexes SPX, and a small impact on import IM. During the financial crisis, the role of the monetary policy on industrial output GI, consumer price index CPI, NYSE trading volume, NYSE, and S&P 500 SPX was greatly weakened, which had a great impact on imports and exports.

4 Conclusion

During the period of the non-financial crisis, the American economy developed rapidly, and the monetary policy could regulate the market effectively. However, in the early stage of the financial crisis, the phenomenon appeared that the tight monetary policy could not adequately cool down the overheated economy, and bubbles were continually accumulating. During the financial crisis, the conventional monetary policy was mostly ineffective. In the case of extremely low-interest rates, most of the liquidity injected into the market by the expansionary monetary policy will be absorbed by people's infinite money demand, and the effect of monetary policy will be significantly weakened. The implementation of monetary policy is faced with the risk of over-reaction, over-fast reaction, and too long duration. The monetary policy may lead to more significant financial risks when monetary policy plays its role, and the greater financial risks will lead to higher costs. Although the recession of the American economy stopped, and there were signs that the American economy would get out of the crisis in 2009, the vitality of the economy has not been well recovered, and the American economy is still fragile. In the post-financial crisis period, the effectiveness of the monetary policy gradually recovered, and the economic situation of the United States continued to improve. This article not only from the perspective of macroeconomic considerations for America's monetary policy effectiveness but also to the United States from foreign trade and the stock market judging the effectiveness of the monetary policy. This paper can identify more comprehensively and effectively whether the United States has successfully passed the financial crisis, and overcome the evolution of the macroeconomic situation only limits to determine the single monetary policy effectiveness. Empirical research shows that, from the macroeconomic perspective, in the period of a non-financial crisis, USA monetary policy can effectively promote economic growth and price stability. However, in the period of the financial crisis, the monetary policy fails. From the perspective of foreign trade, in the period of non-financial crisis, the monetary policy of the United States had little impact on the import and export trade, and the import and export trade could develop steadily. However, during the financial crisis, the adjustment of the monetary policy made the import and export trade fluctuate considerably, which affected import and export stability. From the perspective of the stock market, in the period of a non-financial crisis, the monetary policy can effectively stimulate the trade of stock market. However, in the period of the financial crisis, the adjustment of monetary policy cannot effectively help the stock market to get out of trouble.

References

1. Mishkin, F. S. (2009). Is monetary policy effective during financial crises? *American Economic Review: Papers & Proceedings, 99*(2), 573–577.
2. Blanchard, O., & Milesi-Ferretti, G. (2009). Global imbalances: In midstream? IMF Staff Position Note, December 22, 2009.
3. Coibion, O. (2011). Are the effects of monetary policy shocks big or small? NBER Working Paper, No.17034, May, 2011.
4. Ma, H., & Sun, X. (2011). Research on the monetary policy effect of the federal reserve during the financial crisis—An empirical analysis based on the monetary market. *International Finance Research, 2.*
5. Wang, Q., Wang, D., & Chen, G. (2012). Empirical test on the effect of financial crisis rescue policies—Based on the perspective of American capital market. *Financial and Economic Perspective, 7.*
6. Wu, T., & Yang, D. (2019). Research on the effectiveness of us subprime mortgage crisis rescue policy—Based on the dual perspective of finance and trade transmission. *Southwest Finance, 4.*
7. Wu, T., & Wang, T. (2019). Research on us financial crisis relief policies: Domestic impact, international spillover and policy implications. *Southwest Finance, 12.*
8. Lee, H., & Lee, W. (2019). International linkage among Mena financial markets. *Economic Computation and Economic Cybernetics Studies and Research, 53*(3), 297–314.

Research on the Impact of Margin Trading on Corporate Investment Efficiency

Hui Zhang and Xuemeng Guo

Abstract This paper examines the impact of margin trading on corporate investment efficiency by using the difference-in-differences model. First, I examine whether margin trading can improve corporate investment efficiency. Second, I investigate whether information transparency and agency costs are mediating factors for margin trading to improve the corporate investment efficiency. The results indicate that margin trading can improve corporate investment efficiency, and its motivation comes from the short selling effect of securities lending trading. In addition, information transparency and agency costs play an intermediary role for margin trading in improving corporate investment efficiency. Further examination found that the effect of margin trading on improving investment efficiency is more obvious in firms with weak governance mechanism, which indicates that margin trading has an external governance role. The results in this paper have certain reference significance for the security regulatory authorities, firms, and investors.

Keywords Margin trading · Investment efficiency · Information transparency · Agency costs

1 Introduction

In 2010, China officially launched the pilot of margin trading. Margin trading is a credit trading, which is divided into financing trading and securities lending trading. Financing trading refers to the trading behavior of borrowing funds to buy securities, while securities lending trading refers to the trading behavior of borrowing securities to sell them. In March 2010, 90 stocks were announced as margin trading pilots for the first time. After the fifth expansion, 950 stocks have become the underlying stocks for margin trading. The introduction of margin trading has promoted a series

H. Zhang (✉) · X. Guo
School of Economics and Management, Beijing Jiaotong University, Beijing, China
e-mail: 19120680@bjtu.edu.cn

X. Guo
e-mail: 19113062@bjtu.edu.cn

M. Li et al. (eds.), *IEIS 2020*, https://doi.org/10.1007/978-981-33-4363-4_18

of related papers. From the perspective of capital market, margin trading can increase the information content of stock prices [1] and improve the pricing efficiency of the underlying stocks [2]. From the perspective of corporate governance, margin trading has suppressed managers' earnings management behavior [3].

Inefficient investment is caused by information asymmetry [4] and agency problems [5]. In addition, the corporate governance level is a major factor influencing investment efficiency [6]. Some scholars have introduced the effect of margin trading on the impact of corporate investment efficiency. When margin trading exists, investor sentiment will be reflected in the stock prices in time, which might cause managers to change their investment strategies to cater for changes in stock prices [7]. Short selling pressure brought by securities lending trading can improve the corporate internal governance and reduce excessive investment, thereby improving investment efficiency [8]. Reference [8] only focused on the effect of securities lending trading without distinguishing between financing trading and securities lending trading. What's more, it lacks empirical evidence. Therefore, this paper studies the impact of margin trading on corporate investment efficiency, and distinguishes the different effects of financing trading and securities lending trading on corporate investment efficiency. The results show that margin trading inhibited the corporate inefficient investment. Its motivation comes from the short selling effect of securities lending trading. Furthermore, this paper explores the channels of margin trading in affecting investment efficiency: agency costs and information transparency. In addition, a series of robustness tests were also conducted in this paper. The results show that the conclusion of this paper is robust.

The contribution of this paper is mainly reflected in the following three aspects. First, this paper examines whether the effect of financing trading and securities lending trading on corporate investment efficiency is the same in the case that the volume of securities lending trading in China is much smaller than the volume of financing trading, which will be a certain reference for the regulatory and policy making. Second, this paper studies the channels of margin trading in influencing corporate investment efficiency, and provides methods for improving investment efficiency. Third, the research in this paper validates the external governance role of margin trading, which has a certain reference value for policy makers in expanding the scope of margin trading pilots in the future.

The remainder of this paper is organized as follows. Section 2 is literature review and research hypothesis. Section 3 is sample selection and research design. Section 4 reports the main empirical results. Section 5 is the extensive test. Section 6 is the robustness test, and Sect. 7 concludes.

2 Literature Review and Research Hypothesis

2.1 The Influencing Factors of Investment Efficiency

Existing research found that information asymmetry and agency problems are the reasons for inefficient investment. Managers' adverse selection and moral hazard problems caused by information asymmetry will affect corporate investment efficiency [5]. High-quality financial statements can alleviate the information asymmetry between the firm and the market, reduce external financing costs, and thus reduce the possibility of corporate investment level deviating from expected investment level [9]. After the financial report disclosed the internal control deficiencies, the corporate investment efficiency has improved significantly [10]. Agency problem is another important cause of corporate inefficient investment [4]. Establishing a perfect corporate governance mechanism can effectively restrict managers' pursuit of private interests, thereby alleviating corporate inefficient investment [11]. Internal governance mechanisms can effectively improve corporate investment efficiency [10].

2.2 Margin Trading and Investment Efficiency

Combined with the real market conditions of stock trading, the effects of financing trading and securities lending trading may be different. Therefore, this section separately analyzes the impact of securities lending trading and financing trading on corporate investment efficiency.

From the perspective of securities lending trading, short selling mechanism mainly affects the corporate investment efficiency from the aspects of improving information environment and reducing agency costs.

2.2.1 Information Environment

First, due to the higher costs, the information held by investors is more authentic in securities lending trading [12]. Short selling mechanism provides investors with a way to tap the corporate negative information for profit, which increases the difficulty and risk of managers hiding negative information, and thus prompts managers to disclose company information more timely and accurately [13]. Second, the implementation of securities lending trading can increase the number of analysts tracking [14], and significantly reduce the optimistic bias of analysts' earnings forecasts [15], which makes the information environment more transparent.

2.2.2 Agency Theory

First, securities lending trading, as an informed trading, can make the behavior of managers more subject to the supervision of shareholders or external investors [3]. The deterrent effect of securities lending trading will force managers to be more cautious when they make investment decisions [16]. Second, the disclosure of negative corporate information not only harms the interests of shareholders, but also has an impact on the interests of managers themselves. Listed firms will circumvent the downside risk of short-term stock prices by issuing more stocks or options incentives to managers, so that the interests of managers and firms tend to be consistent [17].

From the perspective of financing trading, financing trading may have an impact on the corporate investment efficiency in both positive and negative aspects, specifically as follows.

2.2.3 Positive Promotion

First, financing trading will magnify investors' judgments and trading position, and cause abnormal fluctuations in stock prices, which is not conducive to the corporate market value management [18]. However, information disclosure is an important means for corporate management to influence stock prices. Therefore, company management will achieve the objective of stabilizing market value through accurate information disclosure. Second, financing trading can help those good firms to realize the rapid rise of stock price, and the gains from the rise of stock price will further encourage managers to constantly improve the level of corporate governance [19]. In order to avoid the high cost of falling stock price, the management might improve the corporate information environment and reduce opportunistic behavior.

2.2.4 Negative Inhibition

First, financing trading is usually related to irrational behaviors in the market, and the credibility of its information may be low [20]. In China, the participants of financing trading are more active. The more active the financing transaction is, the more likely the interference factors in the stock price will be, which will aggravate the information asymmetry. Second, managers tend to keep increasing stock prices to achieve performance goals [21]. Speculative financing trading is likely to cause the rise of stock prices, so shareholders get good feedback. In this case, managers usually face great pressure to maintain the rise of stock prices [22]. In order to maintain the good news, managers may have some improper behavior, which will worsen the corporate agency problem.

In summary, securities lending trading can effectively reduce the information asymmetry and agency costs, thereby improving corporate investment efficiency. Financing trading may increase investment efficiency or reduce investment efficiency.

Therefore, the impact of financing trading on corporate investment efficiency is two-way and the impact may not be significant in the end. Consequently, this paper believes that the relationship between margin trading and corporate investment efficiency is mainly affected by securities lending trading. Based on this, I propose the following hypothesis.

Hypothesis 1: Margin trading improves corporate investment efficiency, and its motivation lies in securities lending trading.

The foregoing analysis shows that information asymmetry and agency problems are the main reasons for inefficient investment. However, margin trading is conducive to reducing the degree of information asymmetry and agency costs. Therefore, this paper predicts that information transparency and agency costs are the two important channels for margin trading to affect investment efficiency. Hypothesis 2 proposed accordingly is as follows.

Hypothesis 2a: Information transparency is a mediating variable between margin trading and corporate investment efficiency. Margin trading suppresses the corporate inefficient investment by improving information transparency.

Hypothesis 2b: Agency costs is a mediating variable between margin trading and corporate investment efficiency. Margin trading suppresses the corporate inefficient investment by reducing the agency costs.

3 Sample Selection and Research Design

3.1 Sample Selection

This article selects 2009–2018 A-share listed firms as the research sample. In the selection process, the sample was screened as follows: (1) excluding financial firms; (2) excluding ST and *ST firms; (3) removing the samples with missing data; (4) winsorizing all continuous variables at the upper and lower 1% level. The data comes from CSMAR database.

3.2 Variable Definition

3.2.1 Dependent Variable

The dependent variable is *Ineffinv*, which refers to corporate investment efficiency. I use the following equation to calculate *Ineffinv* [6].

$$
\begin{aligned}
Inv_{i,t} &= \alpha_0 + \alpha_1 Growth_{i,t-1} + \alpha_2 Lev_{i,t-1} + \alpha_3 Cash_{i,t-1} \\
&\quad + \alpha_4 Age_{i,t-1} + \alpha_5 Size_{i,t-1} + \alpha_6 \mathrm{Ret}urn_{i,t-1}
\end{aligned}
$$

$$+\alpha_7 Inv_{i,t-1} + \sum Year + \sum Ind + \varepsilon_{i,t} \quad (1)$$

Inv represents new investment. *Growth* is the growth rate of operating income. *Lev* is the ratio of total debt to total assets. *Cash* is the ratio of cash and cash equivalents to total assets. *Age* is the natural logarithm of corporate listing time. *Size* is the natural logarithm of total assets. *Return* is the rate of return on stocks considered for cash dividend reinvestment. The absolute value of the residual is *Ineffinv*. When the residual value is greater than 0, it means that the firm has overinvested. When the residual is less than 0, it means that the firm is underinvested. The greater the value of *Ineffinv* is, the lower the investment efficiency is.

3.2.2 Independent Variable

The main independent variable is *Treat*. *Treat* is an indicator variable defined as 1 if firm *i* is included in the target firm of margin trading in year *t*, and 0 otherwise. In addition, in order to prove that the effect of margin trading on investment efficiency comes from securities lending trading, this paper adds the independent variable *Buy*, which represents financing trading [23]. *Buy* is measured by two indicators: (1) the ratio of financing purchases to circulation market value, *SBR*; (2) the ratio of financing balances to circulation market value, *SBYR* [14].

3.2.3 Mediating Variable

Mediating variables include information transparency and agency costs. Information transparency is expressed in *Opaque*, which refers to the sum of the absolute value of discretionary accruals in the past 3 years [24]. In order to ensure that *Opaque* has the same effect direction on investment efficiency as margin trading, this paper multiplies the sum of the absolute value of discretionary accruals in the past 3 years by −1 to perform regression analysis. The agency costs is expressed in *Mfee,* which refers to the ratio of management fees to income [25]. In order to ensure that *Mfee* has the same effect direction on investment efficiency as margin trading, this paper takes the reciprocal of the ratio of management fees to income to perform regression analysis.

3.2.4 Control Variable

X is the vector of control variables that are commonly used in corporate investment efficiency literature [26]. *Top* is the ratio of shares held by the largest shareholder to total shares. *Indratio* is the ratio of the number of independent directors to the total number of directors. *Dual* is an indicator variable defined as 1 if the chairman and general manager is the same person, and 0 otherwise. *Rmasha* is the ratio of executive shares to total shares. Soe is an indicator variable defined as 1 if the firm is a state-owned, and 0 otherwise. In addition to the above variables, the remaining control variables are *Size, Lev, Cash, Roa, and Growth.*

3.3 Research Model

This paper uses the difference-in-differences method and the mediating effect test method to construct the following equations to test whether the margin trading improves corporate investment efficiency and the mediating role played by information transparency and agency costs [27].

$$Ineffinv_{i,t} = \beta_0 + \beta_1 Treat_{i,t} + BX_{i,t} + \sum Year + \sum Ind + \varepsilon_{i,t} \quad (2)$$

$$Opaque_{i,t}/Mfee_{i,t} = \beta_0 + \beta_1 Treat_{i,t} + BX_{i,t} + \sum Year + \sum Ind + \varepsilon_{i,t} \quad (3)$$

$$Ineffinv_{i,t} = \beta_0 + \beta_1 Treat_{i,t} + \beta_2 Opaque_{i,t}/Mfee_{i,t} + BX_{i,t} + \sum Year + \sum Ind + \varepsilon_{i,t} \quad (4)$$

In the above equations, the subscripts *i* and *t* refer to the *i* company in year *t*, and ε is the random disturbance term. The first step is to regress (2) to test whether margin trading can improve corporate investment efficiency. The second step is to regress (3) to test whether margin trading can increase information transparency and reduce agency costs. The third step is to regress (4) to check whether agency costs and information transparency play an intermediary role.

According to the previous theoretical analysis, margin trading improves corporate investment efficiency due to the short selling pressure effect exerted by securities lending trading, and financing trading does not have a significant effect on it. In order to test whether this hypothesis is true, this paper adds *Buy* to (2) to identify whether margin trading improving corporate investment efficiency is caused by financing trading or securities lending trading. The equation settings are as follows.

$$Ineffinv_{i,t} = \gamma_0 + \gamma_1 Treat_{i,t} + \gamma_2 Buy_{i,t} + YX_{i,t} + \sum Year + \sum Ind + \varepsilon_{i,t} \quad (5)$$

Table 1 Summary statistics

Variables	n	Mean	Std Dev	P1	P99
Ineffinv	19,519	0.0305	0.0308	0.0004	0.1672
Treat	19,159	0.2706	0.4443	0	1
Size	19,159	22.1982	1.2892	19.5550	26.0969
Lev	19,519	0.4526	0.2101	0.0564	0.9409
Cash	19,519	0.1528	0.1153	0.0090	0.5723
Roa	19,519	0.0355	0.5705	−0.2243	0.1955
Growth	19,519	0.2085	0.5364	−0.5822	3.8082
Top	19,519	0.3478	0.1497	0.0845	0.7489
Indratio	19,519	0.3735	0.0533	0.3333	0.5714
Dual	19,519	0.2312	0.4216	0	1
Rmasha	19,519	0.0713	0.1399	0	0.5779
Soe	19,519	0.4335	0.4956	0	1

3.4 *Summary Statistics*

Table 1 reports summary statistics of full sample. The average value of *Ineffinv* is 0.0305, indicating that the difference between the actual investment efficiency and the expected investment efficiency of the sample is 3.05%. The minimum value of *Ineffinv* is greater than 0, indicating that all the sample firms have inefficient investment. The average value of *Treat* is 0.2706, indicating that 27.06% of the sample are the target firms of margin trading.

4 Empirical Results

4.1 *Testing of Hypothesis 1: The Impact of Margin Trading on Corporate Investment Efficiency*

Table 2 reports the result of examining the impact of margin trading on corporate investment efficiency. In Model (1), the regression coefficient of *Treat* is negative at the 1% significance level, indicating that margin trading can suppress corporate inefficient investments. In Model (2) and (3), a further grouping test found that both the regression coefficients of *Treat* in over-investment group and under-investment group were negative at the 1% significance level, indicating that margin trading can suppress over-investment and under-investment, thereby improving investment efficiency.

The regression result in Table 2 shows the impact of margin trading on corporate investment efficiency, but it cannot distinguish the difference between the effects of

Table 2 Margin trading and investment efficiency

Variables	*Ineffinv*	*Overinv*	*Underinv*
	(1)	(2)	(3)
Treat	−0.0029*** (−4.85)	−0.0064*** (−4.67)	−0.0019*** (−4.07)
Size	−0.0006** (−2.45)	−0.0007 (−1.37)	−0.0013*** (−6.58)
Lev	−0.0062*** (-4.31)	−0.0006 (−0.93)	−0.0072*** (−6.45)
Cash	−0.0034 (-1.62)	−0.0245*** (-4.67)	0.0132*** (8.20)
Roa	−0.0124*** (−2.84)	0.0074 (1.04)	−0.0287*** (−8.52)
Growth	0.0063*** (15.14)	0.0001*** (2.71)	0.0017*** (4.57)
Top	0.0006 (0.36)	0.0003 (0.08)	0.0035*** (2.85)
Indratio	0.0091** (2.23)	0.0146 (1.61)	0.0074** (2.25)
Dual	0.0005 (0.97)	0.0004 (0.30)	−0.0001 (−0.22)
Rmasha	0.0056*** (3.09)	0.0034 (0.86)	0.0058*** (3.96)
Soe	−0.0053*** (−10.03)	−0.0091*** (−7.52)	−0.0025*** (−5.78)
Year	Yes	Yes	Yes
Industry	Yes	Yes	Yes
Constant	0.0510*** (9.49)	0.0610*** (5.14)	0.0557*** (13.02)
N	19,519	7526	11,993
R-squared	0.0584	0.0636	0.0605

*, **, *** Indicate statistical significance at the 10%, 5%, and 1% levels, respectively

financing trading and securities lending trading. Table 3 shows the result whether margin trading improving corporate investment efficiency is the impact of financing trading or securities lending trading. In Model (1) and (2), the regression coefficient of *Treat* is still negative at the 1% significance level, while the regression coefficient of *SBR* and *SBYR* is almost 0 and not significant. Therefore, financing trading has no significant impact on corporate investment efficiency. According to the fact that the regression coefficient of *Treat* is significantly negative, it is confirmed that securities lending trading is the driving force for margin trading to improve investment efficiency.

Table 3 Impact of financing trading or securities lending trading

Variables	*Ineffinv*	
	(1)	(2)
Treat	−0.0024*** (−3.07)	−0.0034*** (−3.63)
SBR	−0.0000 (−0.96)	
SBYR		0.0000 (0.65)
Size	−0.0006*** (−2.57)	−0.0006** (−2.23)
Lev	−0.0061*** (−4.30)	−0.0062*** (−4.32)
Cash	−0.0035 (−1.63)	−0.0034 (−1.61)
Roa	−0.0126 (−2.89)	−0.0122*** (−2.77)
Growth	0.0063*** (15.16)	0.0063*** (15.10)
Top	0.0004 (0.26)	0.0007 (0.46)
Indratio	0.0091** (2.21)	0.0092** (2.23)
Dual	0.0005 (0.98)	0.0005 (0.98)
Rmasha	0.0056*** (3.09)	0.0056*** (3.10)
Soe	−0.0053*** (−9.99)	−0.0053*** (−10.04)
Year	Yes	Yes
Industry	Yes	Yes
Constant	0.0519*** (9.52)	0.0503*** (9.06)
N	19,519	19,519
R-squared	0.0584	0.0584

*, **, *** Indicate statistical significance at the 10%, 5%, and 1% levels, respectively

In summary, margin trading can improve corporate investment efficiency, and its motivation lies in securities lending trading. Hypothesis 1 is confirmed.

4.2 Testing of Hypothesis 2: The Mediating Role of Information Transparency and Agency Costs

Table 4 reports the result of testing the mediating effect of information transparency. In Model (1), I first verified whether margin trading can improve corporate investment efficiency. This hypothesis has been verified above. In Model (2), the regression coefficient of *Treat* is positive at the 5% significance level, indicating that margin

Table 4 Mediating effect test of information transparency

Variables	*Ineffinv*	*Opaque*	Ineffinv
	(1)	(2)	(3)
Treat	−0.0029*** (−4.85)	0.0054** (−2.20)	−0.0027*** (−4.33)
Opaque			−0.0006** (2.29)
Size	−0.0006** (−2.45)	0.0141*** (12.83)	−0.0004* (−1.74)
Lev	−0.0062*** (−4.31)	−0.1378*** (−21.14)	−0.0052*** (−3.50)
Cash	−0.0034 (−1.62)	−0.0666*** (−6.48)	−0.0023 (−0.96)
Roa	−0.0124*** (−2.84)	−0.0992*** (−4.98)	−0.0128*** (−2.81)
Growth	0.0063*** (15.14)	−0.0118*** (−6.31)	0.0060*** (14.03)
Top	0.0006 (0.36)	−0.0366*** (−5.05)	−0.0012 (−0.77)
Indratio	0.0091** (2.23)	0.07805*** (4.11)	0.0092** (2.10)
Dual	0.0005 (0.97)	−0.0053** (−2.05)	0.0010 (1.63)
Rmasha	0.0056*** (3.09)	0.0289*** (2.72)	0.0023 (0.93)
Soe	−0.0053*** (−10.03)	0.0325*** (13.50)	−0.0050*** (-8.94)
Year	Yes	Yes	Yes
Industry	Yes	Yes	Yes
Constant	0.0510*** (9.49)	0.4323*** (17.58)	0.0456*** (8.03)
N	19,519	19,519	19,519
R-squared	0.0584	0.1434	0.0564

*, **, *** Indicate statistical significance at the 10%, 5%, and 1% levels, respectively

Table 5 Mediating effect test of agency costs

Variables	*Ineffinv*	*Mfee*	Ineffinv
	(1)	(2)	(3)
Treat	−0.0029*** (−4.85)	0.0188*** (12.59)	−0.0021*** (−4.41)
Mfee			−0.0183*** (3.34)
Size	−0.0006** (−2.45)	−0.0217*** (−36.20)	−0.0002 (−0.78)
Lev	−0.0062*** (−4.31)	−0.0552*** (−15.54)	−0.0052*** (−3.59)
Cash	−0.0034 (−1.62)	0.0489*** (9.27)	−0.0043** (−2.04)
Roa	−0.0124*** (-2.84)	−0.3213*** (−29.65)	−0.0065 (−1.47)
Growth	0.0063*** (15.14)	−0.0103*** (−9.96)	0.0065*** (15.57)
Top	0.0006 (0.36)	−0.0393*** (−10.25)	0.0013 (0.83)
Indratio	0.0091** (2.23)	0.0502*** (4.93)	0.0082** (2.00)
Dual	0.0005 (0.97)	−0.0035*** (−2.62)	0.0005 (0.86)
Rmasha	0.0056*** (3.09)	−0.0047 (−1.04)	0.0057*** (3.14)
Soe	−0.0053*** (−10.03)	−0.0066 (−5.02)	−0.0052*** (−9.81)
Year	Yes	Yes	Yes
Industry	Yes	Yes	Yes
Constant	0.0510*** (9.49)	0.5815*** (43.55)	0.0404*** (7.18)
N	19,519	19,519	19,519
R-squared	0.0584	0.2929	0.0603

*, **, *** Indicate statistical significance at the 10%, 5%, and 1% levels, respectively

trading can significantly improve information transparency. In Model (3), the regression coefficient of *Treat* and *Opaque* are at least negative at the 5% significance level, and the coefficients of *Treat* is increased from −0.0029 in Model (1) to −0.0027, indicating that information transparency has played a part of the mediating effect in the impact of margin trading on the corporate investment efficiency. In summary, hypothesis 2a is confirmed.

Table 5 reports the result of testing the mediating effect of agency costs. In Model (1), I first verified whether margin trading can improve corporate investment efficiency. This hypothesis has been verified above. In Model (2), the regression coefficient of *Treat* is positive at the 1% significance level, indicating that margin trading can significantly decrease agency costs. In Model (3), the regression coefficient of *Treat* and *Mfee* are negative at the 1% significance level, and the coefficients of *Treat* is increased from −0.0029 in Model (1) to −0.0021, indicating that agency costs has played a part of the mediating effect in the impact of margin trading on the corporate investment efficiency. In summary, hypothesis 2b is confirmed.

5 Additional Test

The main task of corporate governance is to standardize management's behavior. Improving the level of corporate governance can reduce the self-interest behavior of the management and reduce the inefficient investment. As margin trading has a certain restrictive effect on managers' behavior, there may be a substitution effect between margin trading and corporate governance. Based on this, this paper further conjectures that margin trading has a stronger inhibitory effect on inefficient investment in firms with poor corporate governance. Because a single indicator is difficult to measure the corporate comprehensive governance level, this paper uses the principal component analysis method to construct a comprehensive index describing the corporate governance level. The corporate governance index is expressed in *Gov*.

The annual industry average level of *Gov* is used to distinguish the level of governance, and the sample is divided into two groups with high governance level and low governance level. Table 6 Shows the regression results. In Model (1), the regression

Table 6 Margin trading, corporate governance and investment efficiency

Variables	Ineffinv	
	High-*Gov* Group	Low-*Gov* Group
	(1)	(2)
Treat	−0.0024* (−1.96)	−0.0039*** (−2.84)
Control Variables	Yes	Yes
Year	Yes	Yes
Industry	Yes	Yes
Constant	0.0545*** (9.63)	0.0564*** (4.45)
N	19,519	19,519
R-squared	0.0554	0.0584

*, **, *** Indicate statistical significance at the 10%, 5%, and 1% levels, respectively

coefficient of *Treat* is −0.0024, which is significant at the level of 10%. In Model (2), the regression coefficient of *Treat* is −0.0039, which is significant at the level of 1%. The coefficients of *Treat* in both groups are negative, but the significance levels are quite different. In the group with low corporate governance level, the effect of margin trading is more obvious, which shows that margin trading has a more significant external corporate governance role and can make up for the lack of internal corporate governance.

6 Robustness Test

In order to ensure the reliability of the research conclusion, the robustness tests are conducted from the following aspects:

(1) *Propensity score matching*: Use propensity score matching method to reset the control group samples to minimize the impact of sample selection errors on the research results.
(2) *Change measures of variables*: Use small profit (*Sprof*) as a measure of information transparency [28], and use asset turnover (*Aturn*) as a measure of agency costs [25].
(3) *Change the sample processing method*: Rank the absolute values of the residuals from small to large and select last 80% of the samples for empirical analysis to eliminate the model and sample biases when calculating *Ineffinv*.
(4) *Add control variable*: Control the volume of securities lending trading to exclude alternative explanation.

The above test results are consistent with the main regression results, indicating that the research results are robust. Due to space limitations, no robustness test results were reported.

7 Conclusion

This paper has the following conclusions. First, margin trading can improve corporate investment efficiency, and this motivation comes from the short selling pressure effect of securities lending trading. Second, agency costs and information transparency are the mediating variables between margin trading and corporate investment efficiency. Margin trading improves corporate investment efficiency by increasing information transparency and reducing agency costs. In addition, in the sample of low corporate governance level, margin trading has more obvious promotion impact on corporate investment efficiency, which indicates that margin trading plays a certain external governance role.

Based on the above research results, practical suggestions are proposed from the following two aspects. First, as securities lending trading plays a more important role in improving corporate investment efficiency, the restrictions on securities lending trading should be relaxed. Second, the regulatory department should urge the listed firms to fulfill their information disclosure obligations in a timely manner, increase the timeliness and authenticity of the financial market information acquisition, and create a good information environment for investors participating in margin trading.

References

1. Karpoff, J. M., & Lou, X. (2010). Short sellers and financial misconduct. *Journal of Finance, 65*(5), 1879–1913.
2. Pedro, A. C. S., & Sigurdsson, K. (2011). Price efficiency and short selling. *The Review of Financial Studies, 24*(3), 821–852.
3. Massa, M., Zhang, B., & Hong, Z. (2015). The invisible hand of short selling: Does short selling discipline earnings management? *Social Science Electronic Publishing, 28*(6), 1701–1736.
4. Jensen, M. C. (1986). Agency cost of free cash flow, corporate finance and takeovers. *American Economic Review, 76*(2), 323–329.
5. Myers, S. C., & Majluf, N. (1984). Corporate financing and investment decisions when firms have information that investors do not have. *Journal of Financial Economics, 13*(2), 187–221.
6. Richardson, S. (2006). Over-investment of free cash flow. *Review of Accounting Studies, 11*(2–3), 159–189.
7. Polk, C., & Sapienza, P. (2008). The stock market and corporate investment: A test of catering theory. *Review of Financial Studies, 22*(1), 187–217.
8. Jin, Q., Hou, Q., Li, G., & Xie, Y. (2015). Deregulation on short selling constraint, corporate investment and valuation. *Jing Ji Yan Jiu, 50*(10), 76–88. (in Chinese).
9. Biddle, G. C., Hilary, G., & Verdi, R. S. (2009). How does financial reporting quality relate to investment efficiency? *Journal of Accounting & Economics, 48*(2–3), 112–131.
10. Cheng, M., Dhaliwal, D., & Zhang, Y. (2013). Does investment efficiency improve after the disclosure of material weaknesses in internal control over financial reporting? *Journal of Accounting & Economics, 56*(1), 1–18.
11. Oliver, H. (1995). Corporate governance: Some theory and implications. *The Economic Journal, 105*(430), 678–689.
12. Boehmer, E., Jones, C. M., & Zhang, X. (2008). Which shorts are informed? *The Journal of Finance, 63*(2), 491–527.
13. Diamond, D. W., & Verrecchia, R. E. (1987). Constraints on short-selling and asset price adjustment to private information. *Journal of Financial Economics, 18*(2), 277–311.
14. Zhang, X., Zhou, P., & Li, C. (2016). Short selling and earnings quality: Evidence from corporate financial restatement, Vol. 08, pp. 175–190. (in Chinese)
15. Li, D., Yuan, C., & Liao, G. (2016). Short selling mechanism and analyst's optimistic bias—Test based on difference-in-difference model. *Kuai Ji Yan Jiu, 000*(009), 25–31. (in Chinese).
16. Christophe, S. E., Ferri, M. G., & Angel, J. J. (2004). Short-selling prior to earnings announcements. *The Journal of Finance, 59*(4), 1845–1876.
17. De Angelis, D., Grullon, G., & Michenaud, S. (2017). The effects of short-selling threats on incentive contracts: Evidence from an experiment. *Review of Financial Studies, 30*(5), 1627–1659.
18. Li, Z., Li, H., Ma, W., & Lin, B. (2017). Information governance effects of short selling and margin trading. *Jing Ji Yan Jiu, 52*(11), 150–164. (in Chinese).
19. Lu, Y., Peng, J., & Feng, J. (2018). Impacts of margin trading and short selling on Chinese listed firms' corporate governance. *Guan Li Ke Xue Xue Bao, 21*(11), 92–111. (in Chinese).
20. Hardouvelis, G. A., & Stavros, P. (1992). Margin requirements, speculative trading, and stock price fluctuations: The case of Japan*. *Quarterly Journal of Economics, 107*(4), 1333–1370.

21. Li, Y., & Zhang, L. (2015). Short selling pressure, stock price behavior, and management forecast precision: Evidence from a natural experiment. *Journal of Accounting Research, 53*(1), 79–117.
22. Goel, A. M., & Thakor, A. V. (2008). Overconfidence, CEO selection, and corporate governance. *the Journal of Finance, 63*(6), 2737–2784.
23. Huang, J., & Gong, G. (2019). Margin trading system and dynamic adjustment of corporate capital structure. *Guan Li Shi Jie, 35*(10), 64–81. (in Chinese).
24. Hutton, A. P., Marcus, A. J., & Tehranian, H. (2009). Opaque financial reports, R2, and crash risk. *Journal of Financial Economics, 94*(1), 67–86.
25. Ang, J. S., Cole, R. A., & Lin, J. W. (2000). Agency costs and ownership structure. *Journal of Finance, 55*(1), 81–106.
26. Han, J., Chen, Z., & Yang, X. (2014). The relationship between accounting conservatism and investment efficiency from top management team's background characteristics. *Kuai Ji Yan Jiu, 12,* 25–31. (in Chinese).
27. Wen, Z., Zhang, L., Hou, J., & Liu, H., Testing and application of the mediating effects. *Xin Li Xue Bao, 5,* 614–620
28. Dechow, P., Ge, R., & Schrand, R. (2010). Understanding earnings quality: A review of the proxies, their determinants and their consequences. *Journal of Accounting & Economics, 50*(2–3), 344–401.

On Industry Tax and the Deadweight Loss in the Context of Industry Development—Three Examples Form China

Lu Yu

Abstract In the market model of microeconomics, the deadweight loss caused by tax burden will lead to incomplete market transaction and inefficient resource allocation. However, from the perspective of the rational man hypothesis, why taxation has continued for thousands of years are inexplicable. With regard to this issue, this paper analyzes the tax burden data of China's finance, medicine and culture industries using micro model and concludes the reasons.

Keywords The deadweight loss · Industry tax · Tax burden

1 Introduction

First of all, it must be stated clearly that the industrial tax in this paper refers to the specific tax rate (aka: tax burden) set by a country for each industry, not the general tax rate for the industry.

And the deadweight loss indicates that the optimal operating configuration of the market can not be achieved, which specifically refers to the loss of consumer and producer surplus.

Under the definitions, a contradiction has emerged between the theory and practice of microeconomics. That is, if the rational person is admitted, then the taxation system, especially the system of proportional taxation, should have disappeared in the long history of mankind. This is because if a system will harm the market transactions in the society, then even if the system was created and its original founders were unable to solve the problem, the continuation of human history should have led to the demise of the institution. Or at least it could be changed, for example, from a proportional tax to a fixed tax, thereby minimizing the deadweight losses. This paper will analyze and explain the issue above with reference to the tax practice of the three industries in China, which the author hopes will contribute to the related research in this field.

L. Yu (✉)
National Academy of Economic Security School of Economics and Management, Beijing Jiaotong University, Beijing, China
e-mail: 18811591360@163.com

M. Li et al. (eds.), *IEIS 2020*, https://doi.org/10.1007/978-981-33-4363-4_19

2 Data

Given that the data of these three industries in China are more convenient to obtain for the author, these three industries are selected as the samples, which have no particularity and cover three different industries of different nature. These three industries come from the heavy industry, the service industry and the light industry respectively. We can intuitively feel a country's people's dependence on them. In fact, any country's industrial data can be utilized to further analyze the issue of this paper. The next step of this study is to compare the industrial tax rate and policy relationship between the similar industries in China and the United States.

The data of the tertiary industry above is particularly picked out for the analysis. T At the national level, unless certain industrial markets are abandoned, the industry will still exist in the country, even if the taxes imposed vary.

3 Analysis

The first question raised at the beginning of the article is that if deadweight loss is inevitable, why the system has continued for centuries in human society according to the rational man hypothesis. This is because even if taxes can not be cancelled, tax rates should be minimized to ensure that more members of society can make deals in the market, which is the decision-making model of rational people hypothesis. Another option is to establish a quota system, which converts the tax into a particular cost that can be imposed on either the supplier or the seller. But the social practice is obviously inconsistent with the above analysis. This phenomenon has nothing to do with national conditions, because except for a very small number of tax-free zones, there is not any notional abandoned taxes.

The issues can be accounted for from two perspectives.

Firstly, government taxation, expenditures and work can be considered as another "public service market" where the demand side is the whole nation and the supply side is the government. In this market, the tax revenue of all industrial markets is the price paid for the demand of the entire population, while government services, such as national defense and infrastructure construction are the products of the supply. Therefore, from this perspective, if the deadweight losses in the industrial market can be directly converted into a certain cost, rather than the theoretical wedge of supply and demand, the deadweight losses will thus become the turnover of the public service market. This explanation is simpler and clearer, because microeconomics claims the role of government taxing is public expenditure. But it can not explain another question: why the government does not replace the proportional tax system with a fixed tax system, thereby reducing the deadweight losses. This problem can be analysed from another perspective below.

Secondly, suppose that the industrial market of a country is a single market, and there is no difference between the suppliers, which means the market is a PCM. In the

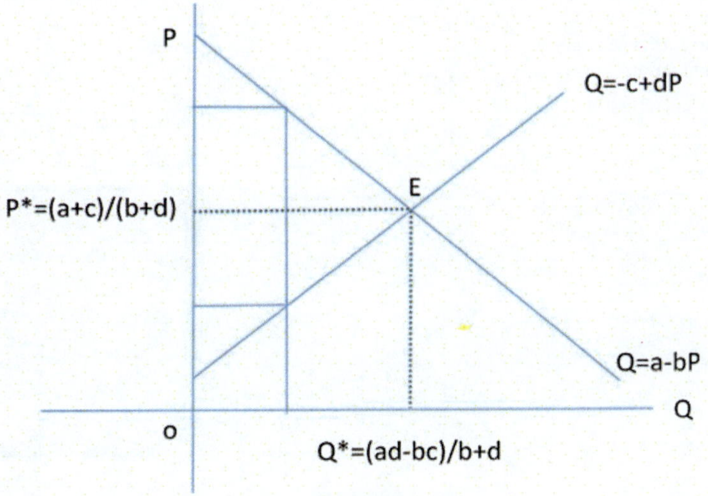

Fig. 1 Linear model of supply and demand market [4]

market model of microeconomics, the smaller the elasticity of supply and demand, the smaller the deadweight loss is.

As shown in Fig. 1, the demand function is Q = a − bp, the supply function is Q = −c + bp, The equilibrium price is $P^* = \frac{a+c}{b+d}$, The equilibrium quantity is $Q^* = \frac{ad-bc}{b+d}$.

Because the less elastic the supply and demand of the market is, wherever the tax wedge is put in, the smaller the area of the deadweight loss is; when the supply and demand reaches full elasticity, the tax burden is meaningless, because there is no wedge at all. In this case, once the tax is increased, the supply and demand sides will immediately leave the market. Therefore, in the field of microeconomics, the less elastic the market demand or supply is, the smaller deadweight loss is. In other words, as shown in Fig. 2, the area inside the dotted line is deadweight loss. The greater the elasticity of the supply and demand sides are, the greater the angle between the supply and demand curves is. It can be illustrated in an expression as below:

$$S_\Delta = \frac{t}{2}(Q^* - Q^t) = \frac{bd}{2(b+d)}t^2$$

On this basis, from the perspective of macroeconomic practice, the supply and demand of a country's industrial market is almost inflexible to the country, and the only alternative is offered in the international market. And the legitimate supply and demand of foreign goods must also enter the domestic market and be integrated into the domestic market supply and demand. Therefore, it can be concluded that as a country's industrial market, the deadweight loss of its tax burden is basically zero. On the other hand, due to the particularity of the industrial market at the national level, the result must be that the deadweight loss is minimal or even infinitely close to

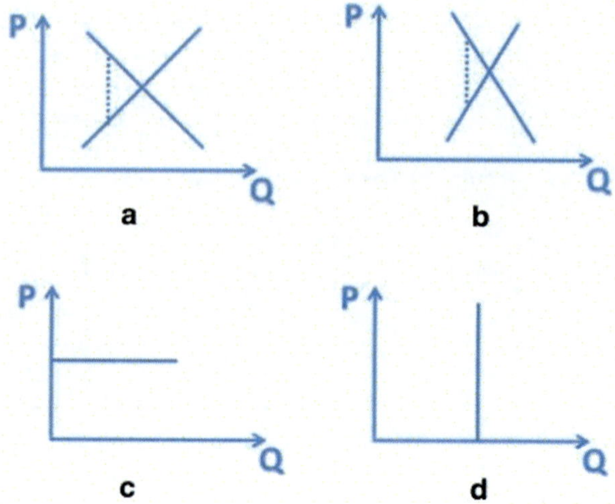

Fig. 2 Four forms of supply and demand market

zero. Of course, it is undeniable that if the tax burden on an industry reaches a certain level, some supply or demand parties will choose to transfer the industry to foreign countries. But this situation will lead to two possible results: First, if a member of both parties abandons the domestic market, it can be regarded as a withdrawal from the domestic market, which means he becomes the supplier or demander of the foreign market, no longer the domestic one. Second, the supplier produces and sells at home abroad. As long as the products are still sold in China and bear the domestic tax burden, the supply still can not deviate from the domestic supply curve, which is almost no different from the above scenario.

Through the analysis of the data of the three domestic industries, we can see that the tax burden rate of the above three industries varies greatly, but no matter which industry it is, there is no tax exemption.

Therefore, it can be seen that the elasticity of both the supply and demand sides in the market is relatively small, for they can not choose to leave the market. Hence, the tax rate has small correlations with the scale of the deadweight loss. Or put differently, at the national market level, the supply-demand relationship of a certain industry is almost inelastic, so the deadweight loss is almost zero as shown in the model.

4 Conclusions

From the above analysis, we can see that the reason why a theoretically unnecessary tax loss system can survive in human society. The reason is that taxes are passed on to the cost of government work as a supply service to another market [5]. On the other hand, from the national macroeconomic level, the demand for industrial supply is almost inelastic, so it is meaningless to simply emphasize proportional tax.

Because consumers in the market have the choice to supply goods to enterprises in the industry, but the demand for the industry itself is inelastic.

In addition, from the supply side, if a country's industry targeted tax rate is too high, it will indeed lead to industrial migration, which means the relevant industrial production will be transferred to foreign countries or completely rely on international import. Therefore, every country should pay close attention to and adjust the industrial tax rate in time to ensure the development of localization of its core industries. And marginal industries can also rely entirely on international supply when needed.

The data in the second chapter of this paper can also support the above conclusions from an empirical perspective. For example, it can be seen from Table 1 that in the construction industry, the tax rate of enterprises invested by the state with foreign capital and enterprises invested by funds from Hong Kong, Macao and Taiwan is significantly higher than state-owned enterprises. According to the theoretical analysis in this paper, it is because the characteristics of construction industry is high upfront cost or sunk costs. So when a company enters the domestic market, it is so hard to leave it. It is no longer necessary to consider its market elasticity, because this industry market is relatively inelastic in terms of the nature of the industry or the national industry. It can also be seen from the financial industry market in Table 2. VAT rates fluctuate widely because of the nature of the financial sector. Under the background of replacing business tax with VAT, the financial industry faces a short-term tax increase and a long-term tax reduction [6]. From the theoretical analysis of this paper, this is because the professional degree of financial employees is so high, so it is difficult to switch industries, so the short-term increase in enterprise tax

Table 1 Tax rate on output profits of construction industry

Tax rate on output profits of construction industry					
	2014	2015	2016	2017	2018
Production profit tax rate (%)	6.8	6.7077	6.6972	6.5	6.6
State-owned business rate (%)	5.5	5.9479	5.3918	5.1	5.8
Enterprises invested by Hong Kong, Macao and Taiwan's rate (%)	6.2	7.5865	8.1447	7.9	7.2
Foreign-invested enterprise's rate (%)	8.6	7.741	8.7554	7.7	8.6

Source National Bureau of Statistics of PRC

Table 2 Income tax in financial industry

Income tax in financial industry [1]					
Month	02/2019	03/2019	04/2019	05/2019	06/2019
Enterprise Income Tax of financial Industry (100 million)	149.27	693.55	1061.52	1337.18	2352.01
Month-on-month growth of enterprise income tax in financial industry (%)	−37.2	130.1	4.2	24.7	−6

Source Ministry of Finance of PRC and State Administration of Taxation

Table 3 Analysis of actual income tax rate of pharmaceutical industry by year

Analysis of actual income tax rate of pharmaceutical industry by year [2, 3]						
Year	2009	2010	2011	2012	2013	2014
Min	0.0097	0.0204	0.0842	0.0492	0.0954	0.0954
Max	0.3286	0.2796	0.2728	0.251	0.5208	0.3266
AVG	0.1380	0.1354	0.1537	0.1500	0.1644	0.1529
SD	0.0547	0.0483	0.0368	0.0392	0.0655	0.0484
Var	0.0030	0.0020	0.0010	0.0020	0.0040	0.0020

Source State Administration of Taxation

burden will not affect the service supply of industry employees. And in the long run, the government needs to attract more money into the country, so it needs to lower the tax burden on financial firms. And Table 3, medicine industry, it can be seen that the historical average rate is relatively stable. This is due to the pharmaceutical industry is a light industry. Compared with the heavy industry and service industry, it is most vulnerable to international competition. So governments should stabilize the tax rate and tax policy. Thus, firms in the industry will stay at home.

Finally, from the perspective of international trade, if a country depends entirely on international supply to meet domestic demand, first of all, the issue of tariffs also needs to be considered. If tariffs are too high to cause international suppliers to abandon export trade, then domestic demand can not be satisfied. Of course, from a microeconomic point of view, the optimal solution is to eliminate trade barriers and establish a free trade market between the two countries, namely, the two markets can be integrated into a larger market to a certain extent. However, as we all know, fully competitive markets (PCM) have always existed only in theory and not in practice, especially for international trade. In the game between countries especially in the market field, there will still be certain Nash equilibrium. That is, if one party uses non-free trade for its own profit to plan the market, then the multi-party country with which it deals must follow only the above-mentioned dominant decision. Because the unilateral opening of free markets represents the abandonment of bargaining power, which will inevitably lead to the outflow of domestic wealt [7].

Therefore, under the premise of ensuring domestic demand, how to reduce trade barriers, adjust industrial tax rates for different industries, and establish a perfect industrial structure according to the domestic situation will be further studied.

References

1. Lu, N., & Yi, J. (2012). The change of China's financial industry tax and its influencing factors—Based on panel data of listed companies in financial industry. *Journal of Shanghai University of Finance and Economics, 14*(04), 80–89.
2. Zhu, Z. (2019). Analysis on the best tax rate and countermeasures of pharmaceutical manufacturing industry in China. Harbin Commercial University.

3. Wang, L., & Wang, X. (2016). A study on the influencing factors of actual tax burden of listed companies in pharmaceutical industry in China. *Business Accounting, 22,* 40–42.
4. Zhang, Z. (2017). The relationship between tax wedge and deadweight loss and tax revenue—Analysis based on linear supply and demand model and elastic theory. *Journal of Henan University of Finance and Taxation, 31*(05), 18–22.
5. Han, F. (2005). Tax policy to promote the development of high-tech industries. *China Finance, 08,* 28–30.
6. Chen, R. (2020). On the Impact of replacing business tax with value-added Tax on financial industry and countermeasures. *Modern Enterprise, 05,* 79–80.
7. Lu, C., & Qizi, Q. (2008). A study on tax policy to promote the development of cultural industry in China. *Shandong Economy* (05), 96–100.

Spot-Futures Market Interaction and the Impact of Arbitrage: An Agent-Based Modelling Method

Xuan Zhou, Qingzuo Kuang, and Honggang Li

Abstract We establish a spot-futures market model based on heterogeneous multi-agent and cross-market traders to study the interaction between stock index futures market and spot market. In our model, spot market and futures market are linked through cross-market traders. The simulation results show that our spot-futures market model reproduces some stylized facts in financial markets such as the price co-movement between the two markets and the fat-tailed distribution of the basis. In addition, by changing the number of arbitrageurs, we study the impact of arbitrage behavior on market liquidity, volatility and price efficiency. The results show that introducing the futures into spot market can decrease volatility, and increase liquidity and price efficiency. However, a further increase in arbitragers proportion will increase volatility and expand spot futures basis, which will hurt the market stability and price efficiency.

Keywords Agent-based modelling · Continuous double auction · Spot-futures market · Arbitrager

---

This work is supported by the National Natural Science Foundation of China under Grant No. 71671017 and the Program of the Co-Construction with Beijing Municipal Commission of Education of China under Grant No. B19H100010.

X. Zhou
School of Economics and Management, Beijing Jiaotong University, Beijing, China
e-mail: zhouxuan@bjtu.edu.cn

Beijing Laboratory of National Economic Early-warning Engineering, Beijing Jiaotong University,Beijing, China

Q. Kuang · H. Li (✉)
School of Systems Science, Beijing Normal University, Beijing, China
e-mail: hli@bnu.edu.cn

Q. Kuang
e-mail: evamail@yeah.net

Q. Kuang
Credit Card Center, China Merchants Bank, Shanghai, China

M. Li et al. (eds.), *IEIS 2020*, https://doi.org/10.1007/978-981-33-4363-4_20

1 Introduction

Stock index futures is a kind of standard futures contract, which take stock index as its underlying asset. The first stock index futures contract was born in the United States in 1982. China launched the first stock index futures contract CSI300 on April 16, 2010. Then, on 12 June 2015, the Chinese stock market crash began and Shanghai index lost about a third of its value in one month. Major aftershocks occurred around the Black Monday on 27 July and 24 August 2015. Shorting stock index futures is accused to be the blasting fuse of this financial crash. It has been controversial for a long time about the impact of the introduction of stock index futures on the financial market. On the one hand, stock index futures has the characters of high liquidity, which can release information into the market and increase stock market's price efficiency and stability. On the other hand, due to the leverage trading in futures market, the excessive speculation in futures market may hurt the market stability and price efficiency.

2 Literature Review

Many empirical studies focus on the impact of the introduction of stock index futures on the underlying asset market, especially its volatility. Among them there are a series of researches work on S&P500 index futures [1–5], or index futures in other mature markets including the markets in the UK, France, Australia and Japan [6–9]. Recently there are also some papers focus on the emerging markets, such as China, Brazil, Russia, India [10, 11]. But these conclusions are not consistent, some papers claim that the introduction of stock index futures can stable the spot markets [1, 8], while some support that the futures will increase spot markets' volatility [2, 3, 6, 7, 10, 11]. Besides, the third group find evidence that there is no significate impact [4, 5, 8, 9]. Gulen and Mayhew [12] studied the financial derivatives markets of 25 countries, and found out that when introduce derivatives into financial market, the volatility significantly increases in stock markets of the United States and Japan, while it does not significantly change in other countries' markets. Antoniou et al. [13] claimed that the inconsistent results about the stock index futures' impact are due to the different proportion of feedback traders in different markets.

Due to the different microstructure of different financial market, no consensus can be reached through empirical research method. Using agent-based simulation method to study financial assets price dynamics is a hot topic in recent years, which provide us an alternative approach. Agent-based modeling is a bottom-up approach based on modern computer technology. In agent-based models, traders can be heterogeneous and bounded rational, which is much closer to the real financial market. Hence, much more stylized facts, such as fat-tailed distribution and volatility clustering, can be reproduced by these computationally oriented models then through theoretically oriented models. Recently, there are several studies on futures and spot markets using the agent-based model. Ohi et al. [14] built a multi-market model and reproduced some statistical properties of Nikkei 225 index futures prices. Xu et al. [15] built up an agent-based model with cross-market structure, compared the simulation data with the real market data and found that the model can reproduce the Chinese financial markets features successfully.

Inspired by the researches of Ohi et al. [14] and Chiarella et al. [16], this work analyzes the interaction between stock index futures market and stock index market in a framework of heterogeneous agents. This paper is organized as follows. Section 3 introduces the multi-agent market model. In Sect. 4, we provide an analysis of the simulation results of the model. Finally, in Sect. 5, we present our conclusion.

3 Model

We build a two-market model based on double auction (CDA) mechanism, which has been widely used in most stock exchanges. In our model, there exists a spot market where the stock index (marked as asset 1) is traded and a futures market where the stock index futures (marked as asset 2) is traded. The spot market and the futures market are linked through the traders investment behaviors. Traders are heterogeneous and bounded rational in our artificial market. We consider each time step t as 5 min, which means that 48 time steps are a trading day (4 h). During each time period t, all traders enter the market randomly and submit orders according to their trading strategies. Traders usually submit limit orders and these orders are executed based on price-time priority and the market price changes at once when transaction happens. Limit orders that cannot be executed immediately will be recorded in the order book. The details of CDA trading mechanism and limit order book can be find in our previous work [17].

3.1 Types of Traders

We classify the traders into five types: fundamentalists, speculators, noise traders, hedgers and arbitragers. Hedges and arbitragers are cross-market traders, namely they can submit orders both in spot market and futures market, while fundamentalists, speculators and noise traders are single-market traders, namely they can only trade in spot market or futures market. The details of heterogeneous trading strategies are described as follows.

(1) Fundamentalist: Fundamentalists believe that assets have their own fundamental value. Suppose the stock's fundamental value $p^*_{1,t}$ follows random walk process:

$$p^*_{1,t} = p^*_{1,t-1} \exp(\sigma_f v_t) \tag{1}$$

where $v_t \sim N(0,1)$ subjects to the standard normal distribution and $\sigma_f \geq 0$ is a given constant. Besides, despite the dividends and convenience yield, the fundamental price of the stock index futures $p^*_{2,t}$ is set according to future parity theorem as:

$$p^*_{2,t} = \mathrm{e}^{r(T-t)} p^*_{1,t} \tag{2}$$

where T is the delivery time of the stock index futures and r is the risk-free rate of interest.

Assume that a fundamentalist i enter the market at time τ $(t < \tau < t+1)$, and the observed fundamental value at this moment is

$$p^*_{a,t\tau} = p^*_{a,t}(1 + \Delta_f z_{t\tau}) \tag{3}$$

where $a \in \{1,2\}$, $z_{t\tau} \sim N(0,1)$. The constant $\Delta_f > 0$ is observation error parameter.

Fundamentalists consider the asset to be overestimated (underestimated) when the market price exceeds (is lower than) the observed fundamental value, and tend to submit sell (buy) orders. Hence, the order direction $H_{a,\,\mathrm{it}\,\tau}(a \in \{1,2\})$ a fundamentalist submit is determined as follow

$$H_{a,\,\mathrm{it}\,\tau} = \mathbf{sgn}(p^*_{a,t\tau} - p_{a,t\tau}) \tag{4}$$

where $p_{a,t\tau}$ is the market price at time τ and **sgn** denotes the sign function. $H_{a,it\tau} = 1\,(-1)$ means that traders tend to submit buy (sell) orders. Furthermore, we assume that the quantity of stock index or stock index futures fundamentalist i wants to trade $\bar{q}_{a,\,\mathrm{it}\,\tau}(a \in \{1,2\})$ is proportional to the spread of observed fundamental value and the asset's current market price:

$$\bar{q}_{a,\,\mathrm{it}\,\tau} = \lfloor |\alpha_f (p^*_{a,t\tau} - p_{a,t\tau})| \rfloor \tag{5}$$

where $\alpha_f > 0$ is a constant measuring the fundamentalist's sensitivity to spread. Similar with the setting in Chiarella et al. [16], the limit price $l_{a,it\tau}(a \in \{1,2\})$ follows uniform distribution between market price and observed fundamental value:

$$l_{a,it\tau} = \begin{cases} U(p_{a,t\tau}, p^*_{a,t\tau}) & (H_{a,it\tau} = 1) \\ U(p^*_{a,t\tau}, p_{a,t\tau}) & (H_{a,it\tau} = -1) \end{cases} \tag{6}$$

(2) Speculator: Speculators in our model make trading decisions based on chartist strategy. They calculate the average of last D_i time periods' close prices, which called moving average prices:

$$m_{a,it} = \frac{\sum_{j=1}^{D_i} p^{\text{close}}_{a,t-j}}{D_i}, \; a \in \{1,2\} \tag{7}$$

where $D_i > 0$ is the individual length of time window. Speculators consider that when the asset's market price $p_{a,t\tau}$ is higher(lower) than its moving average value $m_{a,it}$, the market price will rise(fall) further and they will submit buy(sell) orders. Similarly, the order volume speculators want to submit is assumed to be proportional to the spread between $p_{a,t\tau}$ and $m_{a,it}$. Besides, the limit price is set to be close to asset's current market price. Hence, speculators' trading strategy is determined as follows.

$$H_{a,it\tau} = \mathbf{sgn}(p_{a,t\tau} - m_{a,it}) \tag{8}$$

$$\bar{q}_{a,it\tau} = \lfloor |\alpha_s(p_{a,t\tau} - m_{a,it})| \rfloor \tag{9}$$

$$l_{a,it\tau} = p_{a,t\tau}(1 + |\Delta z_{t\tau}| H_{a,it\tau}) \tag{10}$$

where $a \in \{1, 2\}$, $z_{t\tau} \sim N(0, 1)$ and $\Delta > 0$ is a constant, which describe the aggressive level of speculators.

(3) Noise Traders: We introduce some noise traders with zero intelligence both in spot market and future market. They submit the buy order ($H_{a,it\tau} = 1$) and sell order ($H_{a,it\tau} = -1$) with same probability. The limit volume is random between 3 and 10, and the limit price is close to the current market price:

$$\bar{q}_{a,it\tau} \in \{3, 4, \ldots, 10\} \tag{11}$$

$$l_{a,it\tau} = p_{a,t\tau}(1 + |\Delta z_{t\tau}| H_{a,it\tau}) \tag{12}$$

(4) Hedger: The hedgers are cross-market traders, who believe that the stock index and its futures' prices are driven by the same factors and always move together. Hence, when hedger trade in the spot market, he will conduct reverse operation in the futures market to hedge the risk. We assume that hedgers trade in spot market due to liquidity demand, so they trade in spot market according to random strategy. They submit the buy order ($H_{1,it\tau} = 1$) and sell order ($H_{1,it\tau} = -1$) in spot market with same probability. The volume of limit order is random between 3 and 10, and the limit price is close to the current stock's market price:

$$\bar{q}_{1,it\tau} \in \{3, 4, \ldots, 10\} \tag{13}$$

$$l_{1,it\tau} = p_{1,t\tau}(1 + |\Delta z_{t\tau}| H_{1,it\tau}) \tag{14}$$

Meanwhile, hedgers will conduct reverse operation in the futures market to hedge the risk, hence:

$$H_{2,it\tau} = -H_{1,it\tau} \tag{15}$$

$$\bar{q}_{2,it\tau} = \bar{q}_{1,it\tau} \tag{16}$$

$$l_{2,it\tau} = p_{2,t\tau}(1 + |\Delta z_{t\tau}| H_{2,it\tau}) \tag{17}$$

(5) Arbitrager: Arbitragers attempt to get profit from the basis between futures' market price $p_{2,t\tau}$ and its theoretical prices $\bar{p}_{2,t\tau} = e^{r(T-t)} p_{1,t\tau}$. Specifically, when the futures' market price is higher (lower) than its theoretical price, they will sell (or buy) the futures while buy (or sell) the stock at the same time, and close the positions when the spread disappears to realize profit.

We introduce a threshold δ into our model and assume that only when spread is greater than the threshold, arbitragers will submit limit orders to seize the arbitrage opportunity:

$$H_{1,it\tau} = \begin{cases} -1 & \left(p_{2,t\tau} - \bar{p}_{2,t\tau} < -\delta \bar{p}_{2,t\tau}\right) \\ 1 & \left(p_{2,t\tau} - \bar{p}_{2,t\tau} > \delta \bar{p}_{2,t\tau}\right) \end{cases} \tag{18}$$

$$H_{2,it\tau} = -H_{1,it\tau} \tag{19}$$

Besides, we assume that the desired trading volume is proportional to the spread between futures' market price and its theoretical prices:

$$\bar{q}_{1,it\tau} = \left\lfloor \left| \alpha_a (p_{2,t\tau} - \bar{p}_{2,t\tau}) \right| \right\rfloor \tag{20}$$

$$\bar{q}_{2,it\tau} = \bar{q}_{1,it\tau} \tag{21}$$

and the limit prices of limit orders that arbitragers submit into the spot market and futures market are close to the assets' market prices:

$$l_{1,it\tau} = p_{1,t\tau}(1 + |\Delta z_{t\tau}| H_{1,it\tau}) \tag{22}$$

$$l_{2,it\tau} = p_{2,t\tau}(1 + |\Delta z_{t\tau}| H_{2,it\tau}) \tag{23}$$

It should be noted that when the spread become narrow, namely $\left|p_{2,t\tau} - \bar{p}_{2,t\tau}\right| \leq \delta \bar{p}_{2,\tau}$, the arbitrage opportunity disappears, and the arbitrager submit market orders in both markets to close positions and realize profit.

3.2 Margin Trading and Order Volume

In the spot market, short selling and buying on margin are forbidden, while in the futures market all transactions are executed by margin trading. When traders open positions in futures market, they should pay a corresponding proportion of cash as the margin according to the initial margin ratio. Hence, the actual quantity an agent can submit in spot market $q_{1,it\tau}$ and futures market $q_{2,it\tau}$ are subject to different degrees of wealth restrictions:

$$q_{1,it\tau} = \begin{cases} \min(\bar{q}_{1,it\tau}, \lfloor \frac{C_{it\tau} - \beta_m p_{2,t\tau} |F_{it\tau}|}{l_{it\tau}} \rfloor) & (H_{1,it\tau} = 1) \\ \min(\bar{q}_{1,it\tau}, S_{it\tau}) & (H_{1,it\tau} = -1) \end{cases} \tag{24}$$

$$q_{2,it\tau} = \begin{cases} \max\left[0, \min\left(\bar{q}_{2,it\tau}, \lfloor\frac{C_{it\tau}}{\beta l_{2,it\tau}}\rfloor - F_{it\tau}\right)\right] (H_{2,it\tau} = 1) \\ \max\left[0, \min\left(\bar{q}_{2,it\tau}, \lfloor\frac{C_{it\tau}}{\beta l_{2,it\tau}}\rfloor + F_{it\tau}\right)\right] (H_{2,it\tau} = -1) \end{cases} \quad (25)$$

where $\beta > 0$ and $\beta_m \in (0, \beta)$ are initial and maintaining margin ratio respectively. $C_{it\tau}$ and $S_{it\tau}$ is the amount of cash and stocks that trader i holds at time τ during time period t, and $F_{it\tau}$ is the position held by trader i in futures market.

In futures market, traders' accounts are marked to market on a daily basis, namely the floating profit and loss are accounted at the end of each trading day according to the closing price. If the maintenance margin requirement cannot be met due to the floating loss, the trader is required to close his position.

4 Simulation Results

We conducted simulation of our artificial spot-futures market and observed the agents behaviors and emergent phenomena at macro level. For simplicity, we consider the risk-free interest rate r to be 0, under this assumption, the fundamental value of spot and futures are the same $p^*_{2,t} = p^*_{1,t}$, as well as futures theoretical price and spot's market price, $\bar{p}_{2,t\tau} = p_{1,t\tau}$.

We introduce 1000 heterogeneous traders in our spot-futures market, include 300 fundamentalists (150 in spot market and 150 in futures market), 200 speculators (100 in each market), 200 hedgers, 200 arbitragers and 100 noise traders (50 in each market). The initial cash amount of traders who only trade in futures market (including fundamentalists, speculators and noise traders in futures market) are $C_{i0} \in \{10000, 11000, \ldots, 30000\}$, while other traders who can trade in spot market (include cross-market traders and traders only trade in spot market) are initially hold spot assets randomly between 10 and 30 units, namely $S_{i0} \in \{10, 11, \ldots, 30\}$, besides, their initial cash amount are set according to their initial stock amount as $C_{i0} = 1000S_{i0}$. The initial market price of stock index and its futures are both set to be 1000, while their initial fundamental prices are both set to be 990. Besides, we set the stock index futures' delivery date $T = 1200$ to be the last trading period in simulation, and the spot and futures' market prices in the first 96 time periods are exogenous given randomly. Other parameters used in simulations are listed in Table 1.

4.1 Stylized Facts

Some stylized facts, such as high co-movements between spot and futures prices and fat-tailed distribution of basis, exist in spot and futures markets. These stylized facts are widely observed by empirical studies. The simulation results of our cross-market model can reproduce these price properties perfectly.

Table 1 Parameters used in the simulation

Parameter	Value	Description
N	1000	Number of agents
σ_f	0.001	Volatility of fundamental value
Δ_f	0.0001	Observation error
r	0	Risk-free interest rate
α_f	1	Reaction coefficient for fundamentalists
α_s	1	Reaction coefficient for speculators
α_a	1	Reaction coefficient for arbitragers
D_i	{20, 21, ..., 90}	Length of MA windows
Δ	0.001	Aggressiveness parameter
δ	0.0005	Threshold to submit orders
β	0.2	Initial margin ratio
β_m	$0.75 * \beta$	Maintaining margin ratio

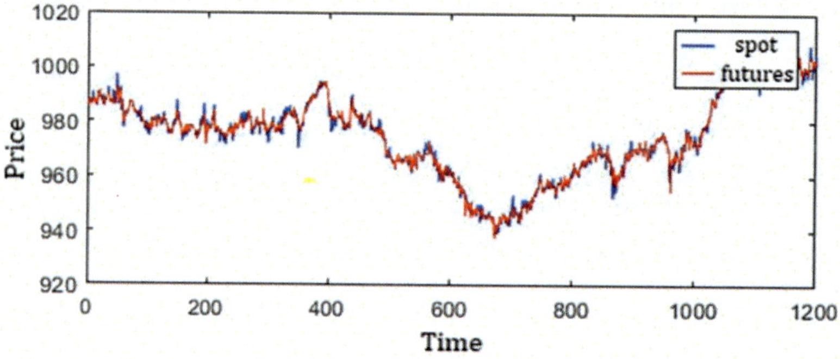

Fig. 1 Co-movement of spot (blue) and futures (red) prices

(1) Co-movement of Prices: Figure 1 shows the variations of the spot and futures' market prices. We can clearly see the co-movement between spot prices and futures prices from this figure. Moreover, the correlation coefficient of the two prices is 0.8269, which further illustrates the high co-movement phenomenon. Further analysis show that the co-movement phenomenon due to two micro level reasons mainly: Firstly, in our model, the fundamental value of spot and futures are closely connected, corresponding to the fact that in the real market the price of spot and futures can be affected by the same factor. Secondly, the investment behavior of the arbitrager will eliminate the prices spread in time, resulting in a close linkage between the spot and futures market prices.

(2) Fat-Tailed Distribution of Basis: Figure 2 shows the distribution of basis, namely the spread between futures price and its theoretical price (equal to spot price when $r = 0$ in our model). Comparing the distribution of basis (blue line) to the normal distribution (red dashed line), we can determine that the basis follow a fat-

Fig. 2 Fat-Tailed Distribution of Basis

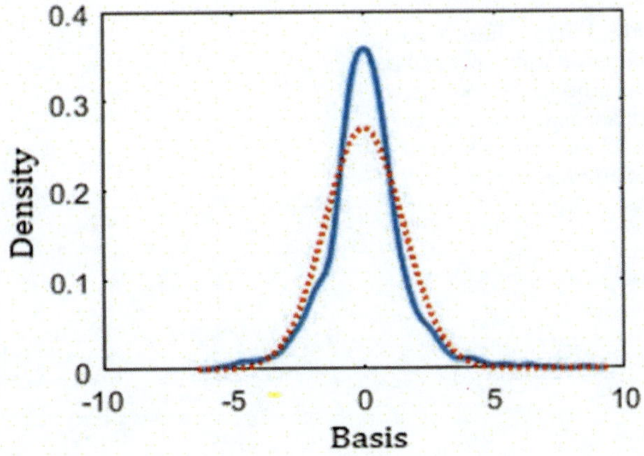

tailed distribution, which is consist with the stylized facts in real financial market. The peak characteristic which means difference between spot prices and futures prices is close to 0 frequently, resulting from arbitrage behaviors of arbitragers, who benefit from basis. On the contrary, speculators and hedgers' trading behaviours may drive the futures price from spot price, which can cause the fat-tailed distribution of basis.

4.2 Arbitragers' Influence on the Markets

Based on this cross-market model, we further study the influence of arbitrage intensity on market quality by increasing the number of arbitragers and observing how the volatility and liquidity of spot and futures markets change. Specifically, we use the bid ask spread S and daily realized volatility RV as metrics of market liquidity and volatility respectively. The calculation details can refer to our earlier work [17].

In order to exclude the influence of other cross-market traders, we only introduce fundamentalists, speculators, noise traders and arbitragers into our cross-market model. As shown in Table 2, the total number of traders in each group of experiments is 1000, and the number of arbitragers N_a is gradually increased from 0 to 200, while the proportion of other traders is reduced proportionally.

Table 2 The number of different type of traders in each simulations

No	Fundamentalists	Speculators	Arbitragers	Noise traders
1	500	400	0	100
2	476	380	50	94
3	450	360	100	90
4	426	340	150	84
5	400	320	200	80

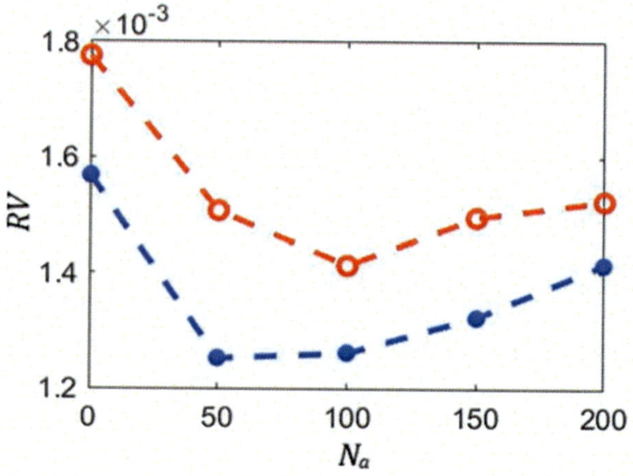

Fig. 3 The results of how mean value of daily realized volatility in the spot market (blue) and futures market (red) change as the number of arbitragers increases

We run each simulations 50 times and record the mean value of liquidity and volatility indicators. From Fig. 3, we can find that the average volatility of the futures market (red) is higher than that of the spot market (blue), which may be due to the margin trading and the compulsory position closing in futures market. Moreover, we can also observe that with the increase of the proportion of arbitragers, the daily realized volatility decreases and then increases both in spot market and the futures market. Arbitrage trading can eliminate mispricing and stabilize markets, but as arbitragers receive homogeneous signals, thus, the introduction of excessive arbitrage will not further increase market stability. On the contrary, too many arbitragers can hurt the market stability and increase market volatility due to the frequent position closing. This result is consistent with the research results of Xiong et al. [18].

Figure 4 shows the bid ask spread of order books in spot and futures markets. As an auxiliary indicator, we also show the change of daily trading volume V in Fig. 5. We can find that the daily trading volume in the futures market is larger than that in the spot market, but its bid ask spread is wider than that in the spot market. This is because margin trading in the futures market can enlarge the supply and demand which can increase trading volume in the futures market, but at the same time, these market orders submitted by traders who close positions will consume the market liquidity and expand bid ask spread of order book in futures market. Further, with the increase of arbitragers' proportion, the bid ask spreads in spot market and futures market both decreases, and the daily trading volumes both increases, indicating arbitrage behaviors can increase liquidities of spot and futures' markets.

At last, in order to study the influence of arbitrage on the basis of futures and spot market, we also calculate the average value of the absolute basis $|B|$ with the increase of arbitrage intensity. The results are shown in Fig. 6. It can be found that the absolute value of the basis decreases first and then increases, indicating that the arbitrage intensity should be controlled; too much or too little arbitrage trading will expand the basis and hurt market price efficiency.

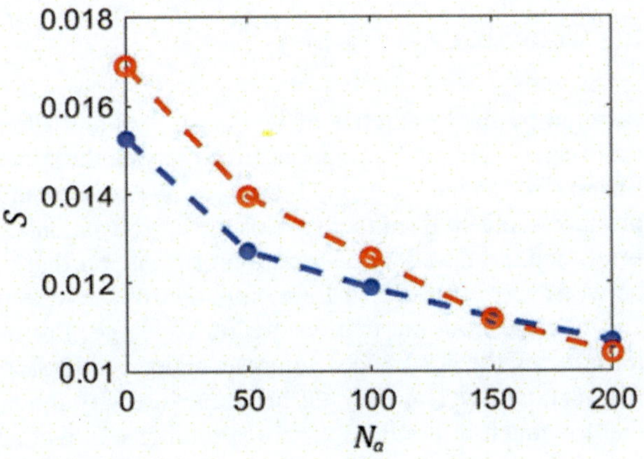

Fig. 4 The results of how mean value of bid ask spread in the spot market (blue) and futures market (red) change as the number of arbitragers increases

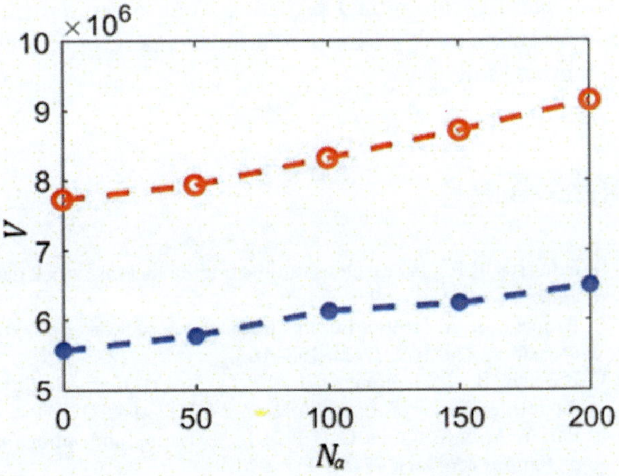

Fig. 5 The results of how mean value of daily trading volume in the spot market (blue) and futures market (red) change as the number of arbitragers increases

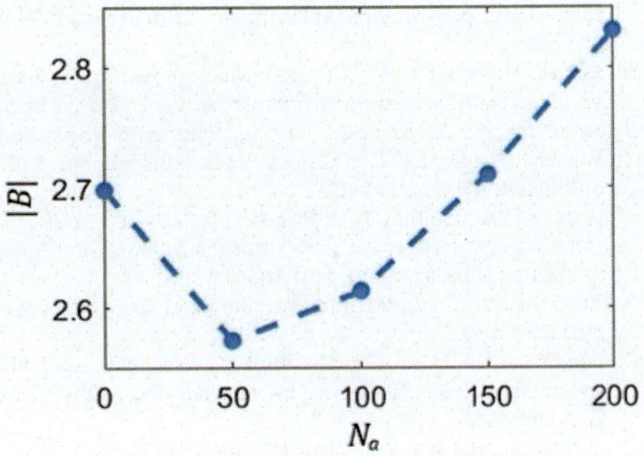

Fig. 6 The results of how average absolute value of basis change as the number of arbitragers increases

5 Conclusions

Inspired by the researches of Ohi et al. [14] and Chiarella et al. [16], we model a multi-agent spot-futures market. In our model, the spot market and futures market are linked through the investors' behaviors. We conduct the simulation of our model and produce some stylized fact such as the co-movement between spot price and futures price, and the fat-tailed distribution of basis. Further, we strengthen the connection between spot market and futures market by increasing the arbitragers' proportion and observe how the market qualities change. From the result, we consider that introducing the stock index futures can decrease spot market volatility and increase market liquidity. However, the arbitrage intensity should be controlled due to the fact that too much or too little arbitrage trading will increase volatility and expand spot futures basis, which will hurt the market stability and price efficiency.

Generally, this model provides some meaningful results and can be extend to conduct more analyses in spot-futures market, such as waterfall effect, circuit breaker and price limit.

References

1. Edwards, F. R. (1988). Does futures trading increase stock market Volatility. *Financial Analysts Journal*, *44*(1), 63–69.
2. Brorsen, B. W. (1991). Futures trading, transaction costs, and stock market volatility. *Journal of Futures Markets*, *11*(2), 153–163.
3. Baldauf, B., & Santoni, G. J. (1991). Stock price volatility: some evidence from an ARCH model. *Journal of Futures Markets*, *11*(2), 191–200.
4. Pericli, A., & Koutmos, G. (1997). Index futures and options and stock market volatility. *Journal of Futures Markets*, *17*(8), 957–974.
5. Darrat, A. F., Rahman, S., & Zhong, M. (2002). On the role of futures trading in spot market fluctuations: perpetrator of volatility or victim of regret? *Journal of Financial Research*, *25*(3), 431–444.
6. Lee, S. B., & Ohk, K. Y. (1992). Stock index futures listing and structural change in time varying volatility. *Journal of Futures Markets*, *12*(5), 493–509.
7. Antoniou, A., & Holmes, P. (1995). Futures trading, information and spot price Volatility: Evidence for the FTSE-100 stock index futures contract using GARCH. *Journal of Banking and Finance*, *19*(1), 117–129.
8. Antoniou, A., Holmes, P., & Priestley, R. (1998). The effects of stock index futures trading on stock index volatility: An analysis of the asymmetric response of volatility to News. *Journal of Futures Markets*, *18*(2), 151–166.
9. Yu, S.-W. (2001). Index futures trading and spot price volatility. *Applied Economics Letters*, *8*(3), 183–186.
10. Fan, X., & De, D. (2017). The spillover effect between CSI 500 index futures market and the spot market: Evidence from high-frequency data in 2015. *China Finance Review International*, *7*(2), 249–272.
11. Tarique, J., Malik, I. R. (2018). The impact of index futures on market efficiency and volatility of spot index: an empirical evidence from emerging economies (BRICS). unpublished.
12. Gulen, H., & Mayhew, S. (2000). Stock index futures trading and volatility in international equity markets. *Journal of Futures Markets*, *20*(7), 661–685.

13. Antoniou, A., Koutmos, G., & Pericli, A. (2005). Index futures and positive feedback trading: Evidence from major stock exchanges. *Journal of Empirical Finance*, *12*(2), 219–238.
14. Ohi, T., Hashimoto, Y., Yu, C., & Ohashi, H. (2011). Simulation of futures and spot markets by using an agent-based multi-market model. *Journal of Advanced Computational Intelligence and Intelligent Informatics*, *15*(2), 204–211.
15. Xu, H.-C., Zhang, W., Xiong, X., & Zhou, W.-X. (2014). An agent-based computational model for China's stock market and stock index futures market. *Mathematical Problems in Engineering*, *2014*, 1–10.
16. Chiarella, C., He, X.-Z., & Pellizzari, P. (2012). A dynamic analysis of the microstructure of moving average rules in a double auction market. *Macroeconomic Dynamics*, *16*(4), 556–575.
17. Zhou, X., & Li, H. (2019). Buying on margin and short selling in an artificial double auction market. *Computational Economics*, *54*(4), 1473–1489.
18. Xiong, X., Liu, J., Haichuan, X., Zhang, W., & Zhang, Y. (2014). Anlyzing the impact on stock index futures market volatility of arbitrage. *Systems Engineering Theory & Practice*, *34*(3), 623–630.

Safety Evaluation Index System of China's Photovoltaic Industry

Xinyi Mei, Chaoxiang Jia, Xuan Lv, and Yanglong Chen

Abstract The safe development of the photovoltaic industry plays an essential role in the structural adjustment of China's new energy industry. China's photovoltaic industry has been the world's most enormous scale. However, there are still a series of issues and fierce disputes for China's solar photovoltaic industry safety. The industrial chain structure is not coordinated, the proportion of domestic and foreign market pattern is unbalanced. The production process is high energy consumption, the strength of scientific research and technology is motionless weak, and the adjustment of industrial policy has a huge impact. All these have troubled the development of China's photovoltaic industry and restricted the industrial safety of the solar photovoltaic industry. This paper uses the theory and analysis tools of economics and management to analyze the photovoltaic industry, puts forward the safety development stage model and dynamic mechanism of the photovoltaic industry, and reveals the law of the safety development of the photovoltaic industry. In this paper, the safety evaluation index system of the photovoltaic industry is constructed. The normalized data value method is used to calculate the 10 indexes, and then the Coefficient of Variation method (CV) is used to determine the weight ratio. The safety evaluation system of the photovoltaic industry in China is helpful to prevent the systemic risk of the photovoltaic industry.

Keywords Photovoltaic industry · Industrial safety · Evaluation index system · Coefficient of variation method

X. Mei (✉) · C. Jia · X. Lv · Y. Chen
Beijing Laboratory of National Economic Security Early-Warning Engineering, Beijing Jiaotong University, Beijing, China
e-mail: meixinyi1234@126.com

C. Jia
e-mail: jcxdante@163.com

X. Lv
e-mail: 18113013@bjtu.edu.cn

Y. Chen
e-mail: chenyanglong@aliyun.com

M. Li et al. (eds.), *IEIS 2020*, https://doi.org/10.1007/978-981-33-4363-4_21

1 Introduction

On June 13, 2014, when presiding over the meeting of the central leading group for the financial and economic affairs, Comrade Jinping Xi proposed a new energy security strategy of "four revolutions and one cooperation." Energy security is an overarching and strategic issue that concerns China's economic and social development. It is crucial to the country's prosperity and development, the improvement of people's lives, and the long-term peace and stability of society. The safe development of the photovoltaic industry plays an essential role in the structural adjustment of China's new energy industry.

2 Literature Review

International research on industrial safety, as early as in the nineteenth century, Henry Charles Carey believed that the American industry needed specific national tariff protection to prevent direct competition with British industrial products. Cardoso, a Latin American economist, put forward the famous dependency theory in the 1960s. Michael Porter, in 1990, argued that When there were high-productivity foreign competitors, industrial development and industrial security would be affected [1].

Research on the safety of the photovoltaic industry abroad, Lesourd advocated reducing the cost of photovoltaic power generation and pointed out that the government should make up for the high cost of power generation with appropriate subsidies [2]. European expert Winfried Hoffmann proposed a technology roadmap and measures to expand the market of solar photovoltaic products [3]. Avanden Heuvel and Van Den Bergh proposed a method to analyze the safety of the solar photovoltaic industry by adopting multi-level diversity. Dusonchet and Telaretti proposed the central supporting policies for promoting the photovoltaic technology in European countries to strengthen the safety development of the photovoltaic industry [4].

Domestic scholar Yu put forward and classified the proposition of industrial insecurity [5]. Jing established a new industrial safety assessment model [6]. Li proposed two kinds of the industrial safety classification system and set up a theoretical system of industrial safety [7]. Jiang emphasized that the technical safety problem is not the product of the open industrial market [8]. He focused on the core of the industrial security concept [9]. Sun proposed the evaluation system of industrial ecological security [10].

In the safety field of the photovoltaic industry, Zhou put forward a theoretical research system for the sustainable development of the photovoltaic industry [11]. Wu et al. proposed to give full play to the role of the government in the transformation of market structure [12]. Sun [13], Yu [14], Zhu [15], and Sun [16] proposed to improve the national support system for photovoltaic industry policy from different perspectives. Yu built the safety evaluation index system of the solar photovoltaic with three first-level indicators, 11 second-level indicators [17].

The above research either focuses on the theoretical research of overall industrial safety or focuses on the policy Suggestions, but lacks comprehensive, rigorous, and in-depth research on the photovoltaic industry, a subdivision of the industry.

3 Index Selection

This paper defines the safety of the photovoltaic industry and establishes a first-level index from three dimensions of the production environment, market environment, and external dependence of the photovoltaic industry. According to the design principle of the industrial safety evaluation index system, ten second-level evaluation indexes corresponding to the first-level indexes are selected to construct the evaluation index system suitable for the safety of China's photovoltaic industry in Table 1. In terms of evaluating the production environment, we select four indicators: capital cost, labor cost, raw material cost, and R&D investment in GDP. In terms of evaluating the market environment, we select four indicators: domestic demand growth rate, domestic market share, international demand growth rate, and international market share. In the evaluation of external dependence, we selected two indicators of external dependence on imports and export of the photovoltaic industry. The safety evaluation index system for the photovoltaic industry based on the coefficient of variation method. According to the evaluation index, the principal component is extracted, the weight ratio of the lowest index is determined, and then the weight ratio of the first-level index can be calculated. The calculation process is carried out by software. Finally, the overall score obtained by multiplying the factor score and the corresponding weight ratio is the safety degree of the photovoltaic industry. In

Table 1 The safety evaluation index system for China's photovoltaic industry

First-level indicators	Second-level indicators
Evaluation of the production environment	Capital cost
	Labor cost
	Raw material cost
	The proportion of R&D investment in GDP
Evaluation of the market environment	The growth rate of domestic demand
	Domestic market share
	The growth rate of international demand
	International market share
Evaluation of the external dependence	Import external dependence
	Export dependence

this way, the safety of the photovoltaic industry can be scientifically evaluated and quantified.

4 The Empirical Analysis

4.1 Evaluation Index Model

The photovoltaic industry safety evaluation model of this project draws lessons from the methods adopted by the most authoritative research institutions in the world, the International Institute for Management Development based in Lausanne, Switzerland, and the World Economic Forum, to integrate the multi-indicator system of the international competitiveness [18].

The photovoltaic industry safety evaluation index model of this project is:

$$S = \alpha X + \beta Y + \gamma Z \tag{1}$$

S—safety degree of photovoltaic industry; X—photovoltaic industry production environment evaluation score; Y—photovoltaic industry market environment evaluation score; Z—the evaluation score of external dependence of photovoltaic industry; α, β, γ are the weight coefficient of each first-level index respectively.

$$X = \sum \alpha_i x_i;\ Y = \sum \beta_i y_i;\ Z = \sum \gamma_i z_i \tag{2}$$

x_i,y_i,z_i are respectively the third-level indicators under the second-level indicators;

α_i, β_i, γ_i, are the corresponding weight value of the second-level indicators; the above formula can be concluded as:

$$S = \alpha X + \beta Y + \gamma Z = \sum \alpha_i x_i + \sum \beta_i y_i + \sum \gamma_i z_i \tag{3}$$

According to the above formula, the safety degree of the whole photovoltaic industry can be quantitatively calculated by normalizing the data value and assigning the weight ratio of each index reasonably.

4.2 Target Assignment

This paper adopts the normalized data value method. Based on the data of 2014, the data of 2015, 2016, 2017, and 2018 were compared with the relevant data of 2014. The final scores of the evaluation index system of each region were compared in each year. The specific method is to first normalize the data. The minimum value of the index is 60 points, and the maximum value is 100 points. The score of the index can be obtained through the formula of the standardized score.

The normalized data value = (the actual value of the indicator − the minimum value of the indicator) / (the maximum value of the indicator − the minimum value of the indicator).

$$\text{Standardized score} = 60 + 40 * \text{normalized value}$$

4.3 *Coefficient of Variation Method for Weight Ratio*

Calculate the standard deviation of each indicator:

$$S_j = \sqrt{\frac{\sum_{j=1}^{m}\left(a_{ij} - \overline{a_j}\right)^2}{m}} \tag{4}$$

The coefficient of variation of each index was calculated:

$$V_j = \frac{S_j}{a_j} \tag{5}$$

V_j represents the coefficient of variation of the first indicator j.

The coefficient of variation was normalized to determine the weight ratio:

$$\overline{W_j} = \frac{V_j}{\sum_{j=1}^{n} V_j} \tag{6}$$

$\overline{W_j}$ represents the weight ratio of the jth index.

By calculation, the weight ratio table can be obtained in Table 2.

4.4 *Statistics of Indicators for Each year*

See Table 3.

4.5 *Determine Industrial Safety*

According to the evaluation value and weight ratio of each secondary index, the evaluation value of each primary index is calculated by the weighted average method.

Similarly, the estimated value of the safety degree of China's solar photovoltaic industry is calculated. The rating table can be obtained by calculation in Table 4.

Table 2 The weight ratio of China's solar photovoltaic industry safety evaluation index

First-level indicators	Primary index weight ratio	Second-level indicators	Secondary index weight ratio
Production environment evaluation	0.24	The cost of capital	0.36
		Labor cost	0.34
		Raw material cost	0.22
		The proportion of R&D investment in GDP	0.08
Market environment assessment	0.16	The growth rate of domestic demand	0.35
		Domestic market share	0.11
		The growth rate of international demand	0.43
		International market share	0.11
Evaluation of external dependence	0.6	Import external dependence	0.11
		Export dependence	0.89

Table 3 Statistics of indicators for each year

Time	2014	2015	2016	2017	2018
The loan interest rate	0.061	0.0518	0.0462	0.0462	0.0462
Labor cost (yuan)	47,241	50,684	54,338	58,049	64,643
Average price of polysilicon (usd/kg)	22.12	18.89	19.09	20.83	18.5
R&D investment in GDP (%)	2.03	2.07	2.12	2.15	2.19
Growth rate of domestic demand	0.273	0.7333	0.3100	0.2308	0.0160
Domestic market share	0.3431	0.4355	0.6325	0.6701	0.5297
Growth rate of international demand	0.2635	0.0300	0.0231	0.3027	0.5031
International market share	0.5023	0.3770	0.2619	0.2640	0.3814
Export external dependence	0.0169	0.0457	0.0370	0.0260	0.0072
External import dependency	0.4332	0.4161	0.4289	0.4027	0.3600

Table 4 The solar photovoltaic industry safety rating scale

Year	Production environment score	Market environment score	External dependency rating	The safety rating of the photovoltaic industry
2014	74	79	87	83
2015	88	77	61	70
2016	91	71	68	74
2017	84	79	80	81
2018	86	82	100	94

4.6 Analysis of Safety Evaluation Results of China's Solar Photovoltaic Industry

According to the calculation, the safety score of China's solar photovoltaic industry in 2018 is 94, the highest score since 2014.

In terms of the production environment score, China's solar photovoltaic industry ranked third in 2018. From the loan interest rate, China's loan interest rate has dropped sharply since 2014, but it has remained at a certain level since 2016, and the loan cost was low in 2018. From the perspective of labor cost, the annual salary of employees in China's manufacturing enterprises above the scale keeps rising. In 2018, the salary of employees was nearly 1.4 times that of 2014, which dramatically increases the production cost of the photovoltaic industry. According to the average price of polysilicon, the spot price of polysilicon in 2018 is the lowest in five years, which can better save the cost of raw materials. From the perspective of R&D investment, the proportion of R&D investment in China keeps increasing every year, and more R&D investment can more effectively improve production efficiency. Therefore, the production environment score of China's photovoltaic industry is at a reasonable level in 2018.

From the perspective of the market environment, China's market environment score has not changed much in the past five years, and the market environment in 2018 is the best. In terms of the growth rate of domestic demand, the growth rate of China's PV installation volume in 2018 is the lowest in five years, and the share of PV installation volume in the total output is also in the middle range. From the perspective of the domestic market, the development of China's solar photovoltaic industry in general. However, in 2018, China's international photovoltaic market has been well developed. In general, the market environment of China's solar photovoltaic industry is generally favorable.

From the perspective of the import and export dependence of the solar photovoltaic industry, the import and export dependence in 2018 is the lowest. It shows that China's photovoltaic industry is less affected by the international market and has a high degree of industrial safety.

In general, the safety degree of China's solar photovoltaic industry has been dramatically improved compared with that of 2014, and it is generally in a relatively safe state.

5 Policy Suggestions

5.1 Expand the Domestic Consumer Market of Photovoltaic Products

The installation and application of photovoltaic products in the downstream links of China's photovoltaic industry are mainly used for export. The severe shortage of domestic photovoltaic system application market leads to the high dependence of the photovoltaic industry on foreign trade. We will expand the domestic consumer market, promote photovoltaic applications, especially photovoltaic construction integration projects, and focus on reversing the severe imbalance between the domestic and foreign markets of downstream products, to improve the situation that trade is heavily dependent on the international market. In the photovoltaic application link, we will increase investment to develop solar photovoltaic power generation and photovoltaic building integration projects.

Opening the domestic photovoltaic application market can also be combined with agricultural science and technology projects and poverty alleviation work. Photovoltaic agriculture is a new field of photovoltaic application, and there is still a large potential market for photovoltaic applications in agricultural production to be developed and popularized. In addition, the photovoltaic (PV) in photovoltaic water pumping technology, solar energy resources were applied to agricultural water conservancy construction, crops and farmland irrigation, the photovoltaic) together with agriculture, very suitable for application in the desert governance, grassland animal husbandry, and agriculture irrigation can also help solve the production and living of defect areas without electricity, actualizing the farmers increase production, the benefit is remarkable. Besides, the application of photovoltaic streetlamps on urban roads can reduce municipal expenditure and expand the domestic consumer market of photovoltaic applications.

5.2 Raise the Technological Level of China's Photovoltaic Industry

The lack of core technology is the fundamental reason for China's photovoltaic industry has been repeatedly subjected to foreign "double reverse". China's photovoltaic industry needs to improve the technical level of each link in the upper, middle and lower reaches, which is the key to improve the international competitiveness of China's photovoltaic products and the trade security of the photovoltaic industry.

In terms of polysilicon production, only by improving the technical level, reducing the overall energy consumption and production cost, and controlling the cost below the market price, can the competitive trade advantage be created. Technological innovation is essential for survival. Only by strengthening the technical transformation, can we reduce the cost, improve the quality, and could compete. Through the introduction of foreign advanced technology and equipment, combined with mature domestic technology, to continuously improve the ability of independent innovation, comprehensive utilization of waste recycling, reduce production costs.

In terms of photovoltaic cell production, the company continuously improves the technology and optimizes the process, improves the conversion efficiency of photovoltaic cells and the average power of battery components, and enhances the product quality advantage. Upgrade the motor structure of photovoltaic cells, reduce the risk of hot spots generated by photovoltaic modules, and improve product performance and life; Master completely independent intellectual property rights to avoid related patent risks. At the same time, pay attention to perovskite battery technology research and development. Decrease the production cost of photovoltaic cells, the formation of core competitiveness in the domestic and foreign markets.

In the field of photovoltaic application, we will increase the investment in the construction and development of solar photovoltaic power generation and photovoltaic building integration projects, including technology development, equipment production, and promotion and application.

5.3 Actively Respond to International Trade Disputes

The trade sanctions against China's photovoltaic industry have brought a significant impact on China's photovoltaic industry. China should actively deal with the international trade disputes, attach importance to the investigation and application of major international photovoltaic enterprises and related industries, and create a functional trade security environment for China's photovoltaic industry.

Strengthen the study of relevant commitment mechanism and WTO rules, familiar with the WTO trade dispute response mechanism and settlement framework, familiar with the WTO rules, regional trade organization trade rules and relevant agreements, flexible response to various potential trade disputes and trade security threats; We will actively participate in global and regional trade consultations and international negotiations, and strive to enhance our right of participation and voice in making rules, standards and dispute settlement mechanisms. We will strengthen the in-depth cooperation between Chinese companies in the photovoltaic industry and major photovoltaic trading countries such as Europe and the United States in the areas of market development, technology research and development, and mutual financial assistance, enhance enterprise trust.

Establish a global information exchange mechanism for the photovoltaic industry so that international counterparts can communicate and coordinate with each other to enhance mutual trust. It has the role of government participation and guidance and non-governmental industry organizations to play the role of communication and coordination, and the timely national adjustment of relevant industrial policies to give full play to the guiding function.

6 Conclusion

In this paper, the safety evaluation index system of China's photovoltaic industry is constructed, the normalized data value method is used to calculate the index, and the coefficient of variation method is used to weight the index ratio. Data related to China's photovoltaic industry from 2014 to 2018 were adopted. Thus, the overall score of safety evaluation of China's photovoltaic industry from 2014 to 2018 is calculated as 83,70,74,81,94. In 2014, the safety state of the photovoltaic industry was relatively stable, and the trend of decline was particularly apparent in 2015 and 2016. In 2017 and 2018, the safety problem of the photovoltaic industry in China began to stabilize again.

The suggestions of the PV industry safety were put forward based on the empirical analysis results: improve the technology level of China's photovoltaic industry, expand the domestic consumer market of photovoltaic products and actively respond to international trade disputes.

References

1. Porter, M. E. (1990). *Competitive advantage of nations*. MY: The Free Press.
2. Lesourd, J.-B. (2001). Solar photovoltaic systems: The economics of a renewable energy resource. *Environmental Modelling & Software, 16,* 147–156.
3. Hoffmann, W. (2006). PV solar electricity industry: Market growth and perspective. *Solar Energy Materials & Solar Cells, 90,* 3285–3311.
4. Dusonchet, L., & Telaretti, E. (2012). Comments on "Economic analysis of different supporting policies for the production of electrical energy by solar photovoltaics in western European Union countries" by Luigi Dusonchet and Enrico Telaretti. *Energy Policy, 48,* 846–849.
5. Yu, X. (2000). Research and countermeasures on industrial protection and industrial safety after China enter into WTO. *Learning and Exploration, 02,* 4–12.
6. Jing, Y. (2006). Research on the industrial safety evaluation index system. *Economist, 02,* 70–76.
7. Li, M. (2008). Research on the industrial organization security theory. *Productivity Research, 2008*(24), 9–11 + 46.
8. Li, M., & Jiang, Z. (2009). Review of the development of industrial economics theory. *Journal of China's Circulating Economy, 23*(4), 30–32.
9. He, W., & Du, P. (2013). Research on the safety evaluation index system of strategic emerging industries. *Management Modernization, 04,* 22–24.
10. Sun, M. (2019). Evaluation of the industrial ecological security. *Rural Economy and Technology, 30*(21), 46–47.

11. Zhou, S., Ma, C., & Li, L. (2007). Theoretical research on sustainable development of solar photovoltaic industry. *Technological Progress and Countermeasures, 07,* 88–90.
12. Wu, Z., & Yu, H. (2008). Research on the evolution law of the market structure of China's emerging industries—A case study of organosilicon industry. *China Industrial Economy, 12,* 134–143.
13. Sun, Y. (2009). Analysis of the current situation and prospect of the competitiveness of China's photovoltaic industry. Fudan University.
14. Yu, L., & Yu, Y. (2012). Evaluation of the photovoltaic industry policy system: Multi-level or diversification. *Reform, 08,* 114–122.
15. Zhu, N., & Tian, L. (2013). Policy research on China's photovoltaic power generation industry—Also on the improvement of China's photovoltaic power generation price system. *Price Theory and Practice, 08,* 93–94.
16. Sun, H., Wang, L. (2013). Analysis of the competitiveness of China's photovoltaic industry. *Technology and innovation management, 34*(06), 569–572 + 591.
17. Yu, T. (2014). Research on the industrial safety of the solar photovoltaic industry in China: theory and practice. Southwest University of Finance and Economics.
18. Wang, L. (2011). Research on early warning system of industrial safety evaluation. Beijing Jiaotong University.

Application Analysis of Big Data in Construction Project Management

Gang Han

Abstract The era background of big data puts forward new requirements for construction project management. This article reviews the literature and analyzes the problems in the application of big data in construction project management. On this basis, this paper establishes the construction engineering safety evaluation index system, and uses the analytic hierarchy process and fuzzy comprehensive evaluation method to evaluate the construction engineering safety. Assuming enough data, this article attempts to use BP neural network to analyze its applicability to construction project management. Finally, this article puts forward suggestions to strengthen the application of big data.

Keywords Big data · Construction engineering · Project management · Application

1 Introduction

The construction industry is an important industry of the national economy and has important significance for China's economic development. Take 2019 as an example, the value-added of the construction industry accounts for 7.16% of GDP in China. The construction industry is an important material production sector of the national economy, and it has very close links with other industries of the national economy, making a huge contribution to the local economy. With the advent of the era of big data and the intensification of market competition, the traditional construction industry is in urgent need of transformation and upgrading. By fully digging out and using big data, the construction industry gradually transforms the traditional extensive management method into a refined and information management method to improve enterprise production efficiency and economic benefits. But for now, the application of big data in the construction industry is still in the early stages and the exploration stage.

G. Han (✉)
Postdoctoral Programme of Management Science and Engineering, Beijing Jiaotong University, Beijing, China
e-mail: hgkj2002@163.com

M. Li et al. (eds.), *IEIS 2020*, https://doi.org/10.1007/978-981-33-4363-4_22

2 Literature Review

The related research of big data technology is mostly in the initial stage, and the lack of cognitive ability of big data is an important factor affecting the decision-making process. According to the analysis of Acikara [1], as early as 1990s, Latham and Egan pointed out that the innovation in the field of engineering projects and the upgrading of the industrial structure were slow, and the project management performance had not been significantly improved [1]. This is related to the ability of enterprises to collect data and convert it into valuable information service decisions.

Big data provides new opportunities for the innovation of engineering project management. Memon [2] believes that the application of big data can provide the information which is critical to the project decision-making process to the project managers and stakeholders. So it is necessary to correctly process and analyze the large amount of data generated during project execution [2]. Zeng [3] analyzed the application of big data mining in engineering project management and designed a project progress control model for real estate projects. Ren [4] studied engineering project management innovation from the perspective of project organization, and established a project management innovation model. Some scholars also proposed innovative countermeasures for engineering project management in data collection, big data decision-making, and big data application .

By combing the literature review, the following conclusions can be drawn. First, big data technology has wide application fields and strong applicability [5]. The project field should be integrated into the era of big data, and rely on more advanced technology and more scientific methods to help companies achieve breakthroughs in management quality and management efficiency [6]. It can promote the transformation of the traditional management model to the digital and information management model, and ultimately help the transformation and upgrade of the engineering project field. Second, the research and application of big data technology in the field of engineering projects in China is relatively lagging behind. Some scholars have explored the innovation of engineering project management under the background of big data, but the research needs to be further deepened.

3 Main Problems in the Application of Big Data in Construction Project Management

3.1 Massive Data Has not Been Attracted Attention

Compared with other types of projects, construction projects have the characteristics of huge data volume, particularly wide coverage, relatively long construction period, and many uncertainties. The construction project has the main stages of initial design and planning, mid-term construction, post-acceptance and quality assurance from project start, planning, and execution to completion. There are huge amounts of data in many process links such as design, bidding, quality, cost, completion acceptance, and quality warranty of construction projects. Government agencies, project participants, project management parties and many other subjects have different data. These huge amounts of data and information are only kept as process data, and most of the data is considered to be worthless or of little value. It is eventually ignored and discarded. It has not fully played its role in the decision-making of construction project management.

For example, in the progress management of construction engineering projects, if it is not possible to convert information such as the order of entry of various raw materials, the time of entry, and the production efficiency of various types of workers per unit of time into data and process the data in a timely manner, it may cause slack work and affect the overall progress of the project. In design management, if you ignore the data from the market, it may cause the dilemma that the design scheme deviates from the market demand.

In addition, the uncertain factors also generate a lot of information and data. If the decision of the construction project management cannot be adjusted in time according to the information and data generated by the emergency, it will bring great challenges to the progress, cost, quality and security of construction projects. Only by paying enough attention to process data and making full use of data calculation and analysis can it truly provide assistance for specific decisions.

3.2 *Lack of Suitable Big Data Application Methods Leads to a Single Construction Project Management Method*

According to a survey in the "Information Development Report 2018 of the Chinese Construction Industry", 46.99% of the respondents believe that the construction industry lacks a big data analysis system, which is a big data application problem. In traditional construction project management, the project department is the core department of the entire project management, and the project managers are the managers who most directly contact the construction project. They often change or advance project plans, control project progress, anticipate project risks, feedback management results, and make corresponding decisions according to their subjective consciousness or collective brainstorming, management experience, behavior preferences. In fact, this approach leads to too many individual subjective components. The understanding of some problems stays at the surface, and the solutions and decisions cannot be guaranteed to be scientific. The simplification of management methods has not been significantly changed. At present, construction project management lacks a set of applicable and targeted big data application methods to guide practice. How to use big data technology and cloud computing in construction project management to deeply dig real data, analyze data, and extract effective data as decision support is an urgent problem to be solved.

3.3 *Lack of Professional Management Teams that Are Proficient in Construction Project Management and Big Data Technology*

In the era of rapid development of information technology, professional management teams are required in order to improve the efficiency of project management and achieve project management goals under constrained conditions. The lack of professional talents makes the actual effect of the combination of big data and construction project management not obvious. The main management personnel of construction projects must keep pace with the times, abandon traditional thinking and management models, continuously improve their comprehensive quality, keep up with the development direction of new technologies, and strengthen the study of big data technology theory and practice. Construction project management personnel have been compound talent who are proficient in construction project management business, but also understand the application of big data technology and promotes the true integration of big data and construction project management.

4 Application of Big Data in Construction Project Management—Taking Construction Project Security Management as an Example

4.1 *Establishment of Construction Safety Evaluation Index System*

Due to the characteristics of long construction activity period, high mobility, high-altitude operations, many cross-links, complicated construction technology, and high labor intensity, it leads to high construction risk and many unsafe factors. Construction safety accidents still occur from time to time. From 2007 to the first half of 2017, a total of 6371 construction safety accidents occurred throughout the country, causing 7804 deaths, which not only caused losses to the development of national economy, but also affected social harmony and stability and the well-being of people.

Human factor is the most important factor for ensuring the safety of building construction. Many accidents are closely related to human behavior. Work license is the most basic condition for evaluating whether construction personnel have construction skills. Some personnel have not obtained the qualifications for induction. They have not systematically learned theoretical knowledge and operational skills. If people without qualifications are allowed to work, they are more likely to have safety accidents. Pre-job safety training mainly includes general knowledge of construction safety, basic knowledge of construction quality, warning of typical cases, professional ethics and other contents. It is also a very important indicator to abide by work discipline and operate in accordance with regulations. Operation by personnel in violation of regulations may lead to accidents. Physical condition is an evaluation indicator of whether the construction personnel are in good health or not. If you conceal your medical history, such as heart and brain-related diseases, danger may occur during construction operations.

Equipment factors also play an important role in construction safety. First, whether the mechanical equipment meets the safety regulations and whether the equipment is in good condition is very important for the safety of the construction process. Second, the safety protection measures at the construction site should be in place. For example, protective railings should be installed at the elevator wellhead and stairwell. A protective shed should be erected at the entrance of the passage. There should also be safety protection facilities near the edge.

The safety system stipulates the rules, regulations, policies and regulations that personnel at all levels should observe when working. The perfect safety system and the implementation of the safety system by the staff can effectively ensure the smooth progress of safety production. Emergency treatment measures are how to organize and implement emergency rescue work in a timely and effective manner after an accident occurs.

Environmental factors include climatic conditions and the working environment of the construction site.

This part selects eleven key indicators from four aspects of personnel factor, equipment factor, management factor and environmental factor to construct a construction safety evaluation index system for construction projects, as shown in Table 1.

Table 1 Index system of construction safety evaluation

Target layer	Index layer	Second-level indicator layer	number
Index system of construction safety evaluation	Human factors $\mathbf{A}_1$	Qualification	$\mathbf{A}_{11}$
		Pre-job safety training	$\mathbf{A}_{12}$
		Follow the rules	$\mathbf{A}_{13}$
		body condition	$\mathbf{A}_{14}$
	Equipment factors $\mathbf{A}_2$	Equipment meets safety regulations	$\mathbf{A}_{21}$
		Construction site safety protection measures	$\mathbf{A}_{22}$
	Management factors $\mathbf{A}_3$	Safety inspection	$\mathbf{A}_{31}$
		Security system	$\mathbf{A}_{32}$
		Emergency measures	$\mathbf{A}_{33}$
	Environmental factors $\mathbf{A}_4$	Climatic conditions	$\mathbf{A}_{41}$
		Construction site operating environment	$\mathbf{A}_{42}$

Table 2 Judgment matrix

	Human factors	Equipment factors	Management factors	Environmental factors
Human factors	**1**	**5**	**6**	**7**
Equipment factors	**1/5**	**1**	**2**	**3**
Management factors	**1/6**	**1/2**	**1**	**3**
Environmental factors	**1/7**	**1/3**	**1/3**	**1**

Table 3 Random consistency index table

N order	3	4	5	6	7	8	9	10
RI value	0.52	0.89	1.12	1.26	1.36	1.41	1.46	1.49

4.2 Analytic Hierarchy Process to Determine the Weight

Project A is located in Tongzhou District, Beijing. Taking Project A as an example, this article consulted five experts in the field. The experts scored the four aspects of Project A, Then the article determined the judgment matrix, and used AHP to determine the weight and judgment. The judgment matrix is shown in Table 2.

The calculated index layer weights are 0.63426, 0.17685, 0.12625, 0.06264.

λ is the maximum eigenvalue. $\lambda = 4.143$.

CI is the consistency index. CI = 0.048.

RI is a random consistency indicator, RI calculation is shown in Table 3.

CR is the consistency ratio.

$$\mathrm{CR} = \frac{\mathrm{CI}}{\mathrm{RI}} < 0.1 \tag{1}$$

CR value is less than 0.1, so it passed the consistency test. Using the same method, it can get the weight of the secondary indicator layer.

4.3 Evaluation of Project A with Fuzzy Comprehensive Evaluation Method

Fuzzy comprehensive evaluation is precisely based on fuzzy mathematics, applying the principle of fuzzy relation synthesis to quantitatively study some unclear boundaries and difficult to quantify factors, based on a multi-factor perspective to synthesize the membership levels of things that need to be evaluated an effective method of evaluation. This article divides the construction safety level into five levels, and establishes a review set as $V = \{V_1, V_2, V_3, V_4, V_5\} = \{\text{very safe, safe, relatively safe, unsafe, extremely unsafe}\}$.

W_1 represents the weight of each secondary index in the personnel factor indicator layer. R_1 represents the membership matrix of the personnel factor.

$$W_1 * R_1 = \begin{pmatrix} 0.127538 & 0.445777 & 0.4 & 0.26685 & 0 \end{pmatrix}$$

W_2 represents the weight of each secondary index in the equipment factor index layer. R_2 represents the membership matrix of the equipment factor.

$$W_2 * R_2 = \begin{pmatrix} 0.3 & 0.5 & 0.2 & 0 & 0 \end{pmatrix}$$

W_3 represents the weight of each secondary indicator in the management factor indicator layer. R_3 represents the membership matrix of management factors.

$$W_3 * R_3 = \begin{pmatrix} 0.389338 & 0.321235 & 0.200002 & 0.063335 & 0 \end{pmatrix}$$

W_4 represents the weight of each secondary indicator in the environmental factor indicator layer. R_4 represents the membership matrix of environmental factors.

$$W_4 * R_4 = \begin{pmatrix} 0.4 & 0.4 & 0.3 & 0 & 0 \end{pmatrix}$$

W represents each index factor in the target layer. R represents the membership matrix of the target layer.

$$W * R = \begin{pmatrix} 0.63426 & 0.17685 & 0.12625 & 0.06264 \end{pmatrix} * \begin{bmatrix} 0.127538 & 0.445777 & 0.4 & 0.26685 & 0 \\ 0.3 & 0.5 & 0.2 & 0 & 0 \\ 0.389338 & 0.321235 & 0.200002 & 0.063335 & 0 \\ 0.4 & 0.4 & 0.3 & 0 & 0 \end{bmatrix} = \begin{pmatrix} 0.243527 & 0.430511 & 0.333116 & 0.177248 & 0 \end{pmatrix}$$

According to the analysis, among the four types of influencing factors, the human factor is the most important for the construction project safety. Whether it is equipment, management, or environmental factors, personnel are the main body of execution, and personnel factors are the prerequisite and foundation for ensuring the safety of construction projects. Of course, this is not to say that other factors are not important. All kinds of factors have an impact on the construction safety of construction projects to a certain extent, and should be taken seriously. According to the principle of maximum membership, Project A is in a safe state.

4.4 Training and Prediction of BP Neural Network Model

BP neural network model has input layer, hidden layer and output layer. The input layer gets the stimulus and passes it to the hidden layer. The hidden layer transfers this stimulus to the output layer according to the weight of the neurons' interconnections and according to the rules. The comparison result of the output layer, if it is not correct, it returns the weight of the adjusted neuron interconnection. So you can train and finally learn. The BP neural network has the characteristics of strong nonlinearity. The number of intermediate layers of the network and the number of neurons in each layer can be arbitrarily set according to the specific situation, and its performance is also different with the difference of the structure. It is suitable for solving complex nonlinear problems of construction engineering project management.

According to the above research, the construction safety project influencing factors are four categories: personnel, equipment, management, and environment, that is, the input values of the BP neural network are 4 categories and 11 indicators.

The number of nodes in the hidden layer can be obtained according to the following formula.

$$n = \sqrt{n_i + n_0} + a \tag{2}$$

Among them, n represents the number of nodes in the hidden layer, n_i represents the number of input nodes, n_0 represents the number of output nodes, a is a constant, and the value range is 1–10.

This formula is an empirical formula. In practice, it is difficult to determine the optimal value of the hidden layer neuron node. Generally, it is determined by trial and error. We first set an initial value, and then gradually increase on the basis of this value. By comparing the prediction performance of each network, the number of nodes with the best performance is selected as the number of hidden layer neurons.

Due to the small number of samples, a simple single hidden layer network structure can meet the needs of this paper. When determining the number of nodes in the hidden layer of the neural network, the number of nodes in the middle layer is generally about three-quarters of the number of nodes in the input layer, where the optimal number of nodes in the hidden layer is 8, 9, and 10.Through the experiments of the program, the optimal number of hidden layer nodes is 10.

The 11 indicators in this paper determine that the number of nodes in the input layer is 11.The output value is the construction safety level of the construction project. There are 5 levels: very safe, safe, basic safe, unsafe, and very unsafe. The number of nodes in the output layer is 5, and the output is (1, 0, 0, 0, 0), (0, 1, 0, 0, 0), (0, 0, 1, 0, 0), (0, 0, 0, 1, 0), (0, 0, 0, 0, 1).

In the process, the score values of the training samples are input into the comprehensive evaluation model to obtain the corresponding safety evaluation value, which can be compared with the actual situation, and the correction weight is adjusted according to the error value. The optimal is achieved when the expected error is 0.005 and below.

Table 4 Sample data

Sample	Index										
	A_{11}	A_{12}	A_{13}	A_{14}	A_{21}	A_{22}	A_{31}	A_{32}	A_{33}	A_{41}	A_{42}
1	83	79	81	85	87	90	82	80	80	85	85
2	70	75	75	85	85	85	80	80	85	80	80
3	88	80	75	87	75	75	70	80	82	84	85
4	83	76	74	85	86	75	83	84	87	85	80
5	90	86	81	75	75	90	84	85	82	95	95
6	94	81	80	75	76	80	73	79	82	90	90
7	87	86	85	85	90	89	85	83	80	80	80
8	82	83	85	86	90	80	75	75	80	85	86
9	91	89	87	84	85	87	81	82	84	87	87
10	89	85	85	80	80	85	85	87	83	80	88

After comprehensive comparative analysis, a 3-layer neural network model is established. The model has 11 input nodes, 10 hidden layer nodes, and 5 output nodes. The activation functions of the hidden layer and the output layer are the logsig function and purelin function, respectively. The maximum number of network trainings is 5000. To converge the neural network, the learning rate is set to 0.01, and the expected error is 0.005.

This article identified 10 samples, of which 1–8 were training samples and 9–10 were test samples. In the process of collecting data, we made a questionnaire. In order to improve the objectivity of the questionnaire, the questions designed are as quantifiable as possible, and the project leader provides the exact data. The 5 experts are experienced in the field of construction projects and are familiar with the 10 samples in this article. Experts give objective scores for each indicator of each project, based on their own understanding of the project and the data of the questionnaire. The data of training samples and test samples is shown in the Table 4.

First, all data needs to be normalized, and then the input data is mapped into 0 to 1 using the formula as below.

$$y = (x - x_{\min})/(x_{\max} - x_{\min})$$

Using MATLAB software to achieve the training and simulation of the neural network model, the error curve of the training sample is obtained, as shown in Fig. 1.

The expected error value changes with the number of training times. When training reaches 1788 times, the network training ends and the error accuracy value set in advance is reached.

Compare the data of the two test samples with the data detected by the neural network model. The judgment results are shown in the Table 5.

Comparing the evaluation results of the two sets of test sample data, the safety level corresponding to the expected value is "safe". The prediction of the network

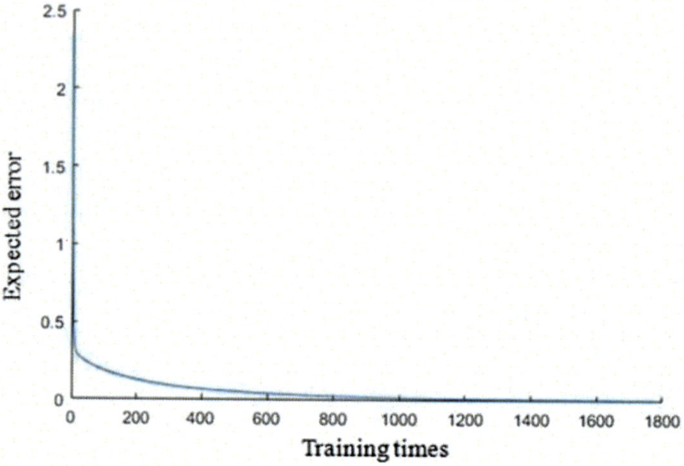

Fig. 1 Error curve

Table 5 Comparison of test data of test samples

Sample No.	Very safe	Safe	Relatively safe	Unsafe	Extremely unsafe	Expected value	Security levels
9	0	0.9914	0	0.1064	0.1041	(0 1 0 0 0)	Safe
10	0	0.8802	0	0.0906	0.0890	(0 1 0 0 0)	Safe

model meets expectations, indicating that the model built can be used for safety evaluation of construction projects.

5 Conclusion and Suggestion

The following conclusions can be drawn from the research in this article.

First, we should collect as much data as possible during the project. The collection, aggregation, processing and effective use of relevant data can transform the data into measurable and valuable information. This information reduces the probability of risky decisions made by project managers and helps project managers make optimal decisions. The more data that can be collected, the more effective information can be extracted, and the more beneficial it is for project managers to provide valuable insights.

Second, AHP, fuzzy comprehensive evaluation and BP neural network methods have strong applicability to the safety of construction projects. Although the sample data in this paper is relatively small, this method has been verified in this paper, especially the feasibility of BP neural network model for the safety prediction of construction projects. Of course, what needs to be improved is that if the regularity of the training samples is good, the more sample data we have, the better results it will get.

In view of the relatively backward application of big data in the construction industry, this article makes the following recommendations.

First, strengthen the idea of big data.

The management should be strengthened by the big data concept and thinking. Managers stand in the strategic perspective of the company's long-term development and conform to the development of the era of big data. Managers should fully understand the value that big data brings to the company, and discuss the specific application path of big data technology to construction project management. Through the data, it can assist decision-making management, improve the effectiveness of corporate decision-making. In the construction of corporate culture, the integration of big data thinking has enabled the company to form an innovative and enterprising working atmosphere.

Second, establish a dedicated information department.

The application of big data and related technologies is not the responsibility of a single department within the company, but a dedicated department at the company level which is responsible for overall planning. Therefore, this article suggests improving the company's organizational structure and establishing a dedicated information department. As the central nervous system of the company, the Information Department is responsible for coordinating the collection, statistics, mining, screening, in-depth analysis and prediction of information and data of various departments of the company, and provides the analysis results to the company's senior management. It provides data support for senior management to make objective and correct decisions.

Third, explore big data technologies and suitable methods for engineering project management.

Scholars have strengthened continuous research on the basic theories, algorithms, and decision analysis of big data in project engineering management, and explored more big data technologies and methods suitable for engineering project management.

Forth, strengthen professional training.

Engineering project managers need to enhance their awareness of informatization, and have a full understanding of big data technology, understand the frontiers of big data technology, and especially know the application situation and future development direction of big data technology at home and abroad in project management.

It is necessary to strengthen the professional training of management personnel, especially the training of mainstream software such as statistical software, engineering cost software, architectural design software, The training can promote the integration of the theoretical knowledge learned with specific projects, and it also can enhance the practical ability of personnel and the application level of information.

References

1. Acikara, T., Kazaz, A., & Ulubeyli, S. (2017). Evaluations of construction project participants' attitudes toward quality management in Turkey. *Procedia Engineering, 196,* 203–210.
2. Memon, S., Changfeng, W., Rasheed, S., et al. (2016). Adoption of big data technologies for communication management in large projects. *International Journal of Future Generation Communication and Networking, 9*(10), 73–82.
3. Hui, Z. (2014). The application of big data mining in engineering project management. *Science and Technology Progress and Countermeasures, 11,* 46–48.
4. Chaoqun, R. (2019). Research on project management innovation based on big data. *Computer Products and Circulation, 06,* 76–77.
5. Shirdastian, H., Laroche, M., & Richard, M. O. (2017). Using big data analytics to study brand authenticity sentiments: The case of Starbucks on Twitter. *International Journal of Information Management*. https://doi.org/10.1016/j.ijinfomgt.2017.09.007
6. Merendino, A., Dibb, S., Meadows, M., Quinn, L., Wilson, D., Simkin, L., & Canhoto, A. (2018). Big data, big decisions: The impact of big data on board level decision-making. *Journal of Business Research, 93,* 67–78.

Study on Influencing Factors of Railway Transportation

Ling Liu, Na Zhang, and Hongchang Li

Abstract There are different degrees of interlinkages between railway transportation and economic and social aspects, mainly reflected in the interaction with industrialization and urbanization. Through panel VAR model and impulse response function, the paper studies the impact of provincial-level railway transportation on industrialization and urbanization in the east, central, and western regions in China, and the reverse impacts the latter two factors have on the former one. Results show that the impact of railway transportation on industrialization and urbanization and the relationship between the latter two on the former are different from region to region. Different regions should put different emphases on giving impetus to urbanization and industrialization by promoting the development of railway transportation.

Keywords Railway transportation · Industrialization · Urbanization · Panel VAR

1 Introduction

Transportation refers to all transportation activities and conditions accompanying the development of industrialization. It is a dynamic process and presents different stage characteristics under different stages of industrialization [1]. Railway transportation refers to the process of changes in the total transportation volume and transportation quality during the development of railway transportation. Since the Industrial Revolution, railways have been used as the main means of transportation for a long time, which has greatly improved all aspects of social and economic life, and still bears the role of cross-regional coordination.

L. Liu (✉) · N. Zhang · H. Li
School of Economics and Management, Beijing Jiaotong University, Beijing, China
e-mail: 18120520@bjtu.edu.cn

N. Zhang
e-mail: zhangna@bjtu.edu.cn

H. Li
e-mail: hchli@bjtu.edu.cn

M. Li et al. (eds.), *IEIS 2020*, https://doi.org/10.1007/978-981-33-4363-4_23

Railway transportation is interlinked with various aspects of the economy and society in different degrees, mainly reflected in the interaction with industrialization and urbanization. There are two main effects of railway transportation on industrialization: the first is the growth effect. The improvement of transportation facilities can directly increase the total output [2], and promote the division of labor specialization [3]. The second is the structural effect, which is manifested by the transfer of industrial structure [4] and the change of the economic spatial pattern [5, 6]. At the same time, social and economic development has also promoted the progress of transportation, and the two have been alternately pushed and pulled [7].

During the interaction between railway transportation and industrialization, urbanization is an important part of the system. Transportation investment promotes enterprise R&D and human capital accumulation, brings economies of scale and promotes long-term economic growth [8, 9]. In particular, the development of high-speed railways in recent years has further compressed time and space, which not only directly promoted economic growth, but also indirectly affected economic growth through aspects such as population, industrial agglomeration and market potential [10]. The development of industrialization can not only promote the progress of railway transportation, but also accelerate the development of urbanization, and urbanization has brought about an increase in the demand for railway transportation, thereby speeding up the construction of transportation [11].

Existing literature has studied the relationship and influence of railway transportation, industrialization and urbanization from different perspectives, but there is no literature that has included the three into the same research framework to quantitatively discuss the dynamic relationship of their interactions. Therefore, this paper aims to empirically analyze the impact of railway transportation on industrialization and urbanization in different regions of China and the reaction of the latter two to railway transportation, and explore their internal relations, so as to provide reference for the future promotion of railway transportation, urbanization and industrialization.

2 Data Selection and Model specification

2.1 Research Methods

The relationship between railway transportation, industrialization and urbanization is shown in Fig. 1. The development of railway transportation, that is, the improvement of the total volume and quality of railway transportation, has played a positive role in technological improvement and efficiency improvement of enterprises along the way, and has promoted the increase of total factor productivity. Reduced interregional transaction costs produce agglomeration effects, and then generate economies of scale through forward and backward linkages. At the same time, various regions are able to participate in the social division of labor based on their own advantageous resource elements, which also promotes the progress of industrialization. On

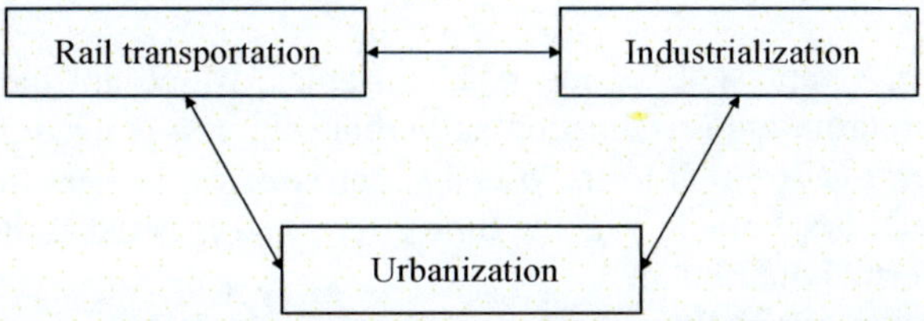

Fig. 1 Correlation between railway transportation and industrialization and urbanization

the other hand, in the process of railway transportation development, population and factors migrate to cities, which in turn affects urban layout and resource allocation [12], which leads to improvement of urbanization, increasement of urban population, expansion of urban area. The travel distances, travel methods and preferences of residents change accordingly, which in turn promotes the growth of transportation demand and the construction of transportation. Therefore, the development of railway transportation has brought about population agglomeration, industrial division of labor, and industrial agglomeration, thereby jointly driving the development of industrialization [13, 14]. At the same time, the socio-economic development's requirements for the transportation speed and quality of railway transportation continue to increase, thereby promoting the progress of transportation. Industrialization has also promoted the development of urbanization on the other hand, thereby indirectly promoting the progress of railway transportation.

Based on the analysis above, there is an interactive relationship among railway transportation, industrialization and urbanization. In order to avoid the estimation error caused by endogenous model, this paper uses panel VAR model to investigate the influence relationship among the three. Due to the vast territory of China and the uneven development of regional development, the interactive relationship between railway transportation and industrialization and urbanization in different regions should also be different. Therefore, this paper will study and compare the data of eastern, central and western provinces (cities and districts), instead of inferring the overall effect.

2.2 *Model Specification*

According to the minimum criteria of AIC, BIC and HQIC, combined with variable time span and impulse response function convergence, after trying multiple lag orders, the models in the eastern, central and western regions all adopt the form of lag 1 order for generalized moment (GMM) estimation. The model is constructed as follows:

$$y_{it} = \sum_{j=1}^{p} \beta_j y_{it-j} + \eta_i + \gamma_t + \mu_{it} \tag{1}$$

Among them, y represents endogenous variables, which are the indicators of railway transportation, industrialization and urbanization. i is for different regions, j is the lag order, t is different years, β is the matrix of parameters to be estimated, η is the individual fixed effect, γ is the time effect, and μ is the random error term following the normal distribution.

Industrialization originally refers to a process in which the proportion of industrial or manufacturing level in national income and employment population gradually changes and increases. The connotation of industrialization in the new era not only includes the expansion and growth of industry in quantity, but also includes the improvement of industrial quality, international competitiveness and modernization level along with the growth in quantity [15]. The essence of the new industrialization with Chinese characteristics is the synchronous and coordinated development of economy, society and humanity [16]. In order to fit the connotation of industrialization in the new era, the per capita GDP of each provincial unit is selected as the measurement index of industrialization development, and the proportion of urban population is selected as the measurement index of urbanization. In consideration of the availability of data, and in combination with the selection of indicators for measuring transportation output factors in relevant literature [17, 18], this paper adopts the converted railway conversion turnover as the measurement index of railway transportation.

The converted railway conversion turnover calculation formula adopts the recommendations of the railway experts of Beijing Jiaotong University as follows:

$$tt = Pt_{rail} + Ft_{rail} \tag{2}$$

The two variables respectively represent the turnover of railway passenger transport and freight transport. The sample data are from China statistical yearbook, China statistical yearbook on the tertiary industry and China statistical yearbook on population and employment, covering the period from 2005 to 2017.

3 Empirical Analysis of Influencing Factors of Railway Transportation

Perform a stationarity test on the data to avoid biased estimation of the model. Based on the data characteristics of the short panel, the same root HT test and different root IPS and PP tests are selected. The results show that the original sequence is non-stationary, and the difference process is further tested again (Table 1). At the significance level of 1%, the per capita GDP (*dlgdp*), total railway transport turnover (*dltt*) and proportion of urban population (*dpup*) after difference are all stationary sequences, that is, there is no unit root problem in panel data regression (Table 2).

Table 1 Model variable setting

Index	Variable symbol	Variable meaning
Industrialization	lgdp	Per Capita GDP
Railway transportation	ltt	Converted total railway transportation turnover (1 passenger volume = 1 ton cargo volume)
Urbanization	pup	Proportion of urban population (%)

Table 2 Panel data stationarity test

Testing method	Variables		
	dlgdp	*dltt*	*dpup*
HT	0.4198***	0.2464***	0.4831***
IPS	−3.0988***	−6.8511***	−5.4427***
PP	160.3945***	415.3601***	353.0305***

$^{*}p < 0.1$; $^{**}p < 0.05$; $^{***}p < 0.01$

Impulse response analysis is used to identify the dynamic effects between variables. Using 500 Monte Carlo simulations to estimate the impulse response function, the result is within 95% confidence interval.

3.1 Impulse Response Analysis in Eastern Region

In the impulse response results (Fig. 2), the dotted line is the impulse function response value, the solid line is the confidence band of plus or minus two standard deviations, the horizontal axis represents the number of lag periods, and the vertical axis represents the degree of mutual impact response between variables.

From the perspective of the interaction between railway transportation and industrialization, Fig. 2 (8) shows that when a positive impact of one standard deviation is given to the growth rate of per capita GDP in the current period, the response value of total railway transportation turnover growth in the current period is 0.0423, but then declines rapidly. When the growth of total railway transport turnover was impacted, the response value of per capita GDP growth lagged behind by one period, and reached the peak of 0.0077 in the first period, and then slowly declined. This shows that the positive effect of the increase in industrialization level on the level of railway transportation is more time-sensitive, and the current force is greater than the reaction force of transportation on industrialization, but the long-term effect is insufficient, and the reaction force of the increase in railway transportation level on the level of industrialization is relatively mild. With a long period of effect, the two form a virtuous circle.

From the perspective of the impact of railway transportation on urbanization, Fig. 2 (4) shows that the response of the first phase of urbanization variable is negative due

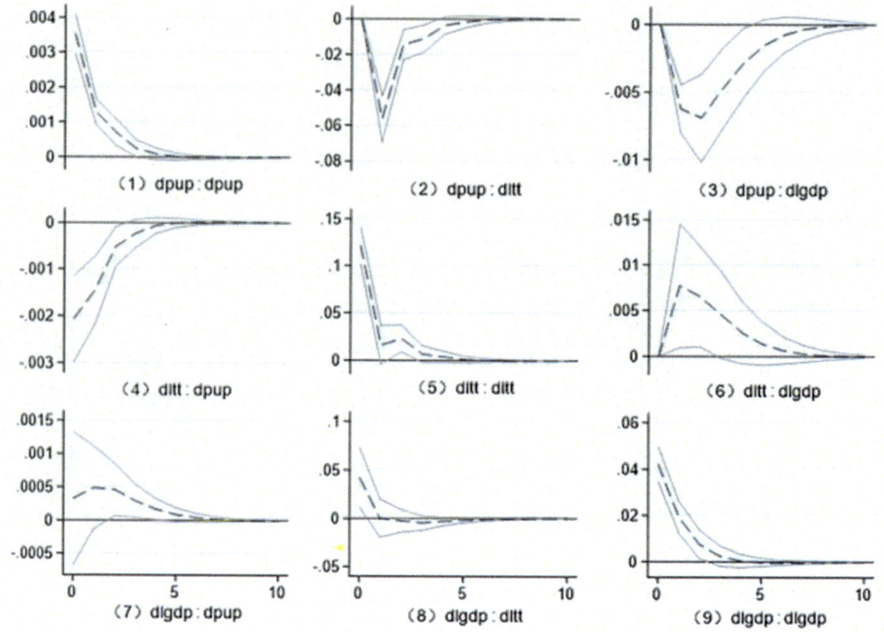

Fig. 2 Impulse response diagram of PVAR model in eastern region

to the impact of the transport variable, and then continues to increase. The response value turns from negative to positive in the sixth phase. The increase of transportation turnover, to some extent, corresponds to the industrial decentralization, that is, the flow from the big cities to the surrounding small and medium-sized cities. For example, after the large-scale investment in transportation and communication infrastructure in South Korea in the early 1980s, the industries previously concentrated in the three largest cities were dispersed in several smaller cities and inland regions [19]. At the same time, due to the improvement of transport accessibility, the service industry is concentrated in big cities, and the population growth that tends to the city due to the agglomeration of service industry gradually exceeds the population growth that moves to the periphery due to the dispersion of industry. Therefore, the corresponding value of urbanization variable changes from negative to positive. From the current situation of China's development, the permanent urban population in China shows an increasing trend from 2010 to 2014, with the increase of municipalities directly under the central government or provincial capitals being higher than that of general cities [20]. Therefore, the dispersion effect of industrial population caused by transportation is actually smaller than the population agglomeration effect caused by other factors.

From the relationship between urbanization and the other two variables, the industrialization and transportation indexes were both reduced and then increased due to the impact of the proportion of urban population, but the responses of the industrialization indexes were all negative in ten years, and the response value of the transportation indexes turned from negative to positive in the seventh period. Therefore, railway transportation indirectly affects industrialization through urbanization, which is negative. Since the urbanization rate in this article refers to the proportion of urban population to the total population, it means that a single increase in the proportion of urban population will result in counterproductive results because there is only "quantity" increase without "quality".

From the perspective of the impact of industrialization on railway transportation, as shown in Fig. 2 (7), when the per capita GDP growth is impacted in the current period, the response value of the proportion of urban population in the current period is positive, and it reaches a peak in the first period, followed by a slow convergence, indicating that industrialization has a long-term positive impact on urbanization. Urbanization has a negative effect on railway transportation, so industrialization has a complex indirect effect on railway transportation.

3.2 Impulse Response Analysis in Central Region

From the perspective of the direct interaction between railway transportation and industrialization, in Fig. 3 (6), the impact on the growth rate of total railway transportation turnover caused the response value of per capita GDP growth in the first phase to reach a peak of 0.0029, followed by a rapid decline and convergence. Figure 3 (8) shows that when the per capita GDP growth rate is impacted, the response of the transportation index in the current period is negative, the peak in the first period reaches 0.0971, and then it converges in a volatile manner. Judging from the magnitude of the mutual impact response of the two, the effect of industrialization on railway transportation is greater than the reaction force of transportation on industrialization. The interaction effect of the two is significant, but it decays fast with time.

From the perspective of the indirect effect of railway transportation on industrialization, in Fig. 3 (4), when the transportation index is impacted, the response value of urban population proportion first decreases and then increases. In the second phase, the proportion of urban population changes from negative to positive and reached a peak, then rapidly declines and tends to converge. In Fig. 3 (3), when the urbanization index is the impact variable, the response value of per capita GDP growth decreases first and then increases, and in the second phase, it turns from negative to positive and reaches the peak value of 0.004. Results show that railway transportation indirectly promotes the development of industrialization by driving the level of urbanization, but this indirect effect lags behind at least three periods.

From the perspective of the indirect effect of industrialization on transportation, as shown in Fig. 3 (7), when per capita GDP growth is impacted, the proportion of urban population to peaks at 0.0017 in response in the first period. From the perspective of the impact of urbanization on transportation, Fig. 3 (2) shows that

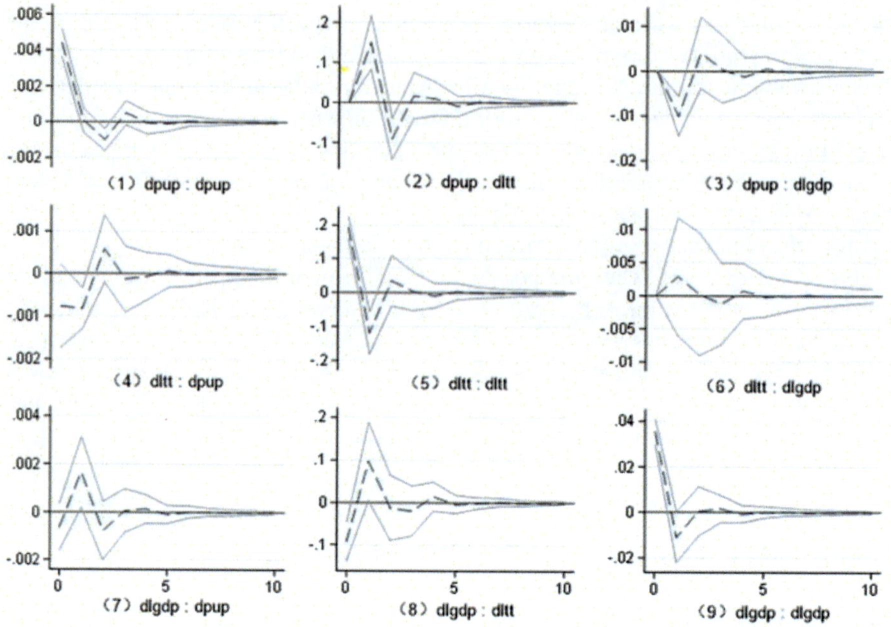

Fig. 3 Impulse response diagram of PVAR model in central region

when the urbanization index is impacted, the response value of the transportation index reaches the peak of 0.1503 in the first period. Therefore, urbanization has a significant effect on railway transportation, and the two promote each other and spiral upward. Industrialization not only directly promotes the development of railway transportation, but also indirectly promotes the development of railway transportation by promoting the progress of urbanization.

The provinces in the central region are located in the interior of China, serving as a bridge between east and west and connecting north and south. Since the implementation of the strategy for the rise of the central region, the central region has shown a strong momentum of economic and social development, with strong growth in transportation demand and historic changes in the construction of transportation infrastructure, which has exerted a far-reaching impact on the social economy and population agglomeration.

3.3 *Impulse Response Analysis in Western Region*

From the perspective of the interaction between railway transportation and industrialization, in Fig. 4 (3), the impact of transportation index causes the growth of per capita GDP to decrease first and then increase. Figure 4 (7) shows that when per capita GDP growth is impacted, the growth rate of total railway transport turnover

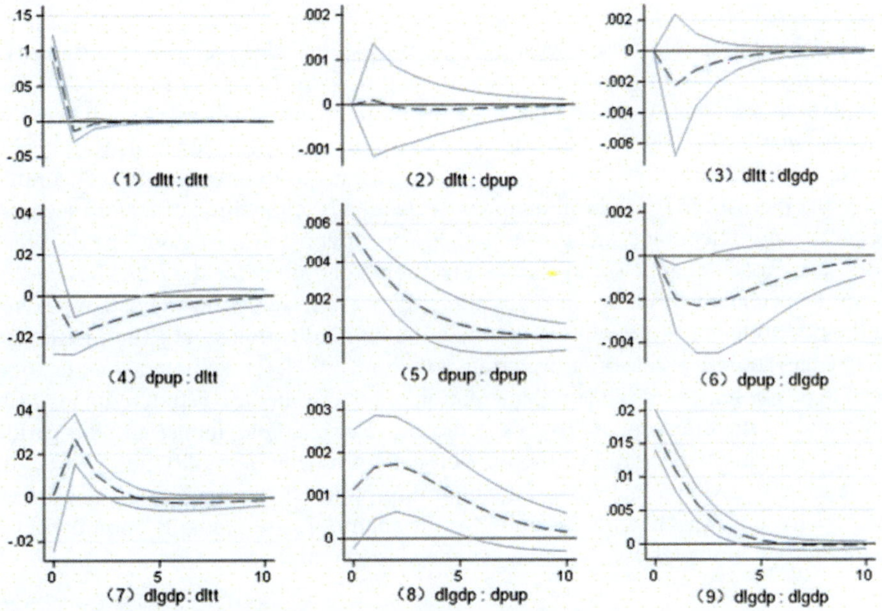

Fig. 4 Impulse response diagram of PVAR model in western region

reaches a peak of 0.0270 in the first phase, and then slowly declines. Therefore, the effect of railway transportation on industrialization is negative, but the negative effect gradually decreases. On the contrary, industrialization has a significant positive effect on railway transportation in the short term, but the effect is not significant in the long run.

From the perspective of the indirect impact of railway transportation on industrialization, Fig. 4 (2) shows that when the transportation index is impacted, the proportion of urban population responds insignificantly. Figure 4 (6) shows that when the proportion of urban population is impacted, the per capita GDP growth rate decreases first and then increases, but the ten-period response values are all negative. Therefore, the indirect effect of railway transportation on industrialization is not significant.

From the perspective of the indirect effect of industrialization on railway transportation, Fig. 4 (8) shows that when the growth of per capita GDP is impacted, the proportion of urban population in the second phase reaches the peak, and the response values in the first ten phases are all positive, indicating that the improvement of industrialization level can significantly promote the development of urbanization. Figure 4 (4) shows that when the proportion of urban population is impacted, the railway transportation index declines in the first two periods and then continues to increase, but the response values are negative.

Analysis above shows that, a single increment of railway transportation or urbanization does not contribute to the improvement of the degree of industrialization, but plays an inhibiting role. for the western region. Western China includes 12 provinces, accounting for 71% of the country's land area; it owns 29% of the country's total population. The western region has a vast area, relatively small population density and backward economy. Due to the terrain factors, the transportation is also relatively blocked. It is an underdeveloped area in China that needs to be developed. Theoretically, the increase in transportation index corresponds to the input of factors such as network, equipment, and manpower, which is conducive to the smooth flow of the "Belt and Road" channels and creates favorable conditions for the free circulation of various element resources in cities along the western region, which is conducive to the development of population urbanization and industrialization. But in fact, to realize the theoretical role of railway transportation, policies, talents and other factors need to be implemented. The implementation of policies in the western region is still restricted by the dilemma of "castles in the air". For example, industrial pilot parks set up in the high-speed rail economic belt cannot operate normally due to lack of related fiscal, taxation and financial supporting policies; transportation trunk lines have been established but the "last mile" traffic conditions are generally poor; barriers to the introduction and use of talents still need resolving.

4 Conclusions

In this paper, by using the PVAR model, the railway conversion turnover is adopted as the index of railway transportation, and the per capita GDP and the proportion of urban population are used as the indicators of industrialization and urbanization. And quantitatively measures the impact of railway transportation on industrialization and urbanization and the reaction force of the latter two on the former in eastern, central and western region Results show that:

- From the perspective of the dynamic relationship between railway transportation and industrialization, the impact of railway transportation on industrialization in eastern and central regions is milder than that of industrialization on transportation, the impact of railway transportation on industrialization in eastern regions is longer, while that in the central region declines rapidly with time. The effect of railway transportation on industrialization in western China is negative, but the negative effect gradually decreases. On the contrary, industrialization has a significant positive effect on railway transportation in the short term, but not in the long run.
- In terms of the dynamic relationship between railway transportation and urbanization, the development of railway transportation in the eastern region has promoted the transfer of urban population to the periphery in a short period of time, while a single increase in the level of urbanization has a negative impact on the railway transportation. The level of railway transportation in the central region and the level of urbanization are mutually reinforcing and spiral upward. The effect of railway transportation in the western region on urbanization is not significant,

while the latter has a negative effect on the former, but the negative effect gradually decreases.

- Railway transportation and industrialization influence each other indirectly through urbanization, but the characteristics of different regions are different. In the eastern region, the indirect effect of railway transportation on industrialization is negative, but the negative effect is gradually weakened, while the indirect effect of industrialization on railway transportation is complicated. In the middle part of China, railway transportation has a significant indirect promoting effect on industrialization, but with a time lag, and the indirect effect of industrialization on railway transportation is also significant. The indirect influence of railway transportation and industrialization on each other is not significant in western China.

The empirical results of this paper indicate that the relationship between rail transportation's impact on industrialization and urbanization and the latter two's reverse impact on the former vary from region to region. The common feature of different regions is that industrialization can significantly promote railway transportation and urbanization at various provincial levels. Therefore, continuing to focus on economic development and industrial optimization and upgrading is an important driving force for the sustainable development of China's railway transportation and urbanization. Different regions should put different emphasis on promoting urbanization and industrialization by promoting railway transportation development.

- In the eastern region where industrial development is relatively complete, the marginal impact of the increase in total railway transportation on urbanization and industrialization has weakened. Therefore, while continuing to promote industrialization, the region should consider expanding the distribution of emerging industries. The improvement of transportation should focus on improving the quality of transportation, strengthening the connection of hubs, multimodal transportation, building a more comprehensive transportation system. Promoting the coordinated development of new-type urbanization, transportation and industrialization, and strengthening the mutual push–pull cycle are the development strategies for the eastern region.
- In the central region where industrialization level is relatively less developed, railway transportation, urbanization and industrialization show a dynamic relationship of "strong demand and support" and significant mutual push and pull. Therefore, the government can vigorously promote "quantity" while pursuing the improvement of "quality".
- In the western region with less developed industrialization level, it is necessary to rationally arrange transportation facilities, expand the total amount of transportation, adjust the railway transportation supply to balance or advance the development of other economic sectors, implement the supporting development of various social and economic factors, Unleash the potential for regional development through multiple measures.

References

1. Rong, C. (1993). *Transportation.* Beijing: China Social Sciences Press.
2. Keynes, J. M. (2018). *The general theory of employment interest and money.* Cham: Springer International Publishing; London: Palgrave Macmillan.
3. Zhang, X., Wang, X., & Wan, G. (2018). A unified framework of road infrastructure's growth effect. *Economic Research Journal, 53*(01), 50–64.
4. Zhang, J. (2018). Research on the impact of transportation infrastructure construction on industrial structure transformation. *Journal of Yunnan University of Finance and Economics, 34*(11), 35–46.
5. Wang, Y., & Ni, P. (2016). Economic growth spillover and spatial optimization of high-speed railway. *China Industrial Economics, 02,* 21–36.
6. Kong, L., Bai, Y., & Jiang, T. (2019). An effect of lanzhou-urumqi high-speed railway on urban tourism economy along the line. *Railway Transport and Economy, 41*(03), 1–5.
7. Zhang, C., & Peng, H. (2012). Study on relationship between regional transportation and the development of social economy based on transportation theory. *Journal of Hunan University (Social Sciences),26*(02), 59–64.
8. Krugman, P. (1991). Increasing returns and economic geography. *Journal of Political Economy, 99*(3), 483–499.
9. Krugman, P. (1980). Scale economies, product differentiation, and the pattern of trade. *The American Economic Review, 70*(5), 950–959.
10. Yao, C., Song, D., & Fan, X. (2019). Research on the mechanism of railway speed-up on economic growth: Based on the perspective of urbanization, industrial agglomeration and market potential. *Inquiry into Economic Issues, 02,* 29–38.
11. Bin, L. (2016). *The evaluation for the mutual influence between the urbanization and development of rail transit.* China: Zhejiang University.
12. Wang, X., & Rong, C. (2014). Research on the interactive development of urbanization and transportation. *Inquiry into Economic Issues, 01,* 52–57.
13. Rong, C. (2011). Re-understanding of comprehensive transportation. *Comprehensive Transportation, 12,* 4–9.
14. Atack, J., Bateman, F., & Haines, M. (2010). Did railroads induce or follow economic growth? Urbanization and population growth in the American Midwest, 1850–1860. *Social Science History, 34*(2), 171–197.
15. Chen, J., & Huang, Q. (2006). Industrial development, the changing national situation and the strategy of economic modernization: An analysis of the national situation for China's growth into a large industrial state. *Social Sciences in China, 02,* 15–26.
16. Kesha, G. (2002). An economic analysis of the relationship between industrialization and urbanization. *Social Sciences in China, 02,* 44–45.
17. Li, T., Cao, X., & Yang, W. (2015). Comprehensive measurement and evolution of regional integrated transport efficiency in China. *Scientia Geographica Sinica, 35*(02), 168–175.
18. Wu, Q., Song, J., & Ju, P. (2015). Spatial distribution of provincial integrated transport efficiency in China. *Economic Geography, 35*(12), 43–49.
19. World Bank State Council Development Research Joint Task Force. (2013). *China in 2030: Building a modern, harmonious and creative society.* Beijing: China Finance and Economics Press.
20. Deng, T., Yan, Y., & Wang, D. (2019). Impact of high-speed rail on urban population size in China. *Finance and Trade Research, 30*(11), 1–13.

A Simulation Research on Innovation-Driving Evolution of Internet Business Organizations

Hui Wan, Jens Mathis Rieckmann, and Guangming Hou

Abstract Taking the theories of organizational innovation and value as the starting point, this paper aims to construct the conceptual framework of an innovation-driven evolution model of Internet commerce organizations combining the characteristics manifest in evolution process of Internet commerce organizations in China. From the perspective of organization system, this paper analyses the mechanism of Internet commerce organizations' evolution, which is driven by innovation, and selects "core ability of competition" and "resource integration ability" as two order parameters. The research analyses the equilibrium point and stability condition of the evolution model of Internet commerce organizations and carries out computer simulation on different evolutionary stages. Additionally the study shows that when the feedback coefficient of the organization system is negative and the increasing rate of core competence is smaller than the elimination rate, so the organization system needs an external force to enter the senior evolving phase transition; while when it is greater, the organizational system enters the positive feedback state and jumps to the advanced order stage. When the self-feedback coefficient of the organizational system is negative and the increase rate of core competence of the organization is greater than the elimination rate, the organization is in a self-stable state at the current programming.

Keywords Innovation driver · Organizational evolvement · Internet commerce organization · Organizational systems

H. Wan · G. Hou
School of Management and Economics, Beijing Institute of Technology, Beijing, China
e-mail: 3120150678@bit.edu.cn

J. M. Rieckmann (✉)
Fraunhofer Institute for Production Systems and Design Technology (IPK), Berlin, Germany
e-mail: mathis.rieckmann@ipk.fraunhofer.de

G. Hou
Department of Management, Beijing Film Academy, Beijing, China
e-mail: guangminghou@bfa.edu.cn

M. Li et al. (eds.), *IEIS 2020*, https://doi.org/10.1007/978-981-33-4363-4_24

1 Introduction

The rapid development of Internet business organizations attracts many scholars. Researchers who focus on theory of organizational learning believe that this development of Internet business organizations is due to their more powerful ability to learn and absorb compared to traditional industry organizations. Some scholars also believe that innovation for the Internet business organization brings a steady stream of development momentum. Innovation includes organizational innovation, such as platform organization [1], small-world network [2, 3], the front-midrange-back construction [4], and the rapid development of information technology [5, 6] on which Internet business organizations rely.

However, in practice, not all Internet business enterprises can grow rapidly or rather survive in the market. Indeed, even in the Chinese market, where market regulations are not as stringent as in other mature markets, the total number of Internet deaths continued to rise in the 11 years from 2005 to 2016, reaching 384 in 2015. About 17% of the companies that closed down in 2017 belong to Internet business organizations [7]. Some nascent entrepreneurs which emphasize on "customer first" and "continuous innovation" are first defeated in the fierce market competition, and partially by the opponent's acquisition. Such kind of anti-management-logical case is constantly repeated, so that it is no longer possible to use common theories like "organizational absorptive capacity", "organizational innovation" or "technological innovation" to explain this phenomenon effectively.

The purpose of this article is to propose a complementary theory that can explain this phenomenon by introducing the concept of "timing" while organizing learning and organizational innovation did play a positive role in organizational performance, which is not to negate the great positive impact of organizational learning and innovation on organizational development. Instead, it analyses when the development of homogeneous Internet business organizations is at the time for organizational learning and innovation, and when it is not suitable to invest a lot of capital in learning and innovation. This discussion is somewhat related to, but not identical to, the formation of organizational duality, which tends to provide advice and guidance on the timing of the development of dual capabilities. It is important to note that the theory and phenomenon discussed in this article have a clear line, that is, when the organization is more suitable for innovation without additional capital injection, when it is better suited to reduce costs and improve efficiency—in short, work hard tenaciously.

Obviously, the cultivation of dual capacity is not unique to Internet business organizations, and traditional industrial organizations will also face this problem. However, unlike traditional industries, Internet business organizations evolve rapidly, and transparency in organizational behavior is not available to other organizations, or is difficult for scholars to obtain. Therefore, the choice of Internet business organization appears to be a short life cycle, it is possible to further refine the study of granular materials, with the organization absorptive capacity and refers to the timing of innovation will emerge. Using the "tissue system of the microscope" in order to find the best time to cultivate ambidexterity.

To accomplish the aforementioned goals this article contains four chapters. It starts with an introduction and a literature review as well as the research logic as a fundament basis, followed by a framework for the dynamic analysis of the organizational evolution. The third part includes the dynamic model and simulation of organizational system. Followed by a conclusions and prospects in the last part.

2 Literature Review and Research Logic

2.1 The Introduction of Time Dimension in Innovation and Organizational Ambidexterity

Since Schumpeter [8] in 1934 proposed in "the theory of economic development", innovation is the reintegration of factors of production and the inclusion of innovation in five forms: the introduction of new products, the adoption of new processes, the opening up of new markets, the use of new raw materials, and the adoption of new organizational methods. The positive effect of innovation on organizational performance has been confirmed by numerous scholars by theoretical derivation or empirical research. Porter [9] divides knowledge, technology, institutions and management as the main factors on which organizational development depends for the national development phase. Innovation is the core of organizational development [10]. This view has become almost the consensus reached by the majority of management scholars. The proposal of organizational ambidexterity further emphasizes the speed and rhythm of organizational innovation under the general direction of organizational innovation, and discusses the contradiction and balance between organization innovation and improving efficiency.

Scholars' research on the organizational ambidexterity mainly focuses on the measurement of the former variables [11, 12], formation mechanism [13, 14] and result variables that affect the formation of organizational ambidexterity [15]. Combining with the existing research conclusions, scholars believe that the former variables that affect the organizational ambidexterity have organizational structure, organization context and executive integration [16], and the research of executive integration is divided into the aspects of executive characteristics and executive social network. The research on the outcome variables of organizational ambidexterity mainly focuses on the relationship between organizational ambidexterity and organizational performance or explores the moderating variables between them. A large number of empirical research has laid an important foundation for exploring the organizational ambidexterity, but from the point of view of theoretical construction, excessive attention to the influence of individual elements or multiple elements in the organization on the ambidexterity is not included. Organizational performance is measured by the length of its lifetime, therefore in success and failure cases of a large number of Internet business organizations, the use of a more sporty and meticulous systematic perspective to explore this problem is suitable. Therefore, this research work introduces the timing of the development regarding the ambidexterity, which is not only the exploration of the theory of organizational ambidexterity in the time dimension, but also the attempt to apply the system science theory to the organization management.

2.2 Systematization of the Organization

Since the 1970s, the system science has made great strides in development, the research category of organizational theory has been expanded gradually, regarding the organization as a system, understanding and analyzing the constituent elements, functions, structures and levels of the organization [17]. The general process of organizational value creation can be abstracted, and the value creation of the organization is liberated from the shackles of the industry and the form of ownership. With the increasing uncertainty of the world organization environment, the multiple organizational phenomenon is becoming more and more complex. Whether it is the flexible trend of organizational structure or the information requirement of organizational management, the researchers of the organization theory have a systematic understanding of the organizational evolution.

Whether it is the innovation ability or the ambidexterity, its essence is the process of the organization system to realize the value of creating this demand, and it is also the result of the cooperative connection of each functional subsystem of the organization [18]. In the initial period of value creation in the organization system, the organization resources are evaluated on the basis of the organization target, and then the required resources are identified. Subsequently the boundary subsystem is responsible for the organization of the material, information and energy exchange between the system and the external environment to prepare the resources for value creation. Finally, the operating subsystem is carried out. In the whole process of the organization system, the control subsystem is responsible for coordination and feedback, constantly helping the system to adjust the environment and adjust itself, so that the organization is close to the organizational target (from the relative initial state). The organizational system relies on organizational capabilities to organize the allocation of resources [19], and ultimately outputs—the value output that condenses the value of organizational labor, as shown in Fig. 1.

Organization resources consist of—which is different in proportion and form—information, material and energy. The components of the value output created by the organizational system are still the three flow resources, but these elements emerge as new values at the overall level of the system through the organization's ability to operate [20]. Therefore, the value creation of organizational system can be presented as:

$$V = G(i, m, e) + F(\alpha, \beta)G(i, m, e) \tag{1}$$

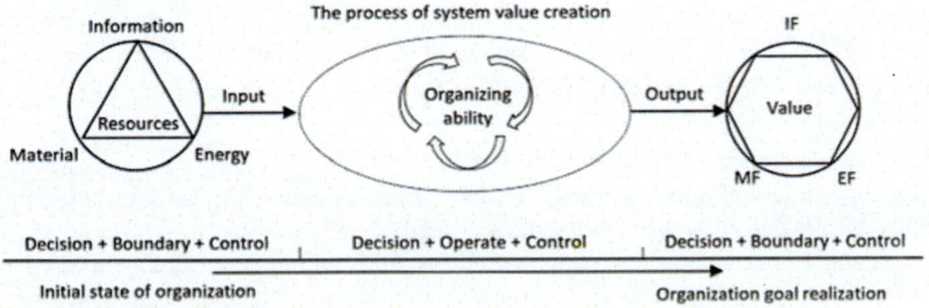

Fig. 1 Value-creation process of organization system

The value of the resource itself (G), which is composed of information flow (i), material flow (m) and energy flow (e), F(α, β) is the ability level of the person in the organization system, α is the ability of human cognition and β is the ability of human being. Since human beings are the foundation of organizational system behavior, organizational capability is determined by the ability of people in the organization. With the improvement of the level of human cognition and the growth of skills, the proportion of resources in the value created by the organization is becoming smaller and smaller, and the contribution rate of organizational ability to value creation has become more and more. When the organizational capability is difficult to create the value needed to achieve the organizational goals, the control subsystem constantly adjusts it in the feedback process. The decision subsystem will carry on the new organizational resource evaluation and the organization target orientation, the boundary subsystem tends to produce more exchange with the environment to obtain more resources, and the operating subsystem will improve the core competence and even carry on the innovative attempt. In this process, the organizational boundaries and organizational processes may change [21], and different organizational systems are chosen based on their existing capabilities.

This study defines the Internet business organization as an organization that participates in Internet business activities. Organizational scale, organizational capabilities and other organizational characteristics do not affect their participation in Internet business activities. The reason for this study is to select the organization evolution related to Internet business as the research object. Compared with the traditional organization, its evolution process is shorter [1], changes more rapidly, the industry involves a wider range of technologies, and the evolutionary impact is deeper [22, 23]. As Internet business organizations rely on the network technology, it greatly expands the organization pattern and organizational structure, making its evolution a microcosm of organizational evolution. Therefore, this study selects the Internet business organization as the point of the dynamic analysis of organizational evolution.

2.3 *Self-Organization and Self-Feedback Characteristics of Internet Business Organization*

The Internet business organization system is an open system [24], which constantly exchanges material flow, information flow and energy flow with the outside world. It has significant self-organization and self-feedback characteristics.

First, the Internet business organization system is an important link between the information technology and the economic system. The typical open system includes the exchange of material, information and energy between the stakeholders of the information technology resources.

Secondly, there are complex non-linear relationships among the elements in the Internet business organization system. In the first instance the information system is organized to collect, transmit, store, excavate and feedback the whole process of information, and it is highly fused with the physical value environment. It is closely integrated with the virtual value environment in the transmission, storage and depth mining of information. This fusion promotes the complexity of the elements of the Internet business organization including a wide range of connections.

Thirdly, there is competition and synergy among Internet business organizations. The core business of organizational unit decision-making, control and boundary subsystem cooperation is to gather the resources information of the external environment as a whole, and promote the sharing of the physical and technical resources by information. The aggregation of information and technology needs to improve the negative entropy of the system of the virtual value environment subsystem. The negative entropy makes the partial subsystem dissipative relaxing. Under the competition of the subsystems, the stochastic fluctuation is formed, and the system develops along the reverse direction of entropy increase and forms the synergy effect, so the order parameter is formed and along with the order parameter finally, a stabled and balanced open system is developed.

Finally, the Internet business organization system is a first-order feedback control system. From the previous analysis, it is obvious that the intersecting depth of the Internet commercial organization and the outside world is deepening, and each organization unit obtains a large amount of information in the process of value delivery, and forms the first order feedback mechanism of the system. Under the synergy system, the system relies on positive and negative feedback mechanism to make the fluctuation constantly enlarge or shrink, to achieve a new stable equilibrium state.

3 Framework for Dynamic Analysis of Organizational Evolution

3.1 Fluctuation Power of Internet Business Organization System

The information technology in the Internet business organization infiltrates into the business domain and collides with the individual business units to form an irregular movement. This process is similar to the Brown movement. Following the Lagrange multiplier method, there is sticky resistance and fluctuation in the infiltration process of Internet business information. The indeterminate impact force.

The critical nature of the evolution of the system to the advanced stage can be presented by the order parameters of each subsystem. They form a set of macro-variables $q = \{q_1, i = 1, 2, \ldots, n\}$, satisfy the equation:

$$\frac{d(q_i)}{dt} = -\gamma_i q_i + r_i(t) \tag{2}$$

r_i represents the viscous resistance of the system; $r_i(t)$ is any random force that is not included in the degree of freedom of $\{q_i\}$.

3.2 Internet Business Organization Synergistic Motivation

The core motivation of the Internet business organization is to maximize the value creation of organizational resources. The new generation of information technology provides a wide space for the interaction and integration of the transformation of organizational capacity and the realization of the value of resources. The competitiveness of the system is mainly embodied in the transformation of organizational ability into the realization of the value of resources and the formation of capacity chain and resources. The organizational value chain formed by chain integration, the bidirectional and alternate transformation of organizational capabilities and resource integration, promote the co-evolution of organizations. It is assumed that the system is only a system composed of an order parameter and a stable module system, which represents the macro variable of the order parameter and the macro-variable of the representative subsystem, which satisfies the equation, respectively.

$$\frac{d(q_1)}{dt} = -\gamma_1 q_1 - a q_1 q_2 + r_1(t) \tag{3}$$

$$\frac{d(q_2)}{dt} = -\gamma_2 q_2 + b {q_1}^2 + r_2(t) \tag{4}$$

As $|y_1| \ll |y_2|$, q_1 is the core ability to drive the organization, q_2 is the integration of organizational resources, γ_1 is the elimination rate of ability, γ_2 is the resistance of q_2, a is the saturation effect of q_1, and b is the synergistic effect of q_1 on q_2.

4 Dynamic Model and Simulation of Organizational System

4.1 *Construction of Internet Business Organization Dynamic Model*

The equilibrium equation of the evolutionary dynamic model of the Internet business organization mainly describes the feedback between the macro-variables and the micro-variables within the single Internet business organization system, and the feedback effect of the system output variables on the input variables, and the system feedback principle is shown in Fig. 2.

The Lagrange equation is established to determine the linear instability point of the evolution of the Internet business organization, while the evolution direction is established by establishing the Hawking model, and the system state equation is established to describe the equilibrium state. Based on three model equations, the evolution model of Internet business organization is obtained.

$$\begin{cases} \frac{d(q_1)}{dt} = (\alpha - \gamma_1)q_1 - aq_1q_2 - \beta {q_1}^2 + r(t) \\ \frac{d(q_2)}{dt} = -\gamma_2 q_2 + b{q_1}^2 \\ \frac{d(s)}{dt} = \eta_1 q_1 + \eta_2 q_2 + \eta_3 q_3 s + \eta_4 q_1 q_2 \end{cases} \tag{5}$$

where q_1 presents the organization core competence, q_2 presents the resource integration ability, α presents the core ability growth rate of the Internet business organization system, γ_1 presents the elimination rate of the core competence, γ_2 presents

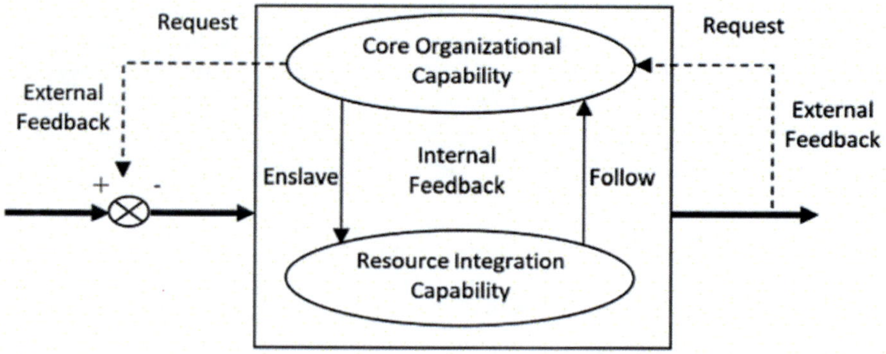

Fig. 2 Dynamic analysis framework of Internet business organization system

the resistance of the resource integration, β presents the influence degree of the core ability on the work of the resource integration ability, a presents the saturation and influence degree of q_1, and b for q_1 pairs q_2 synergy. With the influence degree, η_1 is the influence coefficient of q_1 pairs of organizational system, η_2 is the influence degree of q_2 on the organizational system, η_3 is the feedback coefficient of the system itself, η_4 is the degree of influence on the organizational system between q_1 and q_2, and $r(t)$ is the random influence coefficient. Among them, the first two equations describe the stable state and the approximate adiabatic state of the order parameter servitude when it reaches the advanced order. The last equation describes the comprehensive effect of the stability of the Internet commercial organization system. $\frac{d(q_2)}{dt} = 0$ as relaxation process driven by the core competencies of the organization is much slower than resource integration. The macro-characteristics of the system in the model are expressed by the order parameter, under non damped conditions, $q_1 \neq 0$, stable solution generation. The stable mold system follows the order parameter in an adiabatic way, and the self-organizing state of the subsystems is formed inside the system, which is the regular external behavior of the system. If the organization's own subsystem is in a stable and orderly state, the core capacity driving force and the resource integration ability of the dominant order parameter, which have important influence in the process of evolution, are also unchanged. The system achieves stable equilibrium loading, and the equation satisfies the condition:

$$\frac{d(q_1)}{dt} = 0, \frac{d(q_2)}{dt} = 0, \frac{d(s)}{dt} = 0$$

The model characteristic matrix of the equilibrium point $(0, 0, 0)$ is $\begin{vmatrix} \eta_3 & \eta_1 + \eta_4 q_2 & \eta_2 + \eta_4 q_1 \\ 0 & (\alpha - \gamma_1) - 2\beta q_1 - a q_2 & -a q_1 \\ 0 & 2b q_1 & -\gamma_2 \end{vmatrix}$, and the characteristic equation of the equilibrium point is $\begin{vmatrix} \lambda - \eta_3 & \eta_1 & \eta_2 \\ 0 & \lambda - (\alpha - \gamma_1) & 0 \\ 0 & 0 & \lambda + \gamma_2 \end{vmatrix} = 0$, eigenvalue of the equation is:

$$\lambda_1 = \eta_3, \lambda_2 = \alpha - \gamma_1, \lambda_3 = -\gamma_2$$

According to the differential equation theorem, if all eigenvalues are negative, the equilibrium state of the equation is stable. Since γ_2 is always greater than zero, λ_3 is always negative, that is, the enhancement of resource integration capability will not cause the fluctuation of the system. There are four combinations of values for negligible parameters λ_1, λ_2:

$$\begin{cases} \lambda_1 > 0 \\ \lambda_2 > 0 \end{cases}, \begin{cases} \lambda_1 < 0 \\ \lambda_2 > 0 \end{cases}, \begin{cases} \lambda_1 > 0 \\ \lambda_2 < 0 \end{cases}, \begin{cases} \lambda_1 < 0 \\ \lambda_2 < 0 \end{cases}$$

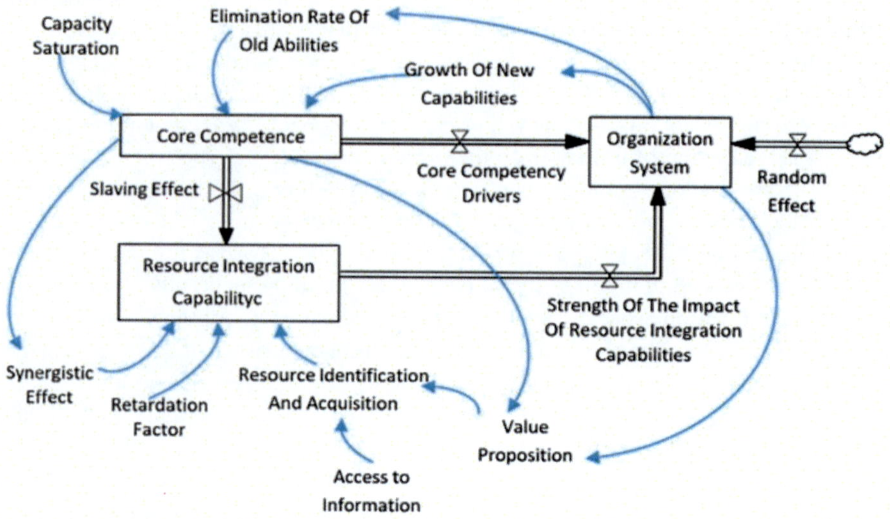

Fig. 3 Evolution flow chart of Internet commerce organization system

4.2 *Model Simulation*

Based on the analysis of MATLAB software and the solution of the Internet business organization system model, Fig. 3 shows that the stability of the system equilibrium state is determined by α, γ_1 and η_3 values. When $\eta_3 = 0$ or $\alpha = \gamma_1$, the organizational system breaks the equilibrium state and enters the instability point, and performs the capability transition under the synergy of the organization's source integration capability and core competence. The transition of organizational ability is influenced by random factors, then $\delta \neq 0$. Assumed parameter assignment as $\eta_1 = 0.5$, $\eta_2 = 0.4$, $\eta_4 = 0.7$, $\beta = 0.6$, $a = 0.8$, $\gamma_2 = 0.3$, $b = 1$, $\delta = 0.2$. Simulation results depict the changing trend of q_1, q_2 and systems.

5 Conclusions and Prospects

5.1 *Conclusions*

Looking at the four combinations of values for negligible parameters $\boldsymbol{\lambda}_1$ and $\boldsymbol{\lambda}_2$ is creates different settings and resulting actions.

1. When $\begin{cases} \lambda_1 < 0 \\ \lambda_2 < 0 \end{cases}$, then $\begin{cases} \eta_3 < 0 \\ \alpha < \gamma_1 \end{cases}$, the self-feedback coefficient of organizational system is negative and the increase rate of organizational core ability is less than elimination rate. The feedback driving force of organizational system is not

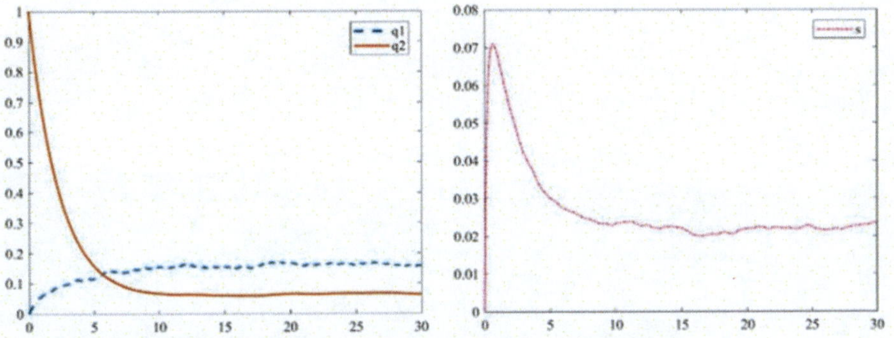

Fig. 4 Trend of q_1, q_2 and s where $\eta_3 < 0, \alpha < \gamma_1, \delta = 0.2$

enough to influence the organization taking the initiative to innovation (Fig. 4). This is because the resources integration ability of organizational system fails to treat the effect of core ability, and the core competence has not been able to improve the process of resource integration and the Internet business organizational system stability. The system needs external stimulation in order to improve the innovation ability, or after a long time to accumulate into the buckling critical point to the high order phase transition. At that time it is a good moment for the organization for incremental innovation.

2. When $\begin{cases} \lambda_1 > 0 \\ \lambda_2 > 0 \end{cases}$, then $\begin{cases} \eta_3 > 0 \\ \alpha > \gamma_1 \end{cases}$, the self-feedback coefficient of the organizational system is positive and the new rate of new organizational core ability is greater than the elimination rate. The organizational system enters a positive positive and the new rate of new organizational core ability is greater than the elimination rate. The organizational system enters a positive feedback state (Fig. 5). The input and output of the organizational system have a strong influence on the evolution of the platform. At the same time, the ability of the organization

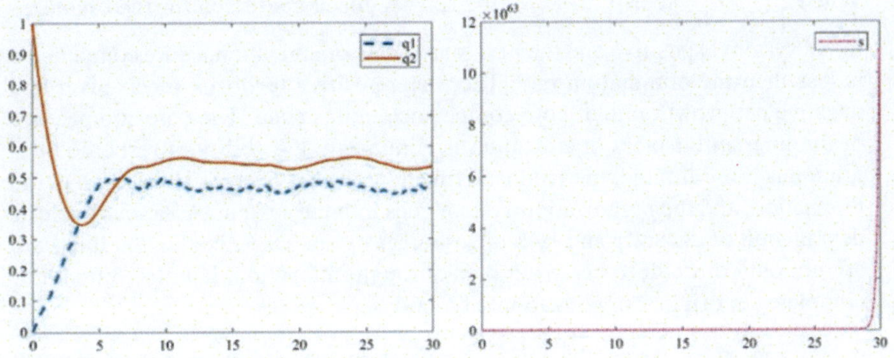

Fig. 5 Trend of q_1, q_2 and s where $\eta_3 > 0, \alpha > \gamma_1, \delta = 0.2$

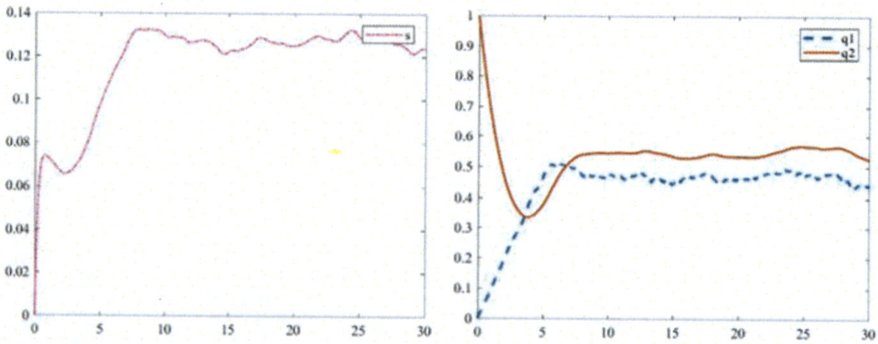

Fig. 6 Trend of q_1, q_2 and s where $\eta_3 < 0, \alpha > \gamma_1, \delta = 0.2$

system increases rapidly, similar to the core competence of the Internet business organization. The integration of resources and resources promotes each other and the organizational system breaks the critical point of instability and rapidly jumps to the advanced order stage, this is a good time for organizations to drive innovations in a sudden way.

3. When $\begin{cases} \lambda_1 < 0 \\ \lambda_2 > 0 \end{cases}$, then $\begin{cases} \eta_3 < 0 \\ \alpha > \gamma_1 \end{cases}$, the self-feedback coefficient of the organizational system is negative and the increase rate of organizational core ability is greater than the elimination rate. The organizational system is in the negative feedback state (Fig. 6). The promotion of organizational core competence is not enough to promote the organization taking innovation in short time although, but the allocation of material flow, energy flow and information flow in the organization resources is continuously improved, and the Internet business organization is maintained in the organization. An advanced state of order in the current period is present. Organizations in this state are better suited to improve the effectiveness of their original functions than to update them.
4. When $\begin{cases} \lambda_1 > 0 \\ \lambda_2 < 0 \end{cases}$, then $\begin{cases} \eta_3 > 0 \\ \alpha > \gamma_1 \end{cases}$, the self-feedback coefficient of the organizational system is positive and the new rate of new organizational core competence is less than the elimination rate. There is a positive feedback mechanism and negative net growth rate of core competence in this state. The core competence of the Internet business organization is eliminated, the ability of resource integration is more limited, the feedback function of the system is stronger than the fluctuation, and the organization system goes into the system. A vicious cycle of degradation of capacity and lack of resources exists. Organizations in this state are not only difficult to carry out breakthrough innovation, but also difficult to complete the original organizational functions.

In addition, the Internet business organization system is in a critical state of equilibrium stability change. Under the organization system each subsystem contract function enhancement, each organization unit mutation type innovation good opportunity.

5.2 Discussion

Colquitt and Zapata-Phelan believe that the contribution of management theory can be examined from two aspects: whether new theories are established and whether existing theories are tested [25]. Based on this, the theoretical contribution and inspiration of this study are mainly in two aspects: the introduction of the time dimension of the study of organizational ambidexterity and organizational innovation. The time dimension is a relative concept for organizations, and not all enterprise organizations have a developed manual indicating the point of time.

As a result, this research paper will—based on the organization as an open complex system, at the stage of system development—explore the core competence of the organization, the new capacity growth, old elimination ability, ability to saturation level, the system synergy, retardation factor, information acquisition, organization identification capability of resource, and obtain the ability and organizational value proposition flow graph relationships. It is concluded that the Internet business organization system of four kinds of organization, describe the relationship between the condition of organizational factors, summarize its characteristics of phase change, and ultimately identify its phases that are suitable for the ambidexterity training mode. That is to provide decision-making support for the timing and mode of organizational innovation. In previous literature, research on organizational ambidexterity have more necessities from the industry, and seize the economic development pattern and the organizational innovation of the industrial structure transformation legitimacy for tissue culture ability of dual macro-direction. However, it is not the timing of organizational innovation and organization mode analysis, and implementation of the strategy.

By using the theory of organization system based on Internet business organization evolution process, in this study the practical way to determine the organization innovation is not only to provide decision support. Thereby helping to know that the cultivation of students ability of innovation in the practice of behavior, thus clear organization stage determination of organizational objectives, help enterprise to try innovation target. Additionally, by simulating the Internet business organization and for example phase organization ability net growth rate, the degree of organization since the feedback, the organization evolution of random influence, or the existence of the specific elements, it can support the organization innovation. Exploring organizational innovation and ambidexterity training of its internal mechanism, it can evade risk and improve efficiency in the organization in the process of innovation and the cultivation of the ambidexterity.

5.3 Prospects

This study on systematic perspective movement, the Internet business organization innovation ability and the ambidexterity in organizing system between the fluctuation and self-organization framework, discusses the Internet business organization as an open complex system. Its core competences are updating the degree of resource conformity ability, contrasting and grouping synergy, using the role and the organization of feedback effects between the power of pulling eventually reach of four kind of organization, and showing on the four kinds of condition of organizational innovation and ambidexterity that training has carried on the theoretical guidance.

Although this paper introduces the timing dimension into organizational research, it is undeniable that the construct measurement mentioned in this paper is still difficult, especially for managers with less management experience. Therefore, young CEOs lacking organizational management experience may need to be more sensitive to observe the pace and rhythm of enterprise organization, so as to seeing the organizational law presented in this paper. Organizational leaders with rich experience in organizational management will have a deeper feeling than inexperienced managers or even scholars who delve into theoretical development. After all, management is about human and organizational behavior and system science linking to interesting cross extension, but this study is only standing on the organizational level of innovation and organization. In case when organizations explore the ambidexterity training, the affected organization and leadership might contribute limited a change to the organization and organizational climate. It is important to consider the impact on the employees to avoid becoming a mechanical prisoners. Therefore, the authors knows that this paper, as the beginning of the interdisciplinary dialogue between organizational theory and system science, can only serve as an introduction. The authors hope this article can arouse the thinking and discussion of scholars from two disciplines, which is the greatest contribution of this article and the authors' original intention.

References

1. Ciborra, C. U. (1996). The platform organization: Recombining strategies, structures, and surprises. *Organization Science, 7*(2), 103–118.
2. Gibbons, R., & Henderson, R. (2012). Relational contracts and organizational capabilities. *Organization Science, 23*(5), 1350–1364.
3. Meng, F., Rieckmann, J. M., & Li, C. (2016). Empirical evidence how social capital effects the internationalisation process of SME in Zhejiang. *Transnational Corporations Review, 8*(3), 196–206.
4. Hansen, K., & Singh, V. (2009). Market structure across retail formats. *Marketing Science, 28*(4), 656–673.
5. Rieckmann, J. M., Wan, H., & Meng, F. (2017). The dynamics of internationalization of the manufacturing industry in China a case study how networks affect SMEs' global strategies. In *Proceedings of the 4th International Conference on Industrial Economics System and Industrial Security Engineering (IEIS)* (pp. 1–6), Kyoto.

6. Rieckmann, J. M., Gao, P., & Meng, F. (2018). The role of social capital during SMEs' internationalization: Build, maintain and use of networks. In *Proceedings of the 2018 International Conference on E-Business and Applications (ICEBA)* (pp. 55–61), Da Nang.
7. China Internet Network Information Center (CNNIC). (2018). *The 44th statistical report on Internet development in China*. Beijing: China Internet Network Information Center.
8. Schumpeter, J. A. (1934). *The theory of economic development: An inquiry into profits, capital, credit, interest and the business cycle* (2017th ed., pp. 57–94). Abingdon: Routledge.
9. Porter, M. E. (1996). Competitive advantage, agglomeration economies, and regional policy. *International Regional Science Review, 19*(1/2), 85–94.
10. Yang, C., & Wei, H. (2016). Organizational transformation as punctuated equilibrium: A literature review and prospects. *Foreign Economics and Management, 38*(4), 36–48.
11. Cao, Q., Gedajlovic, E., & Zhang, H. (2009). Unpacking organizational ambidexterity: Dimensions, contingencies, and synergistic effects. *Organization Science, 20*(4), 781–796.
12. Andriopoulos, C., & Lewis, M. W. (2008). Exploitation-exploration tensions and organizational ambidexterity: Managing paradoxes of innovation. *Organization Science, 20*(4), 696–717.
13. Simsek, Z. (2009). Organizational ambidexterity: Towards a multilevel understanding. *Journal of Management Studies, 46*(4), 597–624.
14. Simsek, Z., Heavey, C., Veiga, J. F., & Souder, D. (2009). A typology for aligning organizational ambidexterity's conceptualizations, antecedents, and outcomes. *Journal of Management Studies, 46*(5), 864–894.
15. Mihalache, O. R., Jansen, J. J. P., van den Bosch, F. A. J., & Volberda, H. W. (2014). Top management team shared leadership and organizational ambidexterity: A moderated mediation framework. *Strategic Entrepreneurship Journal, 8*(2), 128–148.
16. Chang, Y.-C., Yang, P. Y., & Chen, M.-H. (2009). The determinants of academic research commercial performance: Towards an organizational ambidexterity perspective. *Research Policy, 38*(6), 936–946.
17. Hou, G. (2006). *An introduction to organizational systems science*. Beijing: Science Press.
18. Luo, M. (2006). Value constellation: Theoretical explanation and construction of value-creating mechanism. *China Industrial Economy, 1*, 80–89.
19. Luo, M. (2010). Analysis on the evolution and development of organizational design thought. *Foreign Economics and Management, 32*(4), 16–24.
20. Xu, G. (2000). *Systems science and engineering: Theories and applications*. Shanghai: Shanghai Science and Technology Education Press.
21. Kast, F. E., & Rosenzweig, J. E. (2012). *Organization and management: A systems and contingency approach* (3rd ed.). New York: McGraw-Hill.
22. Wan, H., Rieckmann, J. M., Zhang, Q., & Ping, Q. (2020). Research on new energy automobile manufacturing service derivatization based on TRIZ. *Sustainability, 12*(16), 1–11.
23. Wan, H., & Kong, D. (2017). Research on innovation of management concept of urban underground pipelines based on big-knowledge-thinking in the big data era. In *Proceedings of the IEEE International Conference on Big Knowledge (ICBK)* (pp. 220–223), Hefei.
24. Bloodgood, J. M., Hornsby, J. S., Burkemper, A. C., & Sarooghi, H. (2015). A system dynamics perspective of corporate entrepreneurship. *Small Business Economics, 45*(2), 383–402.
25. Colquitt, J. A., & Zapata-Phelan, C. P. (2007). Trends in theory building and theory testing: A five-decade study of the academy of management journal. *Academy of Management Journal, 50*(6), 1281–1303.

Regional Heterogeneity Analysis of Economic Transformation of Resource-Based Cities in China—Based on Quantile Regression

Wei Deng

Abstract This article adopts quantile regression to construct an econometric model with the economic statistics of 68 prefecture-level resource-based cities between 2007 and 2016 according to the fitting relation among the indicators associated with industrial structure upgrading of resource-based cities, to observe and analyze the regional heterogeneity of the indicators' effect significance and influence. The results show: the indicator of technological innovation exerts the most significant influence on transformation and upgrading and shows regional heterogeneity that its influence on the east is greater than the influence on other areas; the indicators of marketization and government behavior play a negative role in overall samples and classified samples, and marketization shall be further improved and government functions shall be transformed to adapt to economic development and industrial structure transformation and upgrading; though environmental regulations show negative influence on the areas of a weak foundation, they promote technological innovation and develop emerging alternative industries on the whole, and thus advance the industrial structure transformation; finance apparently promotes the industrial structure transformation and upgrading; the spillover effect of openness has not been brought into full play; and the tertiary industry contribution rate effectively advances the industrial structure transformation and upgrading.

Keywords Resource-based cities · Economic transformation · Regional heterogeneity · Quantile regression

1 Introduction

In the course of industrialization in China, resource-based cities not only supply resources and energy for national economic construction but also create a lot of jobs for society. Moreover, they extend their influence to neighboring areas, change the economic development pattern and lifestyle in these areas, particularly economically

W. Deng (✉)
School of Economics and Management, Beijing Jiao Tong University, Beijing 100044, China
e-mail: 12113125@bjtu.edu.cn

M. Li et al. (eds.), *IEIS 2020*, https://doi.org/10.1007/978-981-33-4363-4_25

backward areas, drive rapid regional economic growth, and make indelible contributions to the constant and steady development of the national economy [1]. However, rapid economic development necessarily leads to huge resource consumption, so the improvement of economic is accompanied by the gradual increase in resource consumption. Since the 1980s, a great number of resource-based cities have been exposed to numerous economic and social issues such as sluggish economic growth, declining residents' income, rising unemployment rate and deteriorated ecological environment. The traditional economic development pattern that completely depends on resources has become hard to sustain. By 2016, approximately 79% out of the 262 resource-based cities in China had ushered in the stage of maturation or decline and suffered from the imbalance of industrial structure and sluggish economic development, etc. It has become a major strategic issue of national and regional economic development to break the original resource self-circulation mechanism and path dependence, promote the industrial structure transformation and upgrading of resource-based cities, realize industrial coordination and sustainable development and carry out all-round reform of multiple areas, multiple aspects, multiple levels, and multiple perspectives.

In view of resource-based cities' resource endowments and special functions in resources and energy, their transformation is more complex and difficult compared to that of ordinary cities. The transformation and development of resource-based cities have always been hindered by the contradiction between the self-strengthening mechanism of the resource industry and the development of alternatives of the leading industry in resource-based cities, the contradiction between rigid national demand on resource and energy and the resource-based cities' goal of diversified development, the contradiction between the ecological destruction caused by resource development and the resource-based cities' goal of ecological optimization, and the contradiction between the government dependence of resource-based cities during transformation and development and the leading role of the market, etc. The coordination between resource development and economic and social development, and ecological environment protection has not been reached [2]. To cope with the issues in the development of resource-based cities, China had specified the first batch of resource-exhausted cities in the early 1990s. Under the leadership of the Party and through common effort, the transformation of resource-based cities was initiated and carried out steadily with resource-exhausted cities as the starting point. Some cities have completed their industrial structure transformation and achieved breakthroughs upon the financial subsidies and policy support of the government, but most resource-based cities have not been transformed successfully, the economic growth rates of which were lower than the national average level. Moreover, the transformation effect of resource-based cities shows certain regional characteristics.

Based on the current situation of the transformation of resource-based cities, this article adopts quantile regression to construct an econometric model with the economic statistics of 68 prefecture-level resource-based cities between 2007 and 2016, to analyze the important factors affecting the economic transformation of resource-based cities and examine the regional heterogeneity of the indicators' effect significance and influence [3].

2 Determination of the Econometric Model and Evaluation Indicators

At present, China has 126 prefecture-level resource-based cities. Considering the availability and accuracy of data and referring to the classification of resource-based cities in the Sustainable Development Planning of National Resource-based Cities (2013–2020), this article selected the panel data of 68 prefecture-level resource-based cities between 2007 and 2016, trying to cover resource-based cities of all areas and all types to assure comprehensive and representative sampling. The data come from the China Statistical Yearbook and the statistical yearbooks of different provinces and municipalities. This article also divides samples into four regions, the western, northeastern, eastern and central regions, by the regions and geographical environment of these cities. In these samples, Dongying, Jining, Laiwu, Linyi, Longyan, Nanping, Sanming, Tai'an, Zaozhuang and Zibo are in the eastern region; Haozhou, Chenzhou, Chizhou, Chuzhou, Ganzhou, Hebi, Hengyang, Huaibei, Huainan, Jiaozuo, Jingdezhen, Loudi, Luoyang, Ma'anshan, Nanyang, Pingdingshan, Pingxiang, Puyang, Sanmenxia, Shaoyang, Tongling, Xinyu, Suzhou, Xuancheng and Yichun are in the central region; Baiyin, Baoji, Jinchang, Longnan, Pingliang, Qingyang, Shizuishan, Tongchuan, Weinan, Wuwei, Xianyang, Yan'an and Yulin are in the western region; and Anshan, Baotou, Benxi, Chifeng, Daqing, Ordos, Fushun, Fuxin, Hegang, Heihe, Hulun Buir, Huludao, Jixi, Mudanjiang, Panjin, Qitaihe, Shuangyashan, Wuhai, Yichun and Zhangye are in the north-eastern region. By analyzing and studying the common characteristics of the overall samples and heterogeneity of classified samples, this article will examine the regional heterogeneity of the indicators' effect significance and influence.

2.1 Selection of the Econometric Model

This chapter adopts the quantile regression method to analyze the factors affecting the industrial structure transformation and upgrading of resource-based cities and examine the regional heterogeneity of the indicators' effect significance and influence.

Generally speaking, common methods used to appraise influencing factors mainly include principal component analysis, clustering analysis, and factor analysis. With principal component analysis, the comprehensive evaluation of the function meaning may be vague in case there are both positive and negative factor loads in the principal component. Clustering analysis finds it difficult to reach a clustering conclusion against large sample size. Since factor analysis adopts ordinary least square (OLS), its regression results only reflect the structural relations among the data around the mean value, so the results may not be so robust if the sample data fail to satisfy classical assumption as in the cases of peak and heavy tail. Quantile regression (QR) here not only overcomes the average effect but also provides all-round information on the explained variables. The regression results will be more effective and robust if regression is performed against different quantiles to reduce the influence of outliers on the evaluation results to the minimum.

2.2 Selection of Evaluation Indicators

As evaluation indicators will affect the accuracy of analysis results, they shall be selected following the principles of relevance, objectivity, comparability, consistency, importance, systematic science, and dynamic guidance. According to the innovation-driven mechanism for resource-based cities' transformation and the indicatory system of the industrial structure transformation and upgrading, this article chooses technological innovation, tertiary industry contribution rate, marketization, openness, government behavior, finance, urban greening, and three wastes treatment, etc. as the indicators to evaluate the transformation effect of resource-based cities. Details of indicators are as follows:

Explained variable: The industrial structure upgrading coefficient (stru) is the explained variable. Foreign researchers mainly measured this coefficient with Chenery standard structure and Hoffmann coefficient etc. However, these methods are only effective when the data cover a long time span. Considering the quality of data associated with industrial structure upgrading in China and the difficulty to acquire the data, this article refers to the method adopted by Li Fengchun and obtains the measured value of industrial structure upgrading through the weighted operation of the proportions of the industrial output values of different industries in the total output value [4]. See the formula below:

$$stru = \sum_{i=1}^{3} y_1 * 1 + y_2 * 2 + y_3 * 3$$

where yi and i respectively denote the proportion of the output value of the ith industry in the total output value and the weight of the corresponding industry; and stru ranges between 1 and 3, with the closer to 1 suggesting the lower level of industrial structure and the closer to 3 indicating the higher.

Core explanatory variable: Technological innovation (TECH) is the core explanatory variable. Innovation constitutes the major driving force for the transformation and upgrading of the industrial structure, so this indicator is considered as the core explanatory variable. This article regards the proportion of R&D expenditure of a region in the regional GDP as the indicator to measure annual technological innovation. Technological innovation significantly improves industrial efficiency, gathers and attracts a great number of production factors, and promotes the rapid development of the industry and relevant industries [5].

Control variables: Control variables include the tertiary industry contribution rate (CONT), marketization (MKT), openness (FDI), government behavior (GOV), finance (BANK), urban greening (GRN) and three wastes treatment (WST). Their specific economic meanings are explained as follows:

Tertiary industry contribution rate (CONT): The proportion of the tertiary industry in the industrial structure exerts great influence on the economic growth speed, so the proportion of the increment of the tertiary industry in the total increment of GDP is taken as one of the explanatory variables of this analysis model, to analyze the different roles of different factors in economic growth. The price change factor is calculated with the increment of the comparable price.

Marketization: Non-public employment is a frequently used indicator to represent the level of economic marketization in a region. The sum of the numbers of private and non-public employees is adopted as the calculation basis of this indicator [6].

Openness: This indicator is represented by the proportion of FDI in the regional GDP in all regions. FDI can increase capital stock by attracting foreign investment and promote technological progress and regional industrial structure upgrading due to technology spillover [7].

Government behavior (GOV): Local governments can affect the industrial structure by releasing tax, fiscal, and land preferential policies. In this article, the proportion of government expenditure in GDP is considered as government behavior [8].

Finance (BANK): This article uses the BANK loan balance to represent finance and study the influence of finance on industrial structure transformation and upgrading [9].

Urban greenness (GRN): This article adopts the urban per capita green space as the representative of urban greenness, to study the influence of environmental regulations on industrial structure transformation and upgrading [10].

Three-wastes treatment (WST): The three-wastes treatment level is used as the indicator to measure the influence of pollution control on industrial structure transformation and upgrading [11].

2.3 Construction of the Econometric Model

Based on the fitting relation between the industrial structure upgrading coefficient of resource-based cities and 8 explanatory variables, the economic model is constructed as follows:

$$
\begin{aligned}
stru_{it} = {} & c + \alpha tech_{it} + \beta cont_{it} + \beta mkt_{it} + \beta fdi_{it} \\
& + \beta gov_{it} + \beta bank_{it} + \beta grn_{it} + \beta wst_{it} + \varepsilon_{it}
\end{aligned}
$$

Perform logarithmic processing of the model to eliminate heteroscedasticity and assure the data stability. Then, we get the following model:

$$lnstru_{it} = c + \alpha lntech_{it} + \beta lncont_{it} + \beta lnmkt_{it} + \beta lnfdi_{it} + \beta lngov_{it} + \beta lnbank_{it} + \beta lngrn_{it} + \beta lnwst_{it} + \mu_{it}$$

3 Empirical Analysis of Industrial Structure Transformation and Upgrading of Overall Samples

As demonstrated by the F test and the Hausman test of the overall samples' regression results, the fixed-effect model is more rational. Which is shown in Table 1, at 5% significance level.

In the fixed-effect model, the variables of tertiary industry contribution rate X2 and GOV, BANK and GRN show significant influence on the dependent variable, where the influence of GRN on the dependent variable is negative, and the influence of other factors is positive.

With a quantile of 0.1, TECH, CONT, GOV, BANK, GRN and WST have a significant influence on the dependent variable, where GOV exerts a negative influence on the dependent variable, and TECH, CONT, BANK, GRN and WST show positive influence.

With the quantile equation of 0.25, all factors show significant influence, where GOV exerts negative influence and all other factors exert positive influence.

With a quantile equation of 0.5, CONT, MKT, FDI, GOV, BANK, GRN and WST show significant influence on the dependent variable, where MKT and GOV exert negative influence and CONT, fid, BANK, GRN and WST exert positive influence.

With a quantile equation of 0.75, CONT, MKT, GOV, BANK and GRN show significant influence on the dependent variable, where MKT and GOV exert negative influence and CONT, BANK, GRN and WST exert positive influence.

With a quantile equation of 0.9, CONT, MKT, GOV and BANK show a significant influence on the dependent variable. The influence of TECH and GRN is more significant. Meanwhile, TECH, MKT and GOV exert negative influence and CONT, BANK and GRN exert positive influence.

According to the regression results, technological innovation significantly promotes industrial structure transformation and upgrading significantly with a lower quantile, but shows a restrictive role at a higher quantile, which means that technological innovation plays a big part in the early stage of industrial structure transformation and upgrading, but has not produced a marked effect in the senior stage and shall be further improved. The tertiary industry contribution rate and finance significantly promote industrial transformation and upgrading, having advanced the transformation and upgrading of resource-based cities effectively. Marketization and

Table 1 QR results of 68 resource-based cities in China

Variable	Fixed effect	Quantile				
		0.1	0.25	0.5	0.75	0.9
TECH	3.23E−05	9.82E−05***	5.49E−05***	3.08E−05	1.30E−06	−3.12E−05**
CONT	0.914451***	0.953822***	0.930063***	0.997448***	1.065291***	1.022281***
MKT	0.000133	−8.49E−05	−0.000471***	−0.000665***	−0.000605***	−0.000423***
FDI	4.07E−09	2.29E−08	3.12E−08***	2.14E−08***	9.62E−09*	−1.15E−10
GOV	2.24E−05***	−9.04E−05***	−7.80E−05***	−6.90E−05***	−5.16E−05***	−4.15E−05***
BANK	1.01E−05***	3.04E−05***	3.03E−05***	2.66E−05***	2.03E−05***	1.10E−05***
GRN	−0.000190***	0.000509**	0.000580***	0.000543***	0.000780***	0.000666**
WST	4.50E−09	5.76E−07***	3.74E−07***	2.30E−07***	1.03E−07**	2.09E−08
C	1.856610***	1.920180***	1.965009***	1.970734***	1.955679***	1.987241***
R-Squared	0.963438	0.38251	0.427723	0.467991	0.506064	0.536317
F statistics	46.199581					
Hausman test	111.883125					

Note ***, ** and * respectively represent the significance level of 1%, 5% and 10%

government behavior show a negative influence on industrial structure transformation and upgrading, which suggests that the leading role of the market and guiding role of government have not been completely determined during the transformation and upgrading of China's resource-based cities, and the government has not completely withdrawn from the intervention mechanism. Openness shows significant influence at a lower quantile, proving its positive role in the early and middle stages of the transformation and upgrading of resource-based cities. Despite a negative fixed effect, greenness and three-wastes treatment play a significantly positive role regardless of the quantile, which indicates that environmental regulations show a certain threshold effect on industrial structure transformation and will significantly advance industrial structure transformation and upgrading after stepping over the threshold.

4 Empirical Analysis of Industrial Structure Transformation and Upgrading of Classified Samples

Analyze the industrial structure transformation and upgrading of resource-based cities based on the four major regions. Table 2 provides the fixed effect model of each region and the estimates of the quantile regression model.

The indicator of technological innovation shows a more significant effect in the eastern region and exerts significant influence in all regions at a quantile of 0.1 or 0.25, which means that the promoting effect of technological innovation is significantly greater than it in other regions. The tertiary industry contribution rate significantly drives the industrial structure transformation and upgrading in all regions. Marketization shows a significant negative influence on the western, northeastern and central regions but has an insignificant impact on the eastern region, which suggests that marketization needs to be further improved in all areas to adapt to economic development and industrial structure transformation and upgrading. Openness shows insignificant or even negative effects in all regions, which proves that the spillover effect of foreign direct investment has not been made full use of to promote industrial structure transformation and upgrading. Government behavior significantly exerts negative influence in all regions, particularly in the northeastern region, suggesting that government functions shall be further transformed to adapt to market-oriented industrial structure transformation and upgrading. Finance shows the significant influence on the western region, but a relatively weaker influence on the southeastern and central region; and this situation proves that the Western Development strategy has improved the financial and investment environment in the western region and effectively advanced industrial structure transformation. Urban greenness has a significantly positive influence on the western and central region, weaker influence on the eastern region, and the minimum influence or even negative influence on the northeast, which demonstrates that industrial structure transformation in the northeastern region is not exposed to the influence of environmental regulations.

Table 2 QR results of resource-based cities in four regions

Regions	Variable	Fixed effect	Quantile				
			0.1	0.25	0.5	0.75	0.9
Western China	TECH	−3.47E−05	0.000278***	0.0000556	4.05E−05	−5.67E−05	−6.93E−05**
	CONT	0.887240***	0.669878***	0.673300***	1.029169***	0.927104***	0.889633***
	MKT	0.000678**	−0.003656***	−0.001771	−0.002783**	−0.000701	−0.000513
	FDI	−1.22E−05	−3.23E−05	−9.95E−05***	−0.000163***	−0.000194***	−0.000221***
	GOV	−0.00000299	−0.00000158	−0.00000905	−2.28E−05***	−2.19E−05***	−2.39E−05***
	BANK	0.00000562	5.92E−05***	5.03E−05***	4.41E−05***	2.87E−05***	1.75E−05***
	GRN	0.000240*	0.0000306	0.003861***	0.002675***	0.003015***	0.002366***
	WST	1.02E−07	2.36E−06**	3.13E−06**	−0.0000006	−1.60E−06*	−1.33E−06
	C	1.897374***	1.898136***	1.902197***	1.915620***	1.953705***	2.002322***
	R-squared	0.976219	0.21579	0.351688	0.4534	0.590602	0.661949
Northeast China	TECH	0.000201*	0.000613	0.000774***	3.61E−04	2.18E−04	−0.0000598
	CONT	0.947245***	1.065055***	1.070612***	1.004856***	0.991359***	1.313378***
	MKT	0.0000643	0.002128***	0.002067***	0.001903***	0.000748	−0.000368
	FDI	−2.28E−09	5.32E−08***	3.80E−08***	2.28E−08	6.03E−09	−1.46E−08
	GOV	−2.71E−05***	−9.37E−05***	−0.000107***	−8.26E−05***	−8.32E−05***	−9.24E−05***
	BANK	−0.00000139	7.18E−05**	4.79E−05**	5.31E−05**	0.0000115	−0.0000338
	GRN	−0.0000886	0.000358	0.0000465	0.000137	0.0000572	0.000218
	WST	−2.76E−06***	−0.00000195	−0.00000313	−7.04E−06***	−0.00000467	1.67E−06
	C	1.876623***	1.728705***	1.804709***	1.896420***	2.010737***	2.000917***
	R-squared	0.986718	0.483802	0.533172	0.55806	0.520962	0.531846

(continued)

Table 2 (continued)

Regions	Variable	Fixed effect	Quantile				
			0.1	0.25	0.5	0.75	0.9
Eastern China	TECH	2.44E−05	0.000166***	0.000135***	0.000339**	1.87E−04	0.000202**
	CONT	0.965452***	0.640223***	0.813048***	0.762491***	0.871161***	0.982820***
	MKT	0.0000802	0.000638*	0.000372	0.000307	0.0000637	−0.000101
	FDI	0.00000569	2.64E−05***	0.0000159	0.00000762	−0.000016	−2.89E−05**
	GOV	5.28E−05***	−0.0000234	−3.02E−05***	−1.70E−05*	−2.68E−05***	−3.27E−05***
	BANK	0.00000099	3.30E−05***	3.35E−05***	1.67E−05**	0.0000078	0.00000376
	GRN	−0.000165*	0.001044***	0.000909***	0.000920**	0.000888**	0.000758**
	WST	4.39E−08	6.39E−07***	5.86E−07***	4.42E−07***	5.41E−07	7.00E−07**
	C	1.804643***	1.861338***	1.861367***	1.895244***	1.936931***	1.928411***
	R-squared	0.940364	0.419126	0.396425	0.424545	0.495985	0.546328
Central China	TECH	2.29E−05	0.000604***	0.000523**	−3.31E−05	−3.13E−05	−4.81E−07
	CONT	0.963758***	0.978083***	0.823513***	1.017780***	0.921154***	0.840800***
	MKT	0.00017	1.63E−04	−0.000596	−0.001161***	−0.001329***	−0.000983**
	FDI	−3.72E−05*	−1.29E−05	0.000087	0.000137*	0.0000593	−2.67E−05
	GOV	1.84E−05***	−0.00000535	−0.0000257	−6.07E−05***	−6.27E−05***	−0.0000386
	BANK	0.00000595	−0.0000311	0.0000085	3.48E−05*	4.87E−05***	4.52E−05*
	GRN	0.001488	0.011121***	0.006516***	0.005732***	0.004528**	0.005725
	WST	5.72E−07	9.26E−07	9.21E−07	0.00000104	3.91E−07	−7.53E−07
	C	1.825823***	1.722591***	1.850032***	1.906909***	1.968169***	1.974186***
	R-squared	0.985444	0.588105	0.535045	0.558981	0.639015	0.703227

Three-wastes treatment shows the most significant influence on the eastern region, proving that environmental regulations have effectively promoted industrial structure transformation and upgrading in the eastern region.

The driving factors to industrial structure upgrading vary in different regions mainly due to the reasons below: Since the Reform and Opening Up, the eastern region including Beijing-Tianjin-Hebei, Yangtze River Delta and Pearl River Delta has taken the lead to achieve rapid development [12]. Due to high marketization and openness, the eastern region has seen the influx of capital, technology, innovation and resources from countries all over the world, which have greatly inspired the potential of innovative development and cultivated favorable market economy environment and hi-tech enterprises in the east far better than other regions of China. Therefore, capital, technology and talent have shown great marginal effect and the innovation-driven mechanism has gradually matured, effectively promoting industrial structure transformation and upgrading. The western region, despite low marketization and insignificant advantage in technological innovation, has made full use of the Western Development strategy and realized over-speed development by means of technology simulation and transplantation upon the great fiscal support of the government. In this way, technology has been improved sharply, which has effectively advanced industrial structure upgrading. By comparison, industrial structure transformation and upgrading have been slower in the central region. In the central region with a solid industrial foundation, the traditional leading industry has crowded out new technologies and impeded industrial structure transformation and upgrading. The northeastern region is the old industrial base of China. The slow transformation of local government functions has severely impeded the market to adjust resource allocation and therefore hindered industrial structure transformation and upgrading.

5 Conclusion

The following conclusions can be reached by applying quantile regression to systematically observe the factors influencing industrial structure upgrading of resource-based cities: First, the indicator of technological innovation shows significant regional heterogeneity. In overall samples and regional samples, technological innovation shows a significantly positive effect on industrial structure upgrading, with the effect significance following the order "northeast, central, west, east" (from the insignificant to the significant effect). As to the samples of different development stages, technological innovation significantly exerts a positive influence on all samples except the declining cities, and the influence is the most significant in mature cities. Second, marketization and government behavior show a negative influence on both overall and classified samples. It suggests that marketization shall be further improved and government functions shall be further transformed to adapt to economic development and industrial structure transformation and upgrading. Third, environmental regulations may show a negative influence on the regions in the initial stage of transformation or regions with a weak foundation. However, this indicator will promote technological innovation, develop emerging alternative industries, and thus promote industrial structure transformation in the long run. Fourth, finance significantly promotes industrial structure transformation and upgrading. Fifth, the spillover effect of openness has not been made full use of, as openness shows an insignificant influence on industrial structure transformation and upgrading. Sixth, the tertiary industry contribution rate effectively promotes the transformation and upgrading of the industrial structure.

Resource-based cities shall establish the "innovation-driven" transformation and upgrading mechanism, optimize and expand investment demands, transform government functions, establish and improve the market system, effectively utilize foreign investment, introduce key technologies from foreign countries and thus promote transformation and upgrading of resource-based cities.

With regard to different regions, the eastern region shall play its role as an innovation leader and a growth pole and adopt market-oriented means to phase out and eliminate backward [13–15], ineffective and excess production capacity; make use of its regional advantages to support the export of products of self-owned brands, high added-value and independent intellectual property right and explore emerging markets in all-round and multi-level manner; and observe the principle of differentiated guidance and specialized development to build a long-acting development mechanism. The central region shall facilitate the upgrading of pillar industry technology and the reconstruction of leading industries, and encourage optimization and upgrading of production technology, improvement of production equipment and elimination of backward production capacity, as led by the key enterprises; and well receive the industrial clusters moved from the eastern region, establish scientific and effective industrial access criteria and the negative list of enterprise undertaking, refuse to receive any industry or enterprise of high pollution, high energy consumption and low technological content, promote the economic, social and environmental development of the region at the same time and comprehensively improve the welfare level of the region [16, 17]. The western region was faced with low marketization and severe structural issues, and the region shall draw support from the national strategy of opening the west, carry out fine processing and deep processing of resources based on different resource endowments of different areas and extend the industrial chain; attach importance to the combination advantages in energy and mineral resources, industrial base, professional and technological talents and urban and rural coordination, facilitate the ecological, cyclic and green transformation of the industry, spare no effort to advance the transformation and upgrading of traditional industries, receive industries transferred from Pearl River Delta and cultivate new industries, new business forms and new development modes; and give full play to technology advantages in the war industry, promote the integrated military and civilian development and enhance enterprise competitiveness. The northeastern region shall strengthen practical exploration and innovation of the incentive mechanism, advance the development of the tertiary industry and marketization, and increase investment in innovative and emerging industries and construct the leading industry system that is suitable for resource-based cities in the northeast.

Quantile regression is a robust method that can be applied to perform regression analysis of non-average numbers. When applying this method, we shall study various influence factors and their influence significance at different quantiles; and on the other hand, we shall introduce classified samples to the model to analyze the regional heterogeneity of these influence factors and development stages. By applying this method to analyze the transformation of resource-based cities, we can analyze and study the core influence factors that affect transformation as well as the heterogeneity of these factors from multiple dimensions, and hence reach more robust and targeted analysis conclusions. This article adopts quantile regression to analyze the ten-year panel data of 68 prefecture-level resource-based cities in China, performs regression on different quantiles based on all-round information of the explained variable and different regression coefficients at different quantiles, fully analyzes the influence factors that affect the transformation of resource-based cities of different regions and different types, analyzes the factors influencing the transformation efficiency of resource-based cities and their influence level and scope from the perspectives of geographic space and development stages, studies the common law and characteristics of the development of resource-based cities of different types and different regions and provides targeted suggestions for the transformation of different regions from multiple dimensions according to the evident regional heterogeneity and development stage heterogeneity shown in the transformation results of these resource-based cities.

References

1. Xin, Z. (2011). Exploring the way to change the development model of the regional economy—A summary of the views of 2011 annual meeting of China Association for the regional economy. *China Industrial Economics, 09,* 27–35.
2. Qinglin, G. (2009). Industrial transformation and industrial competitive advantage cultivation in regional industrial structure adjustment. *Contemporary Economic Research, 04,* 43–46.
3. Saiyin, Z., Qiang, L., & Jinwei, W. (2019). Statistical analysis of economic differences in Beijing Tianjin Hebei region based on quantile regression method. *Mathematical Statistics and Management, 38*(05), 761–769.
4. Li, F. (2013). Location and industrial choice of China's foreign direct investment to promote industrial upgrading. *International Economic and Trade Exploration, 2,* 95–102.
5. Romanelli, E., & Khessina, O. M. (2005). Regional industrial identity: Cluster configurations and economic development. In *INFORMS 2005*.
6. Ohlin, B. G. (1933). *Interregional and international trade*. Cambridge, MA: Harvard University Press.
7. Barns, T. J., Britton John, N. H., Coffer, W. J., et al. (2000). Canadian economic geography at the millennium. *The Canadian Geographer, 44*(1), 4–24.
8. Li, H., Long, R., & Chen, H. (2013). Economic transition policies in Chinese resource-based cities: An overview of government efforts. *Energy Policy, 55*, 251–260.
9. Atkinson, G., & Hamilton, K. (2003). Savings, growth and the resource curse hypothesis. *World Development, 31*(11), 1793–1807.
10. Siebert, H. (1977). Environmental quality and the gains from trade. *Kyklos, 30*(4), 657–673.
11. Chung, S. (2014). Environmental regulation and foreign direct investment: Evidence from South Korea. *Journal of Development Economics, 108*, 222–236.

12. Deng, W. (2019). Evaluating transformation efficiency of resource-based coastal cities: An AHP and DEA based analysis. *Journal of Coastal Research, 94*(sp1).
13. Perroux, F. (1971). Note on the concept of growth poles. In L. Livingstone (Ed.), *Economic policy for development: Selected readings* (pp. 278–289). New York, NY: Penguin Books.
14. Liu, Y., Hu, Z., Li, H., & Zhu, H. (2019). Does preemption lead to more leveled resource usage in projects? A computational study based on mixed-integer linear programming. *Economic Computation and Economic Cybernetics Studies and Research, 53*(4), 243–258.
15. Lin, Y., Liu, W., & Wang, Y. (2019). An integrated approach using cross-efficiency and shapley value in performance evaluation. *Economic Computation and Economic Cybernetics Studies and Research, 53*(4), 209–224.
16. Algheriani, N. M. S., Majstorovic, V. D., Kirin, S., & Spasojevic Brkic, V. (2019). Risk model for integrated management system. *Tehnicki Vjesnik-Technical Gazette, 26*(6), 1833–1840.
17. Lee, H., & Lee, W. (2019). International linkage among mena financial markets. *Economic Computation and Economic Cybernetics Studies and Research, 53*(3), 297–314.

Efficiency of PPP Investment in Infrastructure

Sixiang Wu and Xuemeng Guo

Abstract In recent years, the Public–Private Partnership (PPP), as a new channel for providing public services, has received strong support from the Chinese government. This paper establishes a relatively objective evaluation index by combining the super-efficiency DEA method with the Malmquist index. It measures the investment efficiency of PPP projects in 18 provinces in North China, East China, South China and Central China from 2006 to 2018. The results show that, on the whole, the investment efficiency of PPP projects in China's infrastructure sector is not high. And the trend is declining year by year. Meanwhile, with the continuous advancement of PPP mode, the influencing factor of investment efficiency has changed from technological progress to efficiency change. In addition, the investment efficiency of PPP projects has the problem of uneven regional development. For example, compared with the other three regions, central China may be relatively weak in technological progress, which drags down the investment efficiency.

Keywords PPP · Investment efficiency · Super-efficiency DEA · Malmquist

1 Introduction and Literature Review

Public Private Partnership, abbreviated as PPP, refers to a partnership between the public and private sectors that aims to better provide public goods or services. Since its birth in 1992 in the United Kingdom, the PPP model has been rapidly promoted in major developed countries by virtue of its unique advantages. Subsequently, it has been promoted and promoted by many international organizations and gradually spread to developing countries. Against the backdrop of the new economic normal, China's economic growth has shifted from rapid growth to medium–high growth;

S. Wu (✉) · X. Guo
School of Economics and Management, Beijing Jiaotong University, Beijing, China
e-mail: 19120676@bjtu.edu.cn

X. Guo
e-mail: xmguo@bjtu.edu.cn

M. Li et al. (eds.), *IEIS 2020*, https://doi.org/10.1007/978-981-33-4363-4_26

fiscal revenue growth has fallen; and the government's debt crisis has further deepened. At this time, the strength of the government alone cannot meet the growing demand for general infrastructure represented by municipal engineering and transportation. The infrastructure investment and financing model needs urgent reform. Therefore, with its own advantages, the PPP model has become an important direction of reform. As of November 2019, a total of 9399 PPP projects have been deposited in the PPP Center of the Ministry of Finance; the total investment of the project is as high as 14.3 trillion RMB. Among them, a total of 6216 projects have been launched, with an investment of 9.5 trillion RMB; a total of 3616 projects have been started, with an investment of 5.4 trillion RMB.

The existing research on the investment efficiency of the PPP model is divided into two levels.

The first is to theoretically focus on the efficiency source of the PPP model to explore why the PPP model is more efficient than traditional government control. PPP can improve efficiency by reducing the full life cycle cost of the project [1]. When the same contractor implements construction and operation, they will have strong incentives to reduce costs. Through partnerships, PPP can delegate some risks in financing and operations to the private sector [2]. Risk sharing can increase the enthusiasm of the private sector to encourage them to improve their technological innovation capabilities and management level, thereby improving the efficiency of the project to avoid cost overruns or insufficient returns. PPP can also increase competition efficiency by introducing competitive bidding in the public domain; they also believe that PPP is an incentive mechanism that improves the efficiency of the supply of public goods [3].

The other is to compare the efficiency of the PPP model with the traditional model from the perspective of empirical research to verify whether the PPP model is more efficient. Some scholars find that in the three aspects of cost, time and quality, the PPP model has better performance than the traditional model. The main performance is that the PPP model can effectively alleviate the cost overruns and delays in the traditional model. Some scholars also analyze the key influencing factors of PPP project efficiency from the perspective of management. For example, Feng identifies the key influencing factors of the successful operation of PPP projects from the regional level, enterprise level and project level [4]; Ye conducts research using questionnaires [5]. At the same time, some scholars have studied the influencing factors of PPP project efficiency from the econometric method, such as discrete selection model [6], stochastic frontier analysis [7], DEA model [8]. Although the influencing factors derived from different research literatures are not completely the same, most of the literature believes that the political environment, economic environment, public and private experience and capabilities, transparency of project operations, risk allocation, government supervision, trust and communication of all parties are the key factors affecting the efficiency of PPP projects. This points the way for the government and the private sector to improve the performance of PPP projects and give full play to the efficiency advantages of PPP.

Although empirical research on the supply efficiency of the PPP model is rare, there is no lack of evaluation methods for the efficiency of public service supply.

Two-stage and three-stage DEA models are commonly used. This paper uses the combination of super-efficiency DEA method and Malmquist index to establish relatively objective evaluation indicators to measure the investment efficiency of PPP projects from 2006 to 2018 in a total of 18 provinces in North China, East China, South China and Central China.

2 Sample Selection and Research Design

2.1 *Super-Efficiency DEA Model*

The data envelopment analysis method, abbreviated as DEA, was first proposed by the famous American operations researchers Cooper and Charrnes. DEA is a system analysis method based on relative efficiency and based on linear programming to evaluate the relative effectiveness of a decision unit with multiple inputs and multiple outputs (DMU) [9]. The basic models are CCR and BCC. In addition, Andersen and Petersen proposed a super-efficiency DEA model on this basis to evaluate the efficiency of a decision-making unit relative to other units [10]. For an effective decision-making unit, the super-efficiency DEA will be reset, and the final efficiency value is greater than 1; and the invalid decision unit, the efficiency value is the same as CCR. The super efficiency model based on CCR is as follows:

$$
\begin{cases}
\min \omega - \varepsilon(es^- + e^T s^+) \\
\sum_{\substack{j=1 \\ j \neq \mathrm{j}_0}}^{n} \lambda_{\mathrm{j}} \mathrm{x}_{\mathrm{j}} + \mathrm{s}^- = \omega \mathrm{x}_0 \\
\sum_{\substack{j=1 \\ j \neq \mathrm{j}_0}}^{n} \lambda_{\mathrm{j}} \mathrm{y}_{\mathrm{j}} - \mathrm{s}^+ = \mathrm{y}_0 \\
\lambda_{\mathrm{j}} \geq 0, \quad \mathrm{j} = 1, 2, \ldots, n \\
s^+, s^- \geq 0
\end{cases}
\tag{1}
$$

2.2 *Malmquist Index*

In 1953, the Swedish economist Sten Malmquist proposed the Malmquist index to analyze changes in consumption in different periods [11]; and later it becomes an important indicator for measuring changes in productivity, based on distance functions in different periods. Because the Malmquist index can better describe the dynamic change of relative efficiency, the researchers combine it with DEA theory. The basis of the construction of the Malmquist exponent is the distance function, which is exactly the inverse of the efficiency value of the CCR and BCC models in DEA theory.

Let the distance function of (x_t, y_t) in period t be $D_t(x_t, y_t)$, and the distance function in period t + 1 be $D_{t+1}(x_t, y_t)$; let (x_{t+1}, y_{t+1}) be at t. The distance function in the period is $D_t(x_{t+1}, y_{t+1})$, and the distance function in the period t + 1 is $D_{t+1}(x_{t+1}, y_{t+1})$; $M_0((x_t, y_t), (x_{t+1}, y_{t+1}))$ is the productivity change from the base period t to t + 1; that is, the Malmquist index is expressed as:

$$M_0((x_t, y_t), (x_{t+1}, y_{t+1})) = \sqrt{\frac{D_t(x_{t+1}, y_{t+1})}{D_t(x_t, y_t)} \times \frac{D_{t+1}(x_{t+1}, y_{t+1})}{D_{t+1}(x_t, y_t)}} \quad (2)$$

If the Malmquist index is greater than 1, it indicates that the total factor productivity from t to t + 1 is on the rise. If it is equal to 1, it means that the efficiency remains unchanged; if it is less than 1, it means that the efficiency has decreased.

It can be further decomposed into a comprehensive efficiency change index (TEC) and a technology change index (TC) in two periods:

$$\begin{aligned} &M_0((x_t, y_t), (x_{t+1}, y_{t+1})) \\ &= \frac{D_{t+1}(x_{t+1}, y_{t+1})}{D_t(x_t, y_t)} \times \sqrt{\frac{D_t(x_{t+1}, y_{t+1})}{D_{t+1}(x_{t+1}, y_{t+1})} \times \frac{D_t(x_t, y_t)}{D_{t+1}(x_t, y_t)}} \\ &= TEC \times TC \end{aligned} \quad (3)$$

TEC is the efficiency change index under the condition of constant scale returns and free disposal of elements. It measures the degree of catch-up from the production decision unit from time t to t + 1 to the optimal production possibility boundary, so it is called the catch-up effect and measures the change in efficiency. When TEC > 1, it indicates that the production of the decision unit is closer to the frontier of production, and the relative technical efficiency has been improved. TC is the technological progress index, which measures the movement of the technological boundary from time t to t + 1; that is, the frontier-shift of efficiency, also known as the growth effect. When TP > 1, it means that the technology has made progress or innovation, and the production frontier is moving upward.

Because the overall efficiency can be decomposed into the product of pure technical efficiency (PTEC) and scale efficiency (SEC), so:

$$\begin{aligned} &M_0((x_t, y_t), (x_{t+1}, y_{t+1})) \\ &= \frac{D_{t+1}(x_{t+1}, y_{t+1})}{D_t(x_t, y_t)} \times \sqrt{\frac{D_t(x_{t+1}, y_{t+1})}{D_{t+1}(x_{t+1}, y_{t+1})} \times \frac{D_t(x_t, y_t)}{D_{t+1}(x_t, y_t)}} \end{aligned}$$

$$= \frac{S_t(x_t, y_t)}{S_t(x_{t+1}, y_{t+1})} \times \frac{D_{t+1}(x_{t+1}, y_{t+1}/VRS)}{D_t(x_t, y_t/VRS)} \times \sqrt{\frac{D_t(x_{t+1}, y_{t+1})}{D_{t+1}(x_{t+1}, y_{t+1})} \times \frac{D_t(x_t, y_t)}{D_{t+1}(x_t, y_t)}}$$
$$= SEC \times PTEC \times TC \quad (4)$$

If PTEC is greater than 1, it means that there is an improvement in the comparison of the efficiency of the two periods. If the SEC is greater than 1, it means that period t + 1 is closer to constant scale returns or long-term optimal scale relative to period t; conversely, if the scale efficiency change is less than 1, it means that period t + 1 deviates from scale returns or long-term optimal scale relative to period t.

2.3 Determination of Evaluation Indicators

2.3.1 The Input Indicators

First, PPP investment is the most representative index of PPP mode, so it is taken as the input index. Second, considering that private capital can participate in the financing, construction, operation and management of PPP projects, the investment proportion of private capital will affect its participation to some extent. At the same time, it is the index that can distinguish PPP from other investment and financing methods of infrastructure. Therefore, it also serves as an input indicator.

The data are from the World Bank PPI website and are in US dollars. Considering the fluctuation of the exchange rate between the US dollar and the RMB, for the sake of comparability, this article converts the PPP investment data. The exchange rate in each year is the average value of the daily value of the RMB/USD exchange rate in the State Administration of Foreign Exchange.

2.3.2 The Output Indicators

Considering the difficulty of directly measuring the output of government public expenditure, the output index is selected to reflect the supply level of typical public goods. Combined with PPI database of the World Bank, the industry distribution of PPP projects in China. It also takes into account the incompleteness of regional air and water transport data. In this paper, electricity consumption, natural gas supply, railway/highway mileage, railway/highway passenger volume, water supply and daily sewage treatment capacity are selected as output indicators. In addition, as the available data cover all infrastructure projects, there is no separate data to measure the public goods supply level of PPP projects. Therefore, existing data needs to be reduced. Proceed as follows:

1. Use the "China Statistical Yearbook of Fixed Assets" to collect the total fixed asset investment of the three provinces of electricity, gas and water production

and supply, railway transportation and road transportation in each year. Then, by dividing the investment amount of PPP projects in various industries in each province and each year and the total fixed asset investment in the corresponding year and industry of the corresponding province, the reduction ratio of the output indicators corresponding to each industry in each province and each year is calculated.

2. Use the reduction ratio obtained above to reduce the eight output indicators obtained.
3. Then, the reduced output indicators are summarized according to the corresponding 18 provinces in North China, East China, South China, and Central China. The input and output values of the investment efficiency of infrastructure PPP projects in North China, East China, South China, and Central China from 2006 to 2018 can be obtained.

2.4 Data Screening

This paper selects 2006–2018 as the research scope. It can be seen from the PPP data collected from PPI database that there are 901 PPP projects in China during the past 13 years. Excluding the data without investment, 841 projects remain. In this paper, 18 provinces in North China, East China, South China, and Central China are selected as the research scope, and a total of 571 PPP projects are obtained.

3 Results Analysis

3.1 Analysis of Total Factor Productivity by Year

Using DEAP2.1 to calculate PPP a total of 18 provinces in North China, East China, South China and Central China from 2006 to 2018. Analysis of the total factor production index of the project shows the results shown in Table 1.

As can be seen from Table 1, the average total factor productivity in the four regions of North China, East China, South China, and Central China from 2006 to 2018 is 0.901, which is less than 1. This shows that total factor productivity is generally in a downward trend. At the same time, except for 2009–2010, 2012–2013 and 2014–2015, investment efficiency shows a regression. In general, the investment efficiency of PPP projects in the infrastructure sector has shown a downward trend during the 13 years, and it is the smallest in the two years 2010–2012.

To further decompose total factor productivity (referred to as tfpch) into efficiency changes (referred to as effch) and technological changes (referred to as techch) to explore the main reasons for the decline in total factor production efficiency. As can be seen from Fig. 1, before 2013–2014, the fluctuation of tfpch is consistent with the fluctuation shape of the techch curve. While after 2014, the fluctuation of tfpch

Table 1 Table of total production factors in 13 years

Year	effch	techch	pech	sech	tfpch
2006–2007	0.986	1.2	1	0.986	1.184
2007–2008	1.025	1.381	1	1.025	1.416
2008–2009	0.924	0.891	1	0.924	0.823
2009–2010	1.077	1.493	1	1.077	1.608
2010–2011	0.94	0.01	0.946	0.994	0.009
2011–2012	1.063	0.01	1.057	1.006	0.011
2012–2013	1.005	1.273	1	1.005	1.279
2013–2014	0.545	0.591	1	0.545	0.322
2014–2015	1.724	0.992	0.946	1.822	1.711
2015–2016	0.984	0.846	1.057	0.932	0.833
2016–2017	0.777	0.877	1	0.777	0.681
2017–2018	1.392	0.681	1	1.392	0.948
Mean	1.001	0.852	1	1.001	0.901

effch stands for efficiency change; *techch* stands for technology change; *pech* stands for pure technology efficiency change; *sech* stands for scale efficiency change; *tfpch* stands for total factor productivity index

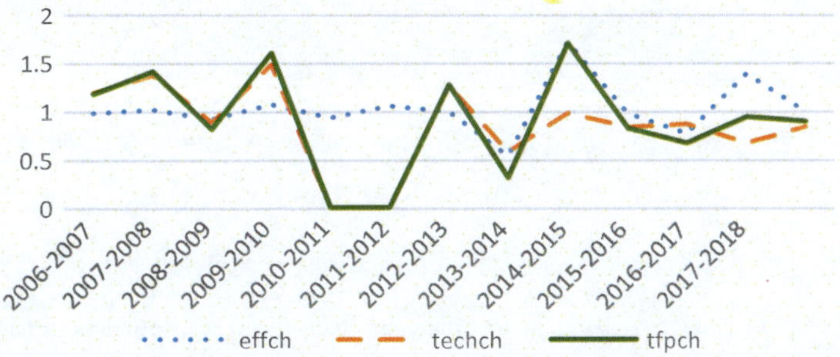

Fig. 1 Line chart of total factor productivity, efficiency change, and technology change

is similar to that of effch curve. This shows that before 2014, technological progress is the main reason for changes in total factor productivity. Since 2014, technological progress has not fluctuated much, and efficiency changes have become the main reason.

From the perspective of techch, the overall volatility is large, showing a downward trend. In 2006–2014, a total of 4 years of techch is greater than 1; while in 2010–2011, techch is very small, only 0.01. Judging from the development process of the PPP model, from 2006 to 2008, it was in the stage of vigorous promotion of PPP projects. The investment in PPP projects as a whole in the society increased, and the technology also progressed, so the project success rate increased as well. From 2009 to 2011, the PPP project was in the low tide. Due to the economic crisis and the implementation of the four trillion stimulus plan, the PPP model was severely hit. Many PPP projects were forced to be suspended, resulting in low overall PPP project investment efficiency. The government then began to attach importance to the importance of private investment, and the PPP model gradually recovered. It can be seen from effch that the overall efficiency is on the rise, indicating that the overall organizational management level of PPP projects has improved. Before 2013, it fluctuated around 1; after 2013, the effch decreased in 3 years. This decrease in efficiency may be due to the reduction in investment scale.

Then we study the reasons for effch from two aspects: pure efficiency change (referred to as pech) and scale efficiency change (referred to as sech). In pech, only the values between 2010–2011 and 2014–2015 are less than 1, and its pech is greater than 1 every next year. It can also be seen from Fig. 2 that the curve of effch and the curve of sech almost coincide. Therefore, the effch less than 1 between 2010–2011 and 2015–2016 is mainly due to the declining input scale. Combining the investment amounts of PPP projects in the four regions in Table 2, it can be seen that the investment amount showed a sudden decrease between 2011 and 2015.

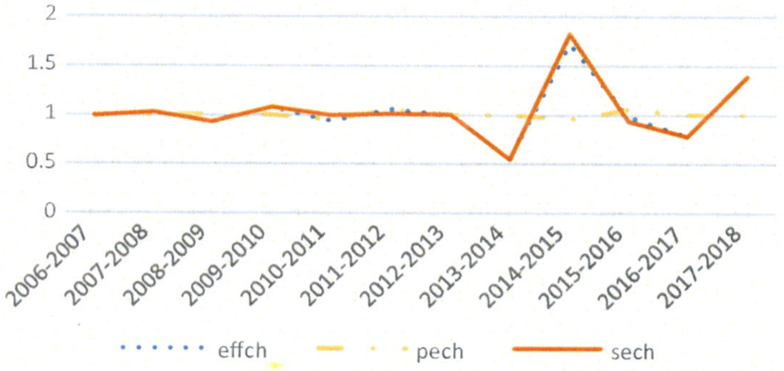

Fig. 2 Line chart of efficiency change, pure efficiency change, and scale efficiency change

Table 2 2006–2018 total factor production index analysis results

Year	effch	techch	pech	sech	tfpch
North China	1.004	0.987	1	1.004	0.998
East China	1	0.652	1	1	0.652
South China	1	3.067	1	1	3.067
Central China	1	0.784	1	1	0.784
Mean	1.001	1.376	1	1.001	1.377

effch stands for efficiency change; *techch* stands for technology change; *pech* stands for pure technology efficiency change; *sech* stands for scale efficiency change; *tfpch* stands for total factor productivity index

3.2 *Analysis of Total Factor Productivity by Region*

Looking at Table 2, it can be found that only the tfpch in South China far exceeds 1, up to 3.067, and the remaining three areas are all less than 1. At the same time, I find that tfpch in various regions is mainly caused by technological progress.

Next, I will analyze the total factor productivity of each region from effch, techch, pech and sech in the 13 years of the four regions.

3.2.1 Efficiency Change

Table 3 is the efficiency change index of each region. It can be seen from the table that the average value of the efficiency change in the four regions is greater than 1, indicating that the organizational management level of the PPP project has improved. Among them, the effch in East China and Central China doesn't change much, fluctuating around 1, indicating that the PPP projects in these two regions have been developing steadily.

As shown in Fig. 3, there have been four times in North China where effch is less than 1. It appeared twice in the four years 2006–2010. This may be because the PPP model was first developed in coastal cities at the time, and it was only promoted in 2002. Therefore, it was only around 2005 that PPP was developed in North China. The factor input level or management level was not perfect, and it was not yet on the frontier of effective production. The other two cases where effch is less than 1 occurred in the last 4 years, that is, between 2014 and 2018. The effch less than

Table 3 Efficiency change index of each region

	North China	East China	South China	Central China
2006–2007	0.947	1	1	1
2007–2008	1.103	1	1	1
2008–2009	0.73	1	1	1
2009–2010	1.369	0.982	1	1
2010–2011	1	1.018	0.766	1
2011–2012	1	1	1.305	0.98
2012–2013	1	1	1	1.02
2013–2014	1	1	0.088	1
2014–2015	0.78	1	11.336	1
2015–2016	1.283	0.732	1	1
2016–2017	0.461	1.366	1	0.578
2017–2018	2.171	1	1	1.73
Mean	1.07	1.008	1.791	1.026

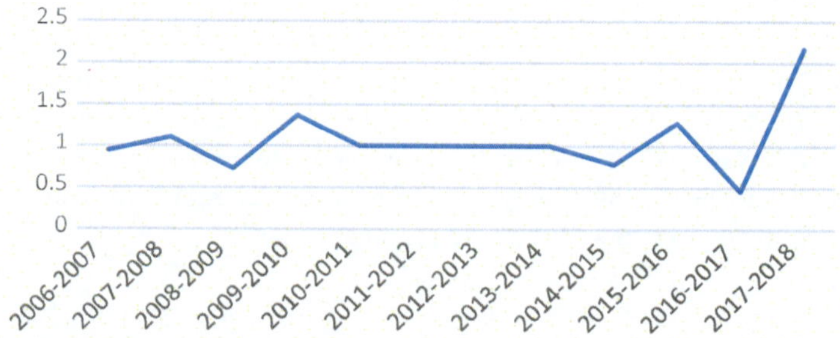

Fig. 3 Efficiency change in North China

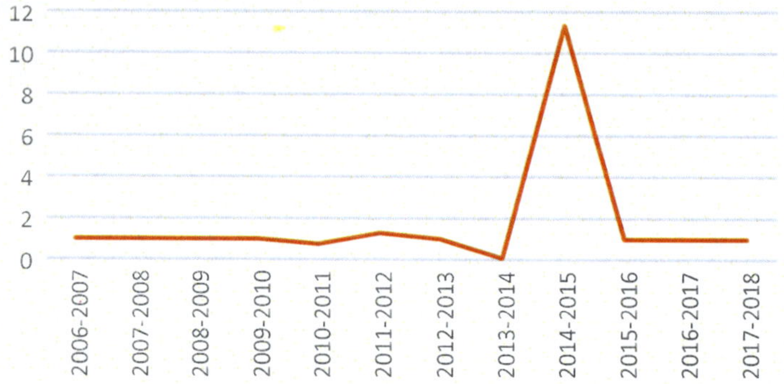

Fig. 4 Efficiency change in South China

1 between 2014 and 2015 was mainly due to the negligence of management level, which decreased by about 20%.

The effch in South China has fluctuated greatly, as shown in Fig. 4. It shows a large fluctuation between 2014 and 2015. Considering that the management level is a relatively stable factor, there isn't be a sudden and particularly large decrease. This shows that the fluctuations in 2014–2015 should be due to the fact that PPP projects in southern China haven't achieve economies of scale, resulting in a low correlation between changes in investment and output. Therefore, the 2015 effch is as high as 11.336 within a reasonable range.

3.2.2 Pure Technical Efficiency

It can be seen from Table 4 that most of the pech are greater than or equal to 1, which shows that the four regions have managed the PPP projects better and have allocated resources reasonably. The government's ability to control and macro-allocate private capital and the ability to allocate various factors of production have reached a good level. Only in North China and South China, the pech did not reach 1 for one year. This may be due to the fact that these years are all years of significant changes in the PPP model, and local governments are too busy reforming to ignore the management of existing projects.

3.2.3 Scale Efficiency

It can be seen from Table 5 that the mean value of sech in all four regions is greater than 1, indicating that the investment scale of PPP projects in these regions is in greater agreement with the production scale on the optimal frontier. Among them, the situation of sech less than 1 in North China in 2006–2010, considering the development of the PPP model, in the early period of North China, PPP has just emerged and the scale of investment was at a relatively small stage.

3.2.4 Technical Progress

The technological progress in North China has been in a fluctuating state, as shown in Fig. 5. Before 2010, it showed a steady growth. This is due to the growth effect caused by technological progress, which not only brought about an increase in short-term output levels, but also brought about the sustainability of economic growth. In 2011 and 2012, there was a lack of innovation in technology. After 2012, it continued to fluctuate, increasing first and then decreasing. This is a catch-up effect of technical efficiency, which only led to the growth of actual output and was not sustainable.

Table 4 Pure technical efficiency values in the four regions

	North China	East China	South China	Central China
2006–2007	1	1	1	1
2007–2008	1	1	1	1
2008–2009	1	1	1	1
2009–2010	1	1	1	1
2010–2011	1	1	0.801	1
2011–2012	1	1	1.249	1
2012–2013	1	1	1	1
2013–2014	1	1	1	1
2014–2015	0.802	1	1	1
2015–2016	1.247	1	1	1
2016–2017	1	1	1	1
2017–2018	1	1	1	1
Mean	1.004	1	1.004	1

As shown in Fig. 6, with 2011–2012 as the boundary, the technological progress in East China is in a state of decreasing first and then increasing. From 2006 to 2011, the majority of techch values were greater than 1, indicating that PPP projects have improved the quality of production factors or improved science and technology. Although it has been growing from 2012 to 2017, it is still less than 1, indicating that there is no obvious innovation in the PPP model in East China in recent years.

As can be seen in Fig. 7, the early stage of technological progress in South China is in a declining stage, but generally greater than 1. This is because South China has become a key area for the promotion of the PPP model in the early stage due to its geographical advantage of relying on the sea. The high level of participation of foreign capital here can better introduce advanced foreign technologies, accelerate the development of the PPP model, and drive economic growth. However, it has fluctuated greatly since 2011 and is less than 1 after 2015. This may be because after 2015, the government focuses on the reform of the PPP model and ignores technological innovation.

Table 5 Scale efficiency values of the four regions

	North China	East China	South China	Central China
2006–2007	0.947	1	1	1
2007–2008	1.103	1	1	1
2008–2009	0.73	1	1	1
2009–2010	1.369	0.982	1	1
2010–2011	1	1.018	0.957	1
2011–2012	1	1	1.045	0.98
2012–2013	1	1	1	1.02
2013–2014	1	1	0.088	1
2014–2015	0.972	1	11.336	1
2015–2016	1.029	0.732	1	1
2016–2017	0.461	1.366	1	0.578
2017–2018	2.171	1	1	1.73
Mean	1.065	1.008	1.786	1.026

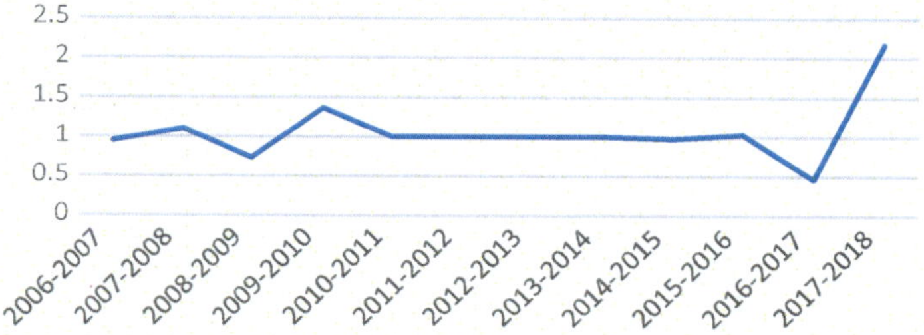

Fig. 5 Line chart of technological progress in North China

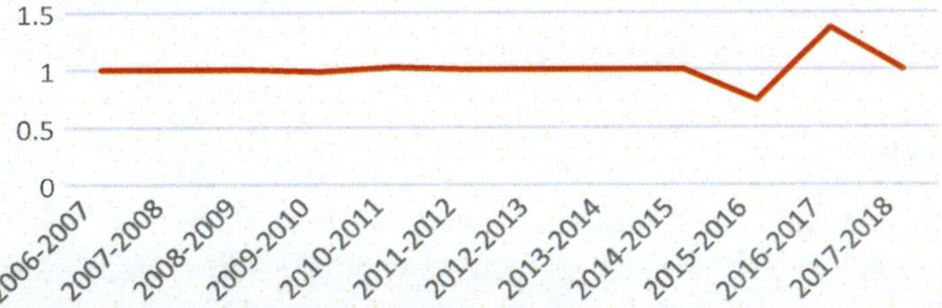

Fig. 6 Line chart of technological progress in East China

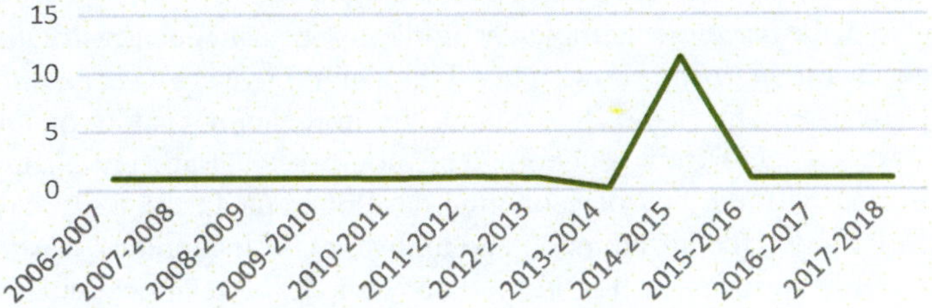

Fig. 7 Line chart of technological progress in South China

The technological progress in Central China is shown in Fig. 8. Although it keeps fluctuating, it is generally less than 1, indicating that there is basically no innovation and new technology introduction in the PPP model in central China. This is in line with the economic and technological development and regional distribution of Central China.

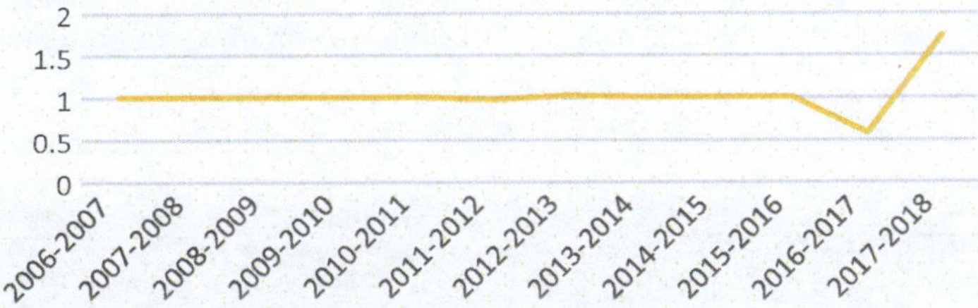

Fig. 8 Line chart of technological progress in Central China

4 Results and Recommendations

4.1 Results

This paper establishes the Malmquist index evaluation system for infrastructure PPP project input to multi-output. Using the super-efficient DEA method, through the calculation of the investment efficiency of infrastructure PPP projects in a total of 18 provinces and cities in North China, East China, South China and Central China from 2006 to 2018, the following conclusions are drawn:

On the whole, the investment efficiency of PPP projects in the infrastructure sector in our country is not high and shows a trend of declining year by year. And with the continuous advancement of the PPP model, the influencing factors of investment efficiency have changed from technological progress to efficiency changes. This change is in line with the law of economic development. In the early days of the rise of the PPP model, the introduction and innovation of technology was the key to the success of PPP projects; while in the later period, the technical conditions are mature, and the importance of organizational management and resource allocation capabilities begins to highlight.

Regional development of PPP projects in my country's infrastructure sector is not balanced. From the perspective of regional distribution, the pure technical efficiency of the four regions is relatively balanced, indicating that the investment scale has reached a moderate level, and the government and society have good management levels. Therefore, efficiency changes are mainly affected by changes in scale efficiency. From the perspective of technological progress, the early technological progress in North China, East China, and South China has been relatively large, and in 2010 or later, there has been a state of technological stagnation and difficulty in progress. In Central China, it has been difficult to break through technological progress.

4.2 Recommendations

Aiming at the above basic conclusions and the existing problems of the PPP model, the following suggestions are made:

Considering that the investment efficiency of the current PPP model is mainly affected by the management level, and the existing PPP model is still in the promotion stage and lacks a comprehensive legal policy system to restrict its management and operation. Therefore, it is first necessary to establish and improve PPP-related policies to fairly divide the rights and obligations of the government and private capital. In addition, a government supervision system should be established, exercise the government's supervision function, supervise the organization and management of PPP projects, and even use some incentives to mobilize the enthusiasm of private capital to improve management and resource allocation capabilities.

The imbalance of regional development in the investment efficiency of the current PPP model may be due to the fact that some local governments have not correctly understood the "PPP model" and believe that the PPP model is only a means of financing. The government requires funds and enterprises require projects.

In addition, it is necessary to strengthen the understanding and emphasis of government departments on PPP projects, increase the initiative of local governments, better promote cooperation with private capital, and prevent rushing PPP projects under immature circumstances.

Aiming at the problem of imbalanced regional development of investment efficiency, it may also be caused by different levels of regional economic and technological development. Therefore, for different regions, the PPP model should be used reasonably on the basis of the original infrastructure. For the more developed regions, because the original infrastructure is already complete, the development can be diverted to some relatively new areas, such as sewage treatment. For less-developed areas, attention should be paid to the improvement of basic infrastructure, such as water supply.

By measuring the investment efficiency of the PPP projects in the four regions, we can find that there is still a lot of room for progress in technology, especially in those regions where economic and technological development is slow and the PPP model is popularized late. Therefore, in the promotion of the PPP model, it is also necessary to continuously explore and test new technologies, reserve excellent talents, and develop a localized PPP model.

References

1. Hart, O. (2003). Incomplete contracts and public ownership: Remarks, and an application to public-private partnerships. *Economic Journal, 113,* C69–C76.
2. Froud, J. (2003). The private finance initiative: Risk, uncertainty and the state. *Accounting Organizations and Society, 28,* 567–589.
3. Peter, R., Colin, D., & Ming, X. (2010). Comparative performance of PPPs and traditional procurement in Australia. *Construction Management & Economics, 28,* 345–359.
4. Feng, Y., Li, N., & Zuo, S. H. (2017). Study on key factors of the success of PPP models. *Public Finance Research,* 51–58 (in Chinese).
5. Ye, X., & Qi, H. (2015). Research on key influencing factors of PPP project cooperation efficiency-based on the perspective of control right. *Project Management Technology, 13,* 9–14 (in Chinese)
6. Liu, Q., & Lu, Y. (2016). Institutional quality, economic environment and efficiency of PPP projects. *Journal of Economics and Management, 30,* 58–65. (in Chinese).
7. Deng, Z., Song, S., & Chen, Y. (2016). Private participation in infrastructure project and its impact on the project cost. *China Economic Review, 39,* 63–76.
8. Zeng, G., & Yin, Y. (2012). Measurement and evaluation of urban infrastructure construction and utilization efficiency in China. *Journal of Zhongnan University of Economics and Law, 5,* 3–8.
9. Charnes, A., Cooper, W. W., & Rhodes, E. (1978). Measuring the efficiency of decision making units. *European Journal of Operational Research, 2,* 429–444.
10. Andersen, P., & Petersen, N. C. (1993). A procedure for ranking efficient units in data envelopment analysis. *Management Science,* 1261–1264.

11. Malmquist, S. (1953). Index numbers and indifference surfaces. *Trabajos De Estadistica, 4,* 209–242.

A Study on Strategic Risk Control Technology of Energy Enterprises in the New Era—A Case Analysis of the State Grid Cooperation of China

Fangcheng Tang, Shuo Sun, Mingqi Duan, Yang Yang, and Mengju Wei

Abstract With the rapid economic development and the acceleration of the globalization process, energy-based enterprises in the new era have also become more and more diversified, and corporate risks have also increased. For energy-based enterprises in the new era, enterprise strategic risk control plays an important role. Taking the State Grid as an example to analyze the current status of strategic risk control of energy-based enterprises in the new period, and to sort out the main methods and models involved in the risk control process, in the future can help companies improve the strategic risk control system and improve corporate strategic risk control technology.

Keywords Energy-based enterprises in the new era · Strategic risk control · Strategic risk assessment

F. Tang · S. Sun (✉) · M. Duan
School of Economics and Management, Beijing University of Chemical Technology, Beijing, China
e-mail: 2019400177@mail.buct.edu.cn

F. Tang
e-mail: fctang@163.com

M. Duan
e-mail: 2019200879@mail.buct.edu.cn

Y. Yang · M. Wei
Energy Development Research Department, State Grid Hebei Economy Research Institute, Shijiazhuang, China
e-mail: yyang8958@126.com

M. Wei
e-mail: jyy_weimj@he.sgcc.com.cn

M. Li et al. (eds.), *IEIS 2020*, https://doi.org/10.1007/978-981-33-4363-4_27

1 Introduction

With the development of economy and the increase of types of enterprises, the diversified development is gradually expanding, which makes the risk level of enterprises intensify and new types of risks increase. From the perspective of strategic management, companies need to recognize potential risks in order to improve their competitiveness and expand market areas. Therefore, it is necessary to effectively prevent the occurrence and development of risks through strategic risk control and related technologies.

Clearly, there is a high demand for strategic risk control techniques by firms. Mark L. Frigo analyzes strategic risks from the internal environment to measure corporate financial risks, governance risks, innovation risks, operational risks, cooperation risks [1]. No matter which kind of risk, it is important for enterprises to control it in a comprehensive and multi-level way. From an internal perspective, effective strategic risk control techniques can promote the economic development of an enterprise and effectively protect its economic interests. At the same time, strategic risk control can prevent problems before they occur to adequate reserves of "power" to cope. From the external point of view, Ruigrok believes that in different environments and development stages, the risks faced by enterprises are different, and their risk control measures should be more targeted [2], improve the external competitiveness of enterprises and the trust of consumers, but also motivation to expand into other areas of business.

This thesis takes State Grid as an example, analyzes the development of State Grid's strategic risk control from the perspective of the principles and purposes of corporate strategic risk control, the current status of corporate risk control and other perspectives, builds the structural framework and development ideas of relevant risk control techniques, organizes and evaluates its commonly used corporate risk control evaluation models, and builds State Grid's strategic risk control system.

2 Purpose and Principles of Enterprise Strategic Risk Control

2.1 Enterprise Strategic Risk Control Purpose

The purpose of strategic risk control in business Ruefli and Sarrazin argued that the environment faced by firms is highly ambiguous and unstable [3]. However, it is possible to standardize and comprehensiveness of the objectives of corporate risk control through the existing systems and framework of the enterprise. From the internal control perspective, the objective is to identify different risk factors in the process and to use risk assessment for risk factors determine the level of risk to take effective measures. In other words, the purpose of risk control is to ensure that the multi-level objectives of the company are achieved and that the strategic environment in which the company operates is safe and legal. Finally, building good corporate relationships with users and creating a core competitive position of the enterprise will help the user and the enterprise as a whole. Strategic decision makes the relationship between users and enterprises from mutual risk transfer to risk sharing. The good relationship between the enterprise and the user can also establish a reputation for cooperation, attract more attention from users to the enterprise, and build the enterprise to compete in the market. The core competitive position in the present.

2.2 Enterprise Strategic Risk Control Principles

In the process of designing risk management and control model, enterprises should follow the principles suitable for their own actual situation. First of all, enterprises should be the principle of legitimacy, the principle of equal rights and responsibilities, the principle of cost-effectiveness and other basic categories of principles; secondly, improve the national grid strategic risk control principles are highly, multi-level and multi-faceted consideration of corporate risk control principles. For example, the principles of reward and motivation, R. S. Kaplan argues that an effective evaluation system motivates leaders and employees into enterprise construction, can effectively stimulate personnel motivation [4]; the principle of sustainable development, National Grid has a unique the nature of the enterprise, in the production process, risk control technology can make the enterprise focus on the principle of sustainability, to ensure that the enterprise to do sustainable and green development; management integration principle, risk control technology can reduce the enterprise power branch too much, the situation of power responsibility is not concentrated.

3 The Current Status of Strategic Risk Control of National Power Enterprises

3.1 Enterprise Overall Risk Control Status

- Building a solid foundation and continuously optimizing the risk control environment. Regulate the scientific control mechanism and integrate the organizational structure of risk management; improve the decision-making mechanism of risk control and make the operation of the risk committee smoother; strengthen risk control. Culture cultivation and the awareness of risk control among all staff is being strengthened.
- Establish rules and regulations, and continuously improve the system of risk control standards. We will promote the construction of system standards in an orderly manner and improve the scientific and applicability of the system; comprehensively construct a framework for classifying major risks in the four sectors of

power grid, industry, finance and international, and unify the language of risk management; standardize the evaluation and supervision of internal control and form risk control operation guidelines.

- Timely response, increasingly mature risk prevention and control mechanisms to carry out KRI analysis and early warning of major risks, and comprehensive study of risk occurrence trends and distribution. Clarify reporting mechanisms, issue risk alert notices, and target response efforts.
- Professional collaboration and joint risk prevention and control are gradually improved. In-depth evaluation of internal control will be carried out to uncover weak management links. At the same time, we will strengthen special auditing and supervision to expand the content and extension of risk control. Optimize the collaborative supervision mechanism and strengthen the comprehensive control of business risks.
- Technical support and the steady advancement of intelligent application of risk control. Online inquiry and improvement of risk control standards, online collection and rectification of problems, online application and analysis of data are realized. According to reasonable normative requirements and effective technical means, the standardization of business operation is enhanced.

3.2 The Current Deficiencies in the Risk Management of Enterprises

Overall, it appears that strategic risk control is progressing rapidly and that relevant regulations and technologies can be applied in a timely manner in risk control to effectively reduce the level of risk. However, in terms of the development process, the strategic risk control of an enterprise is still in the growth stage, compared to large, high-quality enterprises still relatively backward, enterprises should learn from domestic and foreign excellent risk control management system, internalization, to be able to thoroughly transform a set of high standards of risk control.

At the same time, due to the characteristics of the national power grid enterprises, not only belong to the state enterprises, but also related to the daily life of the general public. It also involves different fields, obviously security risk control is more difficult, the nature of the enterprise and its characteristics should be relevant analysis, through effective the ways and means to further improve the enterprise's strategic risk control system, to be able to formulate a complete strategic risk mechanism and regulations, so that the risk the control requirements and objectives are clearer.

4 Research on Corporate Strategic Risk Control Techniques

Based on the above analysis, the factors suitable for the State Grid are selected and identification methods and systems are developed, combining the company's strategy and practice with risk management methods, and the two are fused together to build an operable and directly usable risk control technology process applicable to grid companies.

4.1 *Framing and Thinking of Strategic Risk Control Techniques*

The framework and ideas are outlined below.

Provides an overview of the basic theory of strategic risk control. Introduce the project management theory and risk management methodology. The effect of strategic risk assessment is carried out by combining variables such as the focus characteristics of strategic risk control, the characteristics of risk avoidance, and the judgment of performance completion, carry out strategic risk assessment effect evaluation, and then formulate and propose corresponding strategic risk response strategies. Figure 1 shows the technical approach of this study.

4.2 *Commonly Used Models and Methods for Strategic Risk Control*

Frigo suggested that strategic risk management is based on risk identification and assessment [5]. In the research on methods and models for strategic risk control, the most common methods for risk identification are environmental scanning, self-assessment controls and group discussions, process mapping analysis, risk listing and loss classification, performance analysis, Accident tree method, risk expert survey enumeration method, etc. Each method of risk identification should try to meet the characteristics of comprehensiveness, repetitiveness, diversity of methods, reasonable analysis of resources, abundant experience, and rich data.

The main models and methods of strategic risk evaluation are relatively mature. Based on the results of risk identification, strategic risk evaluation adopts a certain method to assess the identified risks, which mainly includes the probability, the level of importance, and the degree of harm to the project are assessed. There are many ways to evaluate strategic risk, which can be qualitative or quantitative analysis of risk. Richard A. Betis and Vijay Mahajin conduct a multidimensional analysis to determine the level of risk [6]. The current evaluation methods on strategic risk

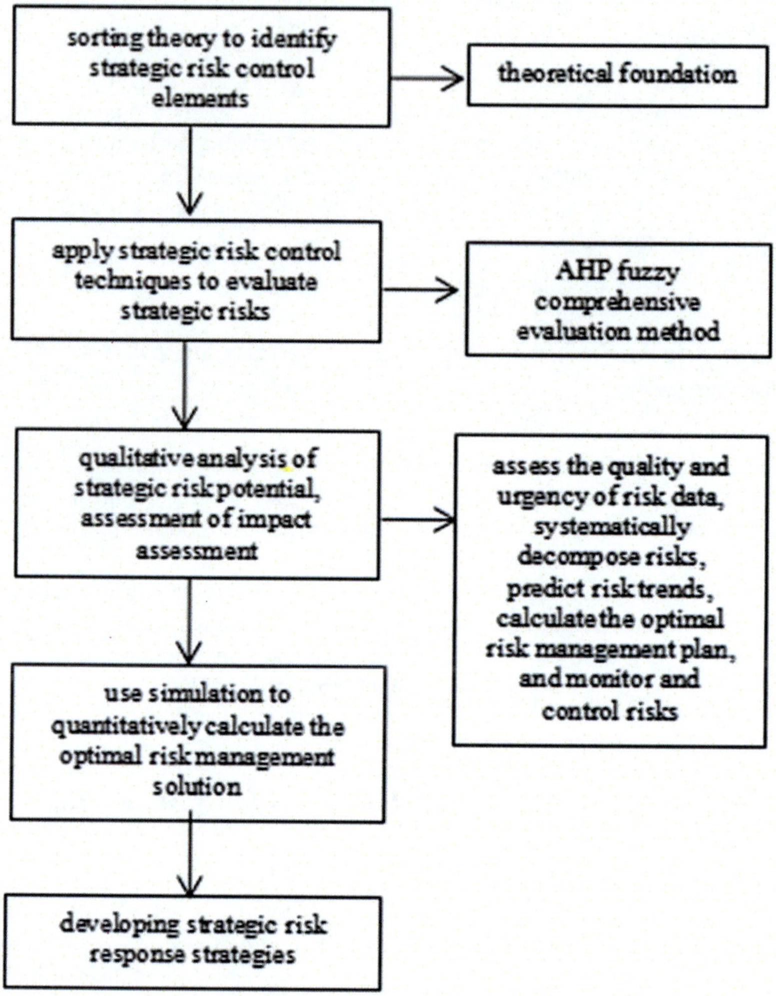

Fig. 1 Technical approach of current study

include the two-dimensional risk assessment method, integrated evaluation method, and hierarchical analysis [7, 8].

In conducting a risk assessment, the impacts to be considered include (among other things) the probability of the risk occurring and the likely impact of the risk, incorporating the various risk is assessed and measured by taking into account external influences. Both qualitative and quantitative analyses are usually applied. Among the methods and techniques for qualitative strategic risk analysis are the assessment of risk likelihood and impact, probability and impact matrices, risk data, and the assessment of external influences. Quality and urgency assessment, risk classification, hierarchical analysis and other methods. Quantitative analysis of risk quantitative analysis of risk methods includes probability distribution, forward projection, sensitivity analysis, probability tree analysis and Monte Carlo simulation methods, etc.

5 National Grid Strategic Risk Control System

Risks resulting from National Grid's increased focus on customers, suppliers, competitors and technology related to the strategy are not attention to the relationship between changes in stock prices on capital markets. As a result, strategic risk assessment began to use a change approach, which exploits the changing characteristics of particular variables, still using the variance of the variables or the most readily available and commonly used variables to estimate the level of strategic corporate risk are return indicators such as ROA and ROE. etc., using changes in returns to assess changes in strategic risk.

Among the many factors influencing strategic risk control, combined with the analysis of the company's actual situation, the factors suitable for the State Grid are selected, and the identification method and system are developed, combined with the company's strategy and practice, and risk management methods, the integration of the two, the construction of a suitable for grid companies, operational and direct use of the risk assessment model. In light of the above analysis, the following strategic risk control system has been developed.

5.1 Qualitative Risk Analysis

The method of establishing appropriate control mechanisms from company's external and internal perspectives, as well as qualitative analysis to investigate and predict risks in order to obtain optimal risk management solutions.

- First of all, building a hierarchical model. When using hierarchical analysis to analyze socio-economic problems, the primary concern is to delve into the nature of the problem and the external environmental variables. Build a basic hierarchical model.
- Second, construct a judgement matrix. The purpose of constructing a judgment matrix is to make a two-by-two comparison of the elements of each layer of the hierarchical model in the first step. The general form of the judgment matrix is as follows:

$$C = \begin{bmatrix} c_{11} & c_{12} & \cdots & c_{1n} \\ c_{21} & c_{21} & \cdots & c_{2n} \\ \vdots & \vdots & \ddots & \vdots \\ c_{n1} & c_{n2} & \cdots & c_{nn} \end{bmatrix} \quad (1)$$

The matrix C has the following properties: $c_{ij} > 0$, $c_{ij} = 1/c_{ji}(i \neq j)$, and $c_{ii} = 1$.

- Third, Solve the judgment matrix and calculate the index weight. Determine the largest eigenvalue of the matrix, and find the corresponding eigenvectors, that is, the corresponding factors in the same scale are compared to the indicators in the previous scale to obtain important weights.
  Set the m-order judgment matrix as:

$$A = \begin{bmatrix} a_{11} & a_{12} & \cdots & a_{1n} \\ a_{21} & a_{22} & \cdots & a_{2n} \\ \vdots & \vdots & \ddots & \vdots \\ a_{m1} & a_{m2} & \cdots & a_{mm} \end{bmatrix} \quad (2)$$

  Normalize the elements in A by row.

$$a_i = a_{ij} \Big/ \sum\nolimits_1^m a_{ij} \quad (i = 1, 2, \ldots, m) \quad (3)$$

$$w_i = a_i \Big/ \sum\nolimits_1^m a_i \quad (i = 1, 2, \ldots, m) \quad (4)$$

- Fourth, the consistency test of the judgement matrix. The judgements need to be consistent and not contradict each other, so as to avoid contradictory judgement results.
- Finally, the hierarchy is ranked singly and overall. The relative importance of a hierarchy factor to the previous hierarchy factor is calculated [9].

5.2 *Quantitative Risk Analysis*

Quantitative analysis of risk means quantifying the identified risk factors and visually expressing them in numerical terms to identify evaluation rating of risk factors. Timely detection and response to threats to the desired strategy from changes and unpredictability of the external environment through the establishment of early warning and crisis systems. The main methods used for the quantitative analysis of risk factors are the expert evaluation method and the fuzzy synthesis method.

1. *Expert evaluation method*: To take advantage of the experts' research in the relevant field, and after talking to the relevant experts, list the risks the project may

face, and the risk factors of the project are classified according to different criteria. Finally, it is referred to experts for investigation and analysis, so as to complete the purpose of evaluating and judging the risk factors of the project. Since there are few historical references for investment risk management in large construction projects, it is important to understand the relationship between project risk management participants and the expert predicts the risk by conducting risk meetings, drawing on the knowledge and experience of the expert, and taking into account the actual situation of the project weighing and evaluating the index weights and ranking of the factors, and then integrating the results of each expert's judgment and using the corresponding methods calculations are made to arrive at a risk evaluation rating and results for the project.

2. *Fuzzy integrated evaluation method*: The specific steps are: firstly, to determine the evaluation factor set. This may include the first-level index set, the second-level index set, the third-level index set, etc.; The second construction weight level can be calculated and determined according to the AHR method. Third, the establishment of an evaluation set. V: $V = (V_1, V_2, \ldots, V_n)$ where V_n denotes the nth evaluation results. Set the evaluation set to $V = (V_1, V_2, V_3, V_4, V_5)$; fourth, evaluate the single factor fuzzy. Establishment of a fuzzy composite evaluation matrix R for the three levels of indicators:

$$R = \begin{bmatrix} r_{11} & r_{12} & \cdots & r_{1m} \\ r_{21} & r_{22} & \cdots & r_{2m} \\ \vdots & \vdots & \ddots & \vdots \\ r_{m1} & r_{m2} & \cdots & r_{mm} \end{bmatrix} \tag{5}$$

Fifth, the fuzzy operation is performed step by step from low to high. The method of fuzzy synthesis is used to perform fuzzy operations step by step, and finally the results of fuzzy operations for the first level indicators are obtained.

$$\mathrm{B_{ij}} = \mathrm{A_{ij}R_{ij}} \tag{6}$$

$$\mathrm{B_{i}} = \mathrm{A_{i}R_{ij}} \tag{7}$$

$$R = (B_1, B_2, \ldots, B_m) = \begin{bmatrix} \mu_{11} & \mu_{12} & \cdots & \mu_{1n} \\ \mu_{21} & \mu_{22} & \cdots & \mu_{2n} \\ \vdots & \vdots & \ddots & \vdots \\ \mu_{m1} & \mu_{m2} & \cdots & \mu_{mn} \end{bmatrix} \tag{8}$$

Finally, the results of the fuzzy composite evaluation are obtained:

$$B = AR(\alpha_1, \alpha_2, \ldots, \alpha_m)\begin{bmatrix} \mu_{11} & \mu_{12} & \cdots & \mu_{1n} \\ \mu_{21} & \mu_{22} & \cdots & \mu_{2n} \\ \vdots & \vdots & \ddots & \vdots \\ \mu_{m1} & \mu_{m2} & \cdots & \mu_{mn} \end{bmatrix} = (b_1, b_2, \ldots, b_n) \quad (9)$$

The research closely follows the strategic planning work of the new era of development of the national grid, based on the optimization of the corporate governance structure and strategic decision-making system to fundamentally realize the company's risk management and control, to achieve a healthy, harmonious and sustainable development and operation. When researching the strategic risk management control model and method, we fully integrate and respond to the head office's proposal to implement new business implementation such as energy storage. Promote cooperation between industrial units and provincial companies to set up joint ventures in areas such as electric vehicles, integrated energy, and commercialization of basic resources. The strategic deployment of the company to build a new platform for innovation and development is the basic starting point for updating the strategic execution evaluation model and methodology. The identification and evaluation of strategic risk control applicable to new business, and can be combined with specific workflows to carry out the corresponding risk control models optimization.

6 Conclusion

For enterprises, the implementation of strategic risk control is very necessary, and an effective and perfect strategic risk control system can improve the enterprises' economic and operational efficiency, the ability to facilitate the achievement of corporate objectives and closely align corporate strategy with risk control, and the ability to build the enterprise strategic risk control system can effectively improve the risk control techniques. Reasonable application of risk control models and methods can effectively help enterprises to improve their enterprise strategic risk control techniques and further expand their advantages to improve business efficiency.

Acknowledgements This research was financially supported by Science and Technology Project of SGCC "Research on Construction of Strategic Goal System and Key Technology of Strategy Implementation for SGCC in the New Era", National Science Fund of China (71532003), Fundamental Research Funds for the Central Universities (buctrc1804), First-class Discipline Construction Funds for BUCT (XK1802-5).

References

1. Paladino, B., Cuy, L., & Frigo, M. L. (2009). Missed opportunities in performance and enterprise risk management. *Journal of Corporate Accounting & Finance, 20,* 43–51.
2. Ruigrok, W., Amann, W., & Wagner, H. (2007). The internationalization-performance relationship at Swiss firms: A test of the S-shape and extreme degrees of internationalization. *Management International Review, 47,* 349–368.
3. Ruefli, T., & Sarrazin, J. (1981). Strategic control of corporate development under ambiguous circumstances. *Management Science, 27*(10), 1158–1170.
4. Kaplan, R., & Atkinson, A. (1996). *Advanced management accounting* (pp. 10–11). Upper Saddle River: Prentice-Hall.
5. Beasley, M. S., & Frigo, M. L. (2007). Strategic risk management: Creating and protecting value. *Strategic Finance, 88,* 24–29, 31, 53.
6. Bettis, R. A., & Mahajan, V. (1990). Risk analysis in corporate performance measurement. In *Risk, strategy and management* (pp. 176–201). Greenwich, CT: JAT Press.
7. Bettis, R. A., & Thomas, H. (1990). *Risk, strategy, and management.* Greenwich, CT: JAT Press.
8. Collins, J. M., & Ruefli, T. W. (1992). Strategic risk: A ordinal approach. *Management Science, 38,* 1707–1731.
9. Jianmu, Y., & Mingran, D. (2006). The dimensional space and flexible control of strategic risk. *Journal of Wuhan University of Technology, 28,* 99–11.

Risk-Return Parity Model for the Broad Assets Based on the Fed Model

Yanglong Chen, Xinyi Mei, and Qingde Wang

Abstract The asset allocation has always been an essential research content of global institutional investors. Based on the Fed model, this paper proposes a risk-return parity model for a wide range of assets to balance the returns and risks of institutional investors. Firstly, this paper reviews the proposal and development of the Fed model. Secondly, this paper constructs a measure of the expected return of assets by different assets of medium and long-term cash flow. The measure could compare valuations of the broad investment assets such as fixed income, stocks, trust, real estate, and maximize the revenue under the regulatory constraints. Meantime, this paper puts forward the model of asset parity and makes empirical retrospection. Finally, this paper introduces the risk adjustment factors of each asset and proposes a risk-return parity model for a large class of assets based on the Fed model. This paper enriches the theoretical system of asset allocation and has widespread practical application value.

Keywords Risk-return parity model · Asset allocation · Fed model

This research is supported by the Program of the Co-Construction with Beijing Municipal Commission of Education of China (Grant No. B20H100020, B19H100010).

Y. Chen (✉)
Beijing Laboratory of National Economic Security Early-warning Engineering, Beijing Jiaotong University, Beijing, China
e-mail: chenyanglong@aliyun.com

X. Mei
National Academy of Economic Security, Beijing Jiaotong University, Beijing, China
e-mail: meixinyi1234@126.com

Q. Wang
National Information Center, Beijing, China
e-mail: 839084971@qq.com

M. Li et al. (eds.), *IEIS 2020*, https://doi.org/10.1007/978-981-33-4363-4_28

1 Introduction

The categories of asset allocation have always been an essential subject for institutional investors. In the field of asset allocation research, the mean–variance model proposed by Markowitz can be regarded as a foundational work. He used the expected return to measure asset return and variance to measure asset risk, created the precedent of quantitative data to evaluate asset return and risk [1]. The model to build the best portfolio could significantly improve the asset risk/reward ratio. The process is now commonly known as the categories of asset allocation. Down the road of risk and return optimization, Sharpe [2], Lintner [3], Mossin [4] established the capital asset pricing model (CAPM) and pointed out that the market portfolio was the best combination of risk assets. Black and Jones [5] and Estep and Kritzma [6] proposed a fixed proportion portfolio insurance strategy (CPPI) and time-invariant portfolio insurance strategy (TIPP) respectively to obtain the portfolio with the fundamental return. Black and Litterman [7] introduced investors' subjective opinions to change the passive characteristics of the previous asset allocation models. Since the minimum square deviation of the mean–variance model is the combination of risk minimization. Many scholars have also explored the path of risk minimization. Choueifaty and Coignard [8] proposed to maximize the diversification portfolio by utilizing the correlation of different asset returns. Bridge water, Qian [9, 10] claimed the risk parity model to pursue the balance of portfolio risk exposure. There are also some researchers to study income maximization. Mills [11], Yardeni [12] successively proposed the GEYR model and the Fed model to utilize the relationship between stock returns and long-term bond yields to maximize benefits. Besides, there are also some configuration models of the economic cycle and subjective judgment. For example, Swensen began making equity-led diversification investments for the endowment since 1985. Greetham and Hartnett [13] used the Merrill lynch clock model formed by the price pass-through of the economic cycle for different assets.

Although the study is very serious and has a specific application value, these models generally face different problems in practice. On the one hand, most of the model is the premise of assumption: financing is convenient, shorting is feasible, market pricing is reasonable, price volatility conforms to specific parameter rules. Nevertheless, this is not the case. On the other hand, many models separate the risks and returns of different assets and do not consider the sources of risks from the perspective of asset appreciation. Based on the Fed model, this paper constructs a comparable income index by considering the income sources of different types of assets and introduces the risk factors of different types of assets to obtain the risk-adjusted income, so as to obtain the risk-return parity model of the broad assets.

2 Theoretical Research and Development of the Fed Model

2.1 The Fed Model

Haycocks et al. found that there was an apparent relationship between the stock price, dividends, and interest rates, stock prices, and bond prices were negative correlation. Mills proposed the Gilt—Equity Yield the wire model (referred to as GEYR), he found the first order co-integration relationship among the long-term bond yields, the stock price, and dividend rate. It was found that the rise of Treasury yields has a significant negative impact on stock prices in the short term. Levin and Wright further proposed that GEYR adjusted by the expected inflation rate, and other factors should be used to make decisions. Brooks and Persand, combined GEYR and area system transformation model, by using the monthly data of the United States and Britain and Germany from 1975 to 1997. Moreover, they found that the trading strategy constructed by this mechanism has a higher yield and lower volatility compared with any other proportion of stock index and static portfolio (buy- and -hold). The Fed model was born in this context [14].

2.2 The Main Contents of the Fed Model

A 1997 report [15] by the federal reserve showed a time-series chart of the expected return on the S&P 500-stock index (the inverse of the expected 12-month forward price-to-earnings ratio, or E_{t+1}/P_t) and the nominal interest rate on the 10-year Treasury. Moreover, Yardeni [12] argued that the two variables were correlated, and proposed the Fed Valuation Model (the Fed's Stock Valuation Model, FSVM), later known as the Fed Model. This Model claimed that stocks and bonds were a competitive asset for investors. There was a competitive equilibrium relationship between stock expected return (E_{t+1}/P_t) and the 10-year Treasury yield (Y_{10t}) when both deviate from the equilibrium state, Market arbitrage forces will drive these two returns to convergence, allowing investors to compare them to make asset allocation decisions.

Yardeni continued to point out that this analysis could be used to calculate the fair market price of the S&P 500-stock index (fair—market price). Through using the listed company earnings divided by the bond yields, ($P_{t,fair} = E_{t+1}/Y_{10t}$) could also guide the investment decisions. Since Yardeni, etc. [16] made a further discussion and put forward the improved version (New Improved FSVM), the improved model has joined the expected profit growth of listed companies and market value, which can be expressed as:

$$\mathrm{FEY} = \mathrm{CBY} - \mathrm{b} \times \mathrm{LTEG}$$

FEY is the expected yield of the S&P 500 index, CBY is the yield to maturity of Moody's A-grade corporate bonds, b is the coefficient, and LTEG is the long-term earnings growth. Nevertheless, they also point out [16] that the simple version has historically worked well because the risk of the corporate bond market (premium) has an average offset the effect of the earnings growth of long-term listed companies.

2.3 Further Study of the Fed Model

The Fed model is simple, intuitive, and has gained lots of support since it was proposed. Firstly, it conforms to the basic logic of investors' investment decisions. Secondly, the yield of medium and long-term Treasury bonds can approximate the risk-free interest rate. With the discount model of stock pricing, there should be a negative correlation between the two. Finally, the empirical data in recent decades also basically support this model, but there are also some doubts about this model.

First, the comparability between the expected yield of stocks and the yield of maturity of bonds is questionable. Second, the two have different risk characteristics, for the significant uncertainty to the realization of the expected profits of listed companies. And in accordance with the "Pecking" Theory of enterprises operating results (Pecking Order going), the shareholders are the last to receive the results of a business, so the investors should be able to demand a certain amount of premium, but the FED model without considering these conditions. Third, the market segmentation theory is valid to some extent. The investors could free switch between the stock market and bond market, and the assumption of homogeneous risk appetite will be subject to some restrictions, so the reality is that the short-term deviations from equilibrium state may often occur.

Scholars have studied these arguments. Asness [17] pointed out that the stock prices and the profits of listed companies would be affected by inflation at the same time. So the division by the result should be a practical value. However, the yield to maturity on the bond is a nominal value, which includes a certain degree of inflation expectations, the two should not be compared directly. They proposed an inverse price-earnings ratio (E/P) that considers volatility in stocks and bonds as a modified Fed model, and then they tested the Fed model. The results showed that for the ten-year-span, the regression coefficient of E/P was significantly positive, while the regression coefficient of E/P-YTM10 was not significant. For the 20-year-span, both had significantly positive regression coefficients. For a one-year-span, when data from 1926 to 2001 or 1881 to 2001 were used, both regression equations had significantly positive regression coefficients, while when data from 1955 to 2001 or 1982 to 2001 were used, neither regression equation had significant explanatory power. They also noted that the two subjects being compared in the Fed model had a relatively low correlation between 1926 and 2001, with a correlation coefficient of 0.18, and a high correlation coefficient of 0.81 between 1965 and 2001.

Koivu et al. [18] used the Vector Error Correction (VEC) model to verify and improve the Fed modeling based on data from 1980 to 2003. They introduced the lag k order term and established the VEC model with the equilibrium correction term. The empirical results showed that the Fed model with RMSE to measure VEC has better prediction ability. Bekaert and Engstrom [19] explored the linkage effect of inflation on stock and bond markets by influencing investors' risk appetite. Clinebell et al. [20], aiming at the Fed model, discussed whether the Fed model was helpful for asset allocation from the perspective of portfolio construction. They pointed out the basic idea of the Fed model is that in the long run, the yield to maturity (YTM, replaced by the yield to maturity of the 10-year Treasury note, YTM10) and the expected yield of the stock market (E/P) should be close. That is, the ratio should be close to 1. If (E/P) > YTM10, the ratio of the two is greater than 1. The Fed model believes that stocks are undervalued relative to bonds and should switch from bonds to stocks for investment. That is, the ratio should be compared with a standard (denoted as λ) to make investment decisions. The traditional strict Fed model declares λ is equal to 1. The authors pointed out that this requirement can be expanded, and λ can be variable. They further modeled the time-varying λ using Kalman filtering, and after trying, found that the random walk model was the most appropriate. They constructed a strategy back-test based on the random walk Fed model. They found that the Fed strategy with the introduction of the random walk had significantly higher effective annualized yield and lower volatility compared with basic Fed strategy.

Moreover, although the Fed model has been created for a long time, the Federal Reserve has never officially recognized or called the Fed model. Instead, the Federal Reserve continued to monitor the relevant indicators that the Fed model considers. For example, in the financial stability report, which was officially released by the Federal Reserve in May 2019. The difference between the expected stock index yield (E_{t+1}/P_t, with a forecast period of the next 12 months) and the real yield of the 10-year Treasury bond (Y_{10t} minus the forecast CPI for the next 10 years) was taken as one of the indicators to reflect the financial stability roughly. Furthermore, there are similar indicators for the real estate market, the Capitalization rate (Capitalization Rates), and real yields on the ten-year United States Treasury note's difference, as one of the rough reaction indexes of financial stability.

3 The Broad Asset Return Parity Model Based on Fed Model

3.1 Building the Broad Asset Return Parity Model Based on Fed Model

The classical Fed model has the advantages of intuitive and straightforward. However, it only considers the two categories of assets, stocks, and bonds. We need to try to generalize the model to the problem of investment decisions with more assets. At present, we mainly consider four types of assets, namely fixed income securities,

Table 1 The income parity of the broad class of assets

Asset classes	Earnings parity	Certainty of return	Expected source of risk	Additional revenue source
Bonds	Yield to maturity	High certainty	The credit risk of the principal issuer	No
Stock	Stock index return (E/P)	Periodicity and waviness	Earnings of listed companies are lower than expected; Stock valuations decline	Earnings of listed companies exceed expectation
Trust	The expected return of risk-adjusted trust products	High probability	The credit risk	No
Real estate	Rental rate of return + expected inflation	High probability	Falling rents; Falling inflation	Rising inflation

stocks, real estate, and trust. First, we will construct a comparable return indicator for these asset classes.

Table 1 briefly lists the construction of comparable income indicators for the four categories of assets.

Bonds do not have returns that exceed expectations, but because the issuing body has the credit risk, it has some default risk. The standardization of creditor's rights assets (referred to as trusts), there is also a potential credit risk. Stocks and real estate could get more than expected earnings, might bear more than expected losses. For stocks, the price can be decomposed into earnings per share by the P/E ratio, respectively, reflect the operating conditions of the enterprise and the valuation level given by the market, and both factors could move in both directions. If the listed company does not perform as expected or investor sentiment declines (led to the lower valuations as a result), then holding stocks can be risky, and vice versa. For real estate, the income mainly comes from the rent received after the purchase, and the rent will rise with the occurrence of inflation, while the purchase cost is fixed. So it would be more realistic to consider the expected inflation rate.

3.2 *Empirical Evidence of Asset Parity Model Based on the Fed Model*

3.2.1 Sample Selection and Data Processing

In this paper, the related indexes are selected for empirical analysis for four types of assets:

Table 2 The comparable index and the conversion index of period income of the return parity model

Asset classes	Earnings parity	Conversion index of period income
Bonds	Chinese corporate bond AA index (1–3 years) yield to maturity	Chinese corporate debt AA index (1–3 years) rise and fall
Stocks	1/ Shanghai composite index PE ratio (TTM)	The Shanghai composite index rose and fell
Trusts	The yield rate of non-securities trust products (2–3 years)	The yield rate of non-securities trust products (2–3 years)
Real estate	Real estate rental return + inflation in first/second-tier cities	Real estate rental rate of return in first- and second-tier cities

Stock: the investment cycle of institutional investors is long, and the stock investment style prefers the blue-chips. Therefore, the inverse of the total market value method of the Shanghai composite index (code: 000001) weighted PE is incorporated into the generalized Fed model for the decision and judgment of asset allocation in broad categories.

Bonds: the maturity of bonds assets of institutional investors is slightly longer, and there are higher requirements for bond investment rating. Therefore, this paper chooses the yield to maturity of the AA index (1–3 years) of Chinese bonds and enterprises as the expected yield of bonds in the broad Fed model.

Trusts: With the consideration of the availability of data, this paper chooses the 2–3 years of non-securities trust product yield as the broad Fed model of trust expected return index.

Real estate: Property has a constant inflow of cash, and rents can fluctuate with inflation after the cost of buying is fixed, long-term rental yields significantly higher than the current rental yields. So this paper uses the "first-tier and second-tier cities' current year rental yields + current year CPI" as the Fed expected rate of return model of real estate assets. Table 2 lists the model selection of indexes and the adjustment method.

All the original data are from Wind. In the above time range, it can be found in Fig. 1 that there are significant differences in the expected returns of the four broad assets.

The expected return of stock assets fluctuates the most, ranging from 5 to 11%, with a fluctuation of nearly 6%. The expected return on trust has remained high, between 6 and 8%; Bond yields came second, with the highest expected yield of 6 percent and the lowest expected yield falling below 4%. The expected return on property (from a rental perspective) has been stable, fluctuating around 4%.

3.2.2 Model Method

According to the analysis of the previous for the generalized Fed model, the four weighted expected return of the asset as the optimal target asset allocation. The

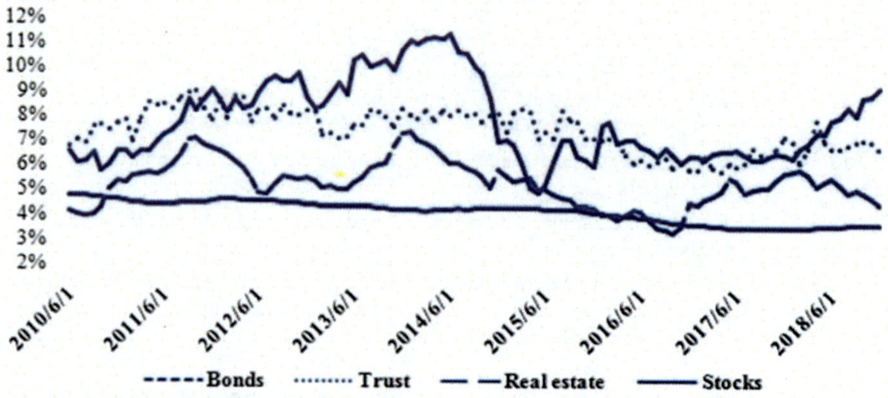

Fig. 1 The trend of the expected return of four categories of assets

regulatory or intra-institutional constraints on the proportion of investment in various assets are taken as the weight constraint condition. According to the expected return of each point of the four categories of assets r_i Optimize the solution for the four categories of asset allocation weights. Then it was held at this optimal weight for one quarter, during which the investment return was calculated every month until the next quarter. The empirical period of the generalized Fed model in this paper was from July 2010 to December 2018.

In order to compare with the income of the real asset allocation, the insurance industry's total investment return is introduced as the benchmark. Because the insurance fund has a wide investment range and a long investment cycle, which is the best benchmark for the asset allocation model test of the broad asset allocation. In the constraint condition, the weight shall refer to the investment restriction of the insurance industry.

$$max\, r = \sum_{i=1}^{4} \omega_i r_i$$

$$st \sum_{i=1}^{4} \omega_i = 1$$

$$\sum_{stock} \omega_i \leq 30\%$$

$$\sum_{non-standard} \omega_i \leq 25\%$$

$$\sum_{real\, estate} \omega_i \leq 30\%$$

$$0 \leq \omega_i \leq 1$$

3.2.3 Empirical Analysis Results

The total investment return of the insurance industry does not include the fluctuation of financial assets available for sale, so it is inappropriate to use the above strategy net value data for comparison directly. In this paper, we calculate the income data of eliminating the floating loss of stocks in the large income parity model. i.e., the stock loss year is uneven; the income is calculated as 0; the profit year is closed; the income is calculated as financial income. Furthermore, the yield of bond assets is changed from the rise and fall of price index to the yield to maturity (i.e., financial yield) at the time of purchase. Table 3 illustrates the income statistics of the vast income parity model.

As is shown in Table 3, two of the three kinds of income parity models are better than the actual investment return in the insurance market. The model is not only original in the model theory but also has the specific application value in real asset allocation.

Table 3 Backtracking of the broad asset return parity model

Year	The total return of the insurance industry (%)	Broad income parity model 1 (%)	Broad income parity model 2 (%)	Broad income parity model 3 (%)
2011	4.85	−3.41	3.09	4.21
2012	3.39	7.60	6.65	4.80
2013	5.04	1.52	3.54	4.52
2014	6.30	22.58	10.31	8.19
2015	7.56	10.88	10.88	9.03
2016	5.66	−0.27	3.42	3.51
2017	5.77	4.86	2.89	3.69
2018	4.30	−3.21	4.16	3.34
Accumulated return	51.75	54.44	54.32	49.49
Annualized return	5.36	5.58	5.57	5.15
Median return	5.35	3.19	3.85	3.95

Table 4 Risk adjustment factors of the broad asset

	Expected return	Risk probability	Risk-adjusted loss
Bonds	AA index (1–3 years) yield to maturity	Bond default rate	Loss on bond defaults
Stock	1/(Shanghai composite price/earnings ratio (TTM))	Probability of ROE decline	ROE decline
Trusts	Return of trusts (2–3 years)	Trusts default rate	Trusts default loss range
Real estate	Real estate rental return + inflation	Historical or expected rent decline probability	The scale of rent decline

4 Risk-Return Parity Model for the Broad Asset Based on the Fed Model

4.1 Risk Adjustment Factors of Each Asset

Different assets have different cash flow characteristics, so different comparable income indicators are constructed. Correspondingly, different assets also have different risks, so this paper constructs risk adjustment factors for different assets (Table 4).

Thirdly, in terms of real estate return and risk adjustment, the downward probability and magnitude of the average residential rent in China's first-tier and second-tier cities since 2008 (cities with too small amounts of data are excluded) are calculated. The rent data is cited from the Eastmoney financial terminal.

Finally, in terms of risk adjustment of stock assets, in order to conduct a more detailed analysis, several representative indexes of the current A-share market are selected. Furthermore, the return on equity (ROE) is used as an alternative index to calculate the downward probability and downward range of each index yield. In order to make the results more stable, each index takes the historical data as long as possible, but there is no uniform data period. The data source of this part of the calculation is still cited from the Eastmoney financial terminal. The above discussion is summarized in Table 5.

4.2 Risk-Return Parity Model for the Broad Asset

According to the expected rate of return of the broad assets at each time point as r_i, Risk probability as p_i, and Risk loss as l_i, Risk Return Parity of the broad assets can be calculated: $RRP = r_i * (1 - p_i * l_i)$. Then, according to the asset allocation constraints of different institutions, the weight of asset allocation can be obtained by optimizing the solution:

Table 5 Assumptions of risk adjustment parameters of each asset

Asset classes	Risk probability (%)	Risk-adjusted loss (%)	Note
Fixed income—deposit	0	0	Bank default risk is not considered in the short term
Fixed income—AAA bonds	0	64	Everbright securities 《Several algorithms for portfolio default rates-bond default rate manual》
Fixed income—AA bonds	2	54	
Trust	1	80	East gold credit rating 《Analysis report on default of creditor's rights trust products in 2018》
Commercial rent	10	5	Annual rent changes in first-tier and second-tier cities since 2008
SZ50 index	50	2	Changes of ROE in each index over the years: (1) downward probability of ROE (2) the downward proportion of average ROE
CSI 100 index	50	2	
CSI 300 index	54	2	
Gem index	38	1	

$$max\ RRP = \sum_{i=1}^{j} \omega_i * r_i * (1 - p_i * l_i)$$

$$st \sum_{i=1}^{j} \omega_i = 1$$

$$0 \leq \omega_i \leq 1$$

4.3 Risk-Return Parity Results for the Broad Assets

According to the indicators of comparable returns of different assets (Table 2) and different risk adjustment factors (Table 4), the risk-adjusted return parity of the broad assets can be calculated. According to the risk-return parity model in Fig. 2, the large-cap blue-chip index has a high return of about 7–9%. The risk-adjusted return of the

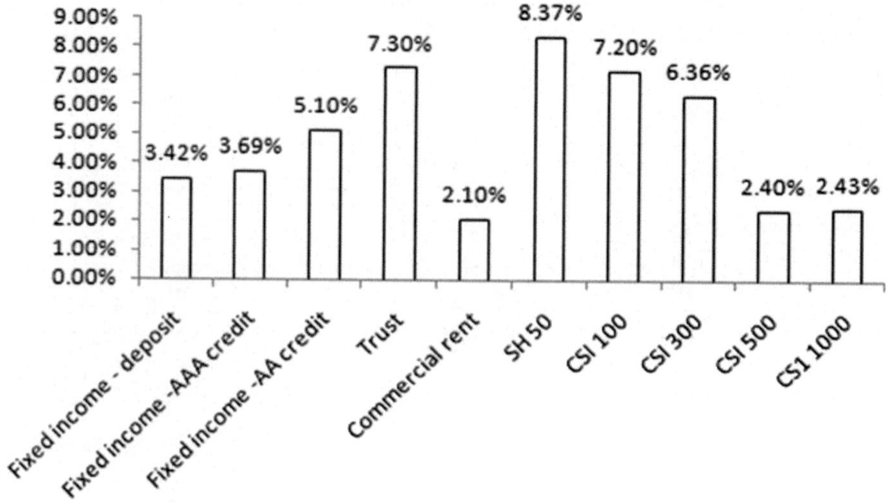

Fig. 2 Risk-adjusted return parity of the broad assets

trust was 7.1%. The risk-adjusted yield on all types of bonds stands at 3–4%, slightly higher than the dividend yield and bank deposits of 2–3%.

It can be seen that the potential returns of the broad asset parity model can be used as an essential reference for institutional investors in asset allocation, due to the asset cash flow and the risk sources.

5 Conclusion and Discussion

This paper extends the classical Fed model and proposes the risk-return parity model for the broad assets. The core idea of the risk-return parity model for the broad assets is to measure the expected return of various assets from the medium- and long-term cash flows of different assets. It introduces the risk adjustment factors of each asset and to maximize the return under regulatory constraints. Compared with the traditional Fed model, which is limited to the allocation of Treasury bonds and stocks, the risk-return parity model for the broad assets constructed in this paper is extended to bonds, stocks, trust, and real estate assets. The Risk-adjusted return parity model is in line with the current institutional investment needs. Compared with other classical models, the risk-return parity model for the broad assets does not rely on the historical volatility of asset prices to measure and evaluate the returns. Nevertheless, the risk-return parity model relies on the future cash flow of assets, which is more consistent with the long-term characteristics of asset allocation and value investment. It also avoids the problem that other classical asset allocation models cannot be optimized without historical price data.

References

1. Markowitz, H. (1952). Portfolio selection. *The Journal of Finance, 7,* 77–91.
2. Sharpe, W. F. (1964). Capital asset prices: A theory of market equilibrium under conditions of risk. *Journal of Finance, 19*(3), 425–442.
3. Lintner, J. (1965). Security prices, risk, and maximal gains from diversification. *Journal of Finance, 20*(4), 587–615.
4. Mossin, J. (1966). Equilibrium in a capital asset market. *Econometrica, 34*(4).
5. Black, F., & Jones, R. W. (1987). Simplifying portfolio insurance. *The Journal of Portfolio Management, 14*(1), 48–51.
6. Estep, T., & Kritzman, M. (1988). TIPP: Insurance without complexity. *The Journal of Portfolio Management, 14*(4), 38–42.
7. Black, F., & Litterman, R. (1992). Global portfolio optimization. *Financial Analysts Journal, 48*(5), 28–43.
8. Choueifaty, Y., & Coignard, Y. (2008). Toward maximum diversification. *The Journal of Portfolio Management, 35*(1), 40–51.
9. Qian, E. (2005). *Risk parity portfolios.* Pan Agora Asset Management Research Paper.
10. Qian, E. E. (2006). On the financial interpretation of risk contribution: Risk budgets do add up. *Social Science Electronic Publishing, 4*(4), 3.
11. Mills, T. C. (1991). Equity prices, dividends and gilt yields in the UK: Cointegration, error correction and 'confidence.' *Scottish Journal of Political Economy, 38*(3), 242–255.
12. Yardeni, E. (1997). Fed's stock market model finds overvaluation. Topical study, 38.
13. Greetham, T., & Hartnett, M. (2004). *The investment clock.* Merrill Lynch, Research Paper, No. 10.
14. Haycocks, H. W., & Plymen, J. (1954). Investment policy and index numbers. *Transactions of the Faculty of Actuaries,* 379–453.
15. Federal Reserve Monetary Policy Report to Congress. (1997). Available at https://www.federalreserve.gov/boarddocs/hh/1997/july/fullreport.htm.
16. Yardeni, E., & Amalia, Q. (2002). *Asset valuation & allocation models.* New York: Prudential Financial Research.
17. Asness, C. S. (2003). Fight the fed model. *The Journal of Portfolio Management, 30*(1), 11–24.
18. Koivu, M., Pennanen, T., Ziemba, W. T., et al. (2005). Cointegration analysis of the fed model. *Finance Research Letters, 2*(4), 248–259.
19. Bekaert, G., & Engstrom, E. (2010). Inflation and the stock market: Understanding the 'fed model.' *Journal of Monetary Economics, 57*(3), 278–294.
20. Clinebell, J. M., Kahl, D. R., Stevens, J. L., et al. (2017). Active asset allocation for retirement funds using the fed model. *Financial Services Review, 26*(2).

Study on Weather Effect of SSE Composite Index

Ying Ren, Yuehui Liu, Peng Liu, and Yingling Tan

Abstract With the rise of behavioral finance, the factors affecting investor behavior have been discovered. Natural environment can not be underestimated among many influencing factors. Therefore, starting from the weather, we propose a new transmission mechanism hypothesis of "weather—investor sentiment—investor behavior—stock market volatility". In this study, with return rate, turnover rate and volatility of SSE Composite Index as the dependent variable and air temperature, air pressure, light, humidity and volatility as the independent variable, the correlation between weather indicators and SSE Composite Index was discussed through EGARCH fitting regression, we finally found that weather can affect SSE Composite Index, and the volatility of SSE Composite Index can reflect the effects of weather on investor behavior more directly. The changes in the weather have stronger influence on the volatility of SSE Composite Index than the weather indicators.

Keywords Weather · Changes in weather · Return rate · Turnover rate · Volatility

1 Introduction

In the 1970s, Fama put forward the efficient market hypothesis and believed that investors were absolutely rational people. In past 30 years, however, some scholars both at home and abroad have found that in violation of the traditional finance theory of financial anomalies, such as equity premium puzzle, the chains of closed-end

Y. Ren (✉) · Y. Liu · P. Liu
School of Economics and Management, Beijing Jiaotong University, Beijing, China
e-mail: renying@bjtu.edu.cn

Y. Liu
e-mail: 18120500@bjtu.edu.cn

P. Liu
e-mail: 14120487@bjtu.edu.cn

Y. Tan
China Development Bank, Beijing, China
e-mail: tanyingling@126.com

M. Li et al. (eds.), *IEIS 2020*, https://doi.org/10.1007/978-981-33-4363-4_29

fund discount, the momentum effect and reversal effect, excessive reaction and the insufficiency, the calendar effect and the effect of seasonal affective disorder, etc. These are the "irrational" behavior of investors as a result.

Daniel Kahneman, a famous American behavioral economist, was awarded the Nobel Prize in economics in 2002, which marked the rise of a boom in behavioral finance in the academic world. Behavioral finance believes that people's behavioral decisions in financial markets can be significantly influenced by cognitive and emotional factors. A large amount of psychological research has shown that people have a more optimistic view of their investment assets when they are in a happy mood than when they are in a low mood, which affects their rational judgment. People's mood is affected by many aspects, the influence of weather conditions on mood should not be underestimated. Numerous studies at home and abroad also show that investor sentiment has a significant impact on the stock market performance [1–5].

Climate change impact on financial markets initially by the American scholar Saunders [6], the paper discusses the influence of cloud cover over Manhattan in on stock market return, the results showed that there is negative correlation between stock returns and clouds, meaning that the lower the cloud cover, the stock market gains will be higher than the average. Keef and Roush [7] found that there was no correlation between cloud cover and the daily return rate of New Zealand stock index, but wind speed had a significant influence on the return rate of New Zealand stock, while temperature had only a slight influence on the return rate of Stock market.

According to the Shanghai Stock Exchange's 2018 statistical yearbook, individual investors accounted for a whopping 99.67% by the end of 2017. And the quality of individual investors in China are relatively high, Central Registration and Settlement Company show that the stock of individual investors in China, 74.7% of the investors have less than a bachelor's degree, that means that some investors are relative lack of stock and stock market theory, its trading behavior is speculation. In addition, the age of individual investors in China is relatively young. Data from the Central Registration and Settlement Company shows that more than 69% of individual investors in China are under 40 years old, and the trend of younger investors' mood instability is more significant. Based on the above analysis, China's stock market is a developing market and its development is not deep enough. There is a serious speculative phenomenon among investors, and the external environment will have a great impact on Chinese stock market investors. From the "weather—investor sentiment—investor behavior—the stock market volatility" the channel comprehensively study on the influence of weather variables and the fluctuation of weather variables on the turnover rate, return rate and volatility of SSE Composite Index maybe can explain the performance and direction of China's stock market.

In this paper, 3293 trading days of SSE Composite Index and weather datas from January 5, 2006 to July 23, 2019 were selected to investigate the influence of weather on the Shanghai composite index. It was found that humidity has a significant influence on turnover rate of SSE Composite Index, and the higher the humidity is, the higher the turnover rate is. The influence of weather and weather variation variables on the return rate of SSE Composite Index is not significant. The weather factors that affect SSE Composite Index include temperature, humidity and air pressure, and these three factors have negative influences on SSE Composite Index. In addition, temperature volatility also affects the volatility of SSE Composite Index. The higher the temperature changes are, the higher the volatility of SSE composite index will be. We also find that the influence of weather change index on volatility is more significant than that of weather factors. Finally, we prove that SSE Composite Index has Monday effect, and the turnover rate, return and volatility of SSE Composite Index on Monday are higher than those of other trading days. To sum up, this study finds that weather does affect the return rate, turnover rate and volatility of SSE Composite Index, especially the more impact on volatility.

Compared with the previous literature, the contribution and innovation of this paper may be in the following aspects: First, this paper selects "the return rate, turnover rate and volatility of SSE Composite Index" as the dependent variables, and a comprehensive selection of the dependent variable indexes can describe the weather in more detail through affecting investor behavior and thus affecting the performance of SSE Composite Index. Secondly, this paper adds the index of weather change. The influence of weather change on investor sentiment is more direct and stronger than the weather itself. So the changes can further improve the transmission channel of weather affecting investor sentiment and then the stock market. Finally, combined with the top five cities of SSE Composite Index in terms of account opening and trading volume, this paper calculates the comprehensive weather and weather change index by weighting, which can reflect the impact of weather on investors of SSE Composite Index more comprehensively and break through the limitation of empirical study based on local weather conditions.

The structure of this paper is as follows: the second part is literature review, the third part is empirical design, the fourth part is empirical results analysis, and the last part is the conclusion of this paper.

2 Literature Review

This paper mainly studied the weather and the influence of the weather changes for the Shanghai composite index, closely related to the literature from the weather, the investor sentiment, investor behavior. So the literature review in this paper focuses on this mechanism.

According to the existing research, different illumination levels will regulate pineal levels in the body, in turn, affect the person's mood: if the sunshine time is not enough, the pineal will be inhibition, and the secretion of melatonin will also be inhibition, and then the adrenaline and thyroid hormone is reduced, thus make a person's cranial nerve tension and the decline in vigor, unresponsive, depression, negative emotions; While sunny weather, the pineal gland level in the body is higher, people's mood is relatively positive and optimistic. In addition, when the air pressure decreases, the oxygen partial pressure in the air also decreases, resulting in people's hemoglobin cannot be saturated with oxygen, resulting in insufficient blood oxygen, heart rate accelerated, shortness of breath and other symptoms of oxygen deprivation resulting in emotional or behavioral loss of control. These studies provide a reliable explanation for the physical effects of weather on mood.

Studies have shown that investor sentiment has a significant impact on stock return and volatility. In terms of investor sentiment on stock return rate, Brown and Cliff [8] construct investor sentiment indicators by means of questionnaire survey, and the results show that investor sentiment will have a significant negative effect on stock market expected return in the next 1–3 years. Schmling [9] investigated the sentiment of institutional investors and found that the sentiment of institutional investors could predict the return rate of stocks. Dragos and Laura [10] use the consumer confidence index to represent individual investor sentiment. Research shows that the consumer confidence index has a positive relationship with the return rate of the stock market, but this effect will be quickly offset by arbitrage. In terms of investor sentiment on stock volatility, Sayim and Rahman [11] by the investor sentiment is divided into rational and irrational factors, found that the change of the rational composition in investor sentiment and negative correlation between Istanbul stock volatility, the irrational factors in the investor sentiment changes and volatility is significantly positive correlation; Da et al. [12] constructed the FEARS index through the search volume data of millions of families on words such as "recession", "unemployment", "bankruptcy", etc., and used the index to reflect investor sentiment. The research results showed that FEARS index could increase stock volatility in the short term. At the same time, domestic scholars have also conducted in-depth studies on the relationship between investor sentiment and the returns and volatility of the stock market. Zhang et al. [13] divided investor sentiment into individual investors sentiment specific emotions and institutional investors to compare, analysis of institutional investor sentiment and individual investor sentiment influence on stock returns shows that the influence of institutional investors on the stock market returns is more significant. Then Zhang et al. [14] used the growth rate of unexpected investor accounts as a proxy for investor sentiment, the empirical results found that stock market value, book market value and other indicators can periodically forecast the stock expected return, but the relationship between beta and stock future earnings is in contrast to the theory. Zhang and Yang [15] did not stop there, they used the market turnover rate, closed fund discount rate and the growth rate of investor accounts three indicators to construct emotional variables indirectly, by correcting DSSW model, using the least square method and GARCH model, finally they found positive and optimistic emotions have a more significant impact on stock returns than negative emotions have on stock market prices, which indicates that the impact of investor sentiment on stock market returns is asymmetric.

In terms of the impact of weather on the stock market, the weather index selected by scholars in the initial study is mostly cloud coverage (light). Saunders [6], an American scholar, who was the first foreign scholar to do the research in this filed, concluded that the higher the cloud cover was, the lower the return of Dow Jones Industrial Index was. Hirshleifer and Shumway [16] took the panel data of major index return rates of 26 countries as the dependent variable and the cloud cover of off-seasonal influence as the independent variable to conduct research. The research results showed that the influence of sunshine on stock market return rates was prevalent and significantly negative in 26 countries. In the following studies, scholars have increasingly selected more and more weather indicators. Keef and Roush [7] mainly selected cloud cover, wind speed and temperature when studying the weather effect of New Zealand stock market. Yi and Wang [17] choose more abundant weather factors, including pressure, temperature, relative humidity and total cloud cover, wind speed and precipitation, etc., the results showed that the temperature, sunshine and humidity have positive effect on the return of SSE Composite index, the influence of wind speed, cloudy and precipitation is negative. Cao and Xu [18] used air pressure, temperature, relative humidity, wind speed, precipitation and sunshine index to comprehensively measure the weather. Only temperature and humidity had significant influences on stock market returns, among which the influence of temperature was negative and the influence of humidity was positive. The weather variables selected by Lu [19] mainly include cloud coverage, temperature, dew point temperature, air pressure, visibility and wind speed, etc. The research conclusion shows that investor sentiment does not lead to the change of the return rate of SSE Composite Index. Only temperature has a positive influence on the return rate, while other weather factors have no obvious influence. Seasonal affective disorder has a close relationship with turnover rate and volatility, which indicates that the weather of proxy sentiment indicator will not directly affect the yield rate, but can directly affect investors' decisions and the liquidity and volatility of the stock market.

To sum up, we can find that most scholars have concluded that weather can significantly affect the return rate and volatility of the stock market, while only a few scholars have concluded that there is no relationship between them or the relationship is weak. On the whole, we can draw the following conclusions: The weather will affect investors' investment behavior by affecting their sentiment, which in turn will affect stock returns and market volatility in the securities market; Moreover, the influence of weather on stock market turnover rate and volatility is more obvious than the influence of weather on stock returns. At the same time, through the comparison of domestic and foreign research results, it can be found that in terms of the impact of weather on the stock market volatility, the domestic developing stock market reacts more strongly than the foreign developed stock market.

We can see that the interception of data year intervals, research methods and research models used in domestic and foreign research literature are relatively comprehensive. But in the selection of dependent and independent variables and is not very comprehensive, independent variables commonly used weather factors, the dependent variable is generally adopted in the stock market returns and volatility. In this study, the return, turnover rate and volatility of the Shanghai Composite Index are selected as dependent variables, and the weather index and weather change index are selected as independent variables. Adding the turnover rate can more vividly reflect investors' behavior, and adding the weather change index is to more clearly see how the weather change directly affects investors' decision-making.

3 Empirical Design

3.1 Data Sources

This study mainly discusses the relationship between weather changes and the volatility of SSE Composite Index.

The sample interval is the daily SSE Composite Index data and weather data from January 5, 2006 to July 23, 2019, with 3293 entries for each variable. We selected the weather data of Beijing, Guangdong, Jiangsu, Shanghai and Zhejiang, including temperature, humidity, dew point, air pressure, visibility, wind speed and light, etc. In addition to the weather of Shenzhen selected by Guangdong, the weather of the capital cities of Jiangsu and Zhejiang were selected, namely Nanjing and Hangzhou. The weather data came from the National Oceanic and Atmospheric Administration.

The data of SSE Composite Index are derived from Shanghai Stock Exchange (SSE), including the daily maximum, minimum, closing price, trading volume, and number of outstanding shares of SSE Composite Index. The return rate, turnover rate and volatility of SSE Composite Index are calculated.

3.2 Definition of Variables

3.2.1 Explained Variables

1. Return Rate

Return rate refers to the rate of change between the stock price of the current day and the stock price of the previous trading day. This paper takes the natural logarithm of the stock price index and conducts a first-order difference to obtain the return rate. The calculation formula is as follows:

$$\mathrm{R} = \ln \mathrm{P_t} - \ln \mathrm{P_{t-1}} \tag{1}$$

2. Turnover Rate

Turnover is a percentage of the total volume of a stock in a given year. It describes how often a stock changes hands in the market over a given period of time. The calculation formula is as follows:

$$\text{Turnover rate} = \text{volume/number of outstanding shares} * 100\% \tag{2}$$

3. Volatility

Volatility is the degree of change in the return on asset investment of the index, the calculation formula is as follows:

$$Volt = \frac{P_h - P_l}{\frac{P_h + P_l}{2}} \tag{3}$$

3.2.2 Explanatory Variables

We use temperature, humidity and pressure, and calculate the corresponding volatility of each weather indicator. In addition, combined with the previous studies of scholars, the stock market has calendar effect, namely January effect and Monday effect, as well as seasonal affective disorder effect, etc. Therefore, dummy variables of autumn and winter, dummy variables of January, dummy variables of Monday and indicators of seasonal affective disorder are introduced to make the analysis more comprehensive.

The definitions of all variables are shown in Table 1.

The fall/winter period is from September 21 to March 20 in the northern hemisphere and From March 21 to September 20 in the southern hemisphere. Though slightly different from the actual situation, let's assume that fall and spring occur from September 21 to March 21, respectively. H_t is defined as the length of time from sunset to sunrise in a place.

$$SAD_t = \begin{cases} H_t - 12 & \text{during autumn and winter} \\ 0 & \text{other trading days} \end{cases} \tag{4}$$

Note that the SAD_t variable only represents autumn and winter, which reflects the variable of night length in autumn and winter minus the annual average night

Table 1 Variable definition

Variable	Definition
R	The return rate of SSE composite index
R_{t-1}	The previous day's return rate SSE composite index
TUR	Turnover rate of SSE composite index
VOL	Volatility of SSE composite index
TEM	The daily average temperature, the unit for degrees Celsius (°C)
HUM	Average daily humidity in percentage (%)
PRE	Daily air pressure, the unit for PASCAL (hg)
FW	The dummy variable of autumn and winter, the interval of which is between 21 September and 20 March, is marked as 1, and the others are marked as 0
JAN	The dummy variable of January, the variable interval between January is marked as 1, and the others are marked as 0
MON	Mon-day dummy variable, marked as 1 for Mondays and 0 for others
SAD	Daily average seasonal affective disorder indicator
TEMV	Daily temperature fluctuation, variance of temperature per hour in a day

length of 12 h. The night length H_t can be obtained by using the spherical triangle standard approximation. In order to calculate the night length of the δ dimension, we first calculate the angular velocity of the sun: λ_t

$$\lambda_t = 0.4102 * \sin\left[\left(\frac{2\pi}{365}\right) * (\text{julian}_t - 80.25)\right] \tag{5}$$

The variables julian_t vary from 1 to 365, representing each day of the year. $\text{julian}_t = 1$ represents January 1st, $\text{julian}_t = 2$ represents January 2nd, and so on. Then we can calculate the night length of each day:

$$H_t = \begin{cases} 24 - 7.72 * \text{arcos}\left[-\tan\left(\frac{2\pi\delta}{360}\right)\tan(\lambda_t)\right] \\ \text{In the southern hemisphere} \\ 7.72 * \text{arcos}\left[-\tan\left(\frac{2\pi\delta}{360}\right)\tan(\lambda_t)\right] \\ \text{In the northern hemisphere} \end{cases} \tag{6}$$

The seasonal mood index SAD_t can be obtained by plugging in H_t.

3.3 *Data Processing*

According to statistical yearbook data issued by the Shanghai Stock Exchange, top five individual investors opening account provinces were Beijing accounted for 24%, followed by 14% of Guangdong, Shanghai and Jiangsu go hand in hand, accounted for 6%, followed by the Zhejiang province accounted for 5%, it can be seen that the top five provinces of opening account has accounted for more than 50%. In addition to the analysis of regional distribution of individual investors to open an account, you also need to analyze the regional distribution of turnover, because our focus is on how the weather affected investor sentiment and influence of investor behavior, with a focus on the investment behavior, although some provinces have more investors to open an account, but have less trading volume, therefore we also need to statistics the provinces in the Shanghai stock exchange transactions (Table 2).

Table 2 Trading volume of five province

Region	Number of the sales department	Trading volume (10 billion RMB)	Proportion (%)
Shanghai	489	2713.57	26.8
Guangdong	703	1928.59	19.05
Beijing	265	1529.75	15.11
Zhejiang	381	636.49	6.29
Jiangsu	363	592.86	5.86
In total	2201	7401.26	73.10

In terms of the transaction volume, Shanghai ranked the first, followed by Guangdong, Beijing ranked the third, and Zhejiang also surpassed Jiangsu to rank the fourth.

Since we focus on the trading behavior of investors, we calculate the corresponding weather indicators based on the proportion of the trading volume in the five cities. The weight of Shanghai's total turnover in the five cities is 36.66%, Guangdong's 26.06%, Beijing's 20.66%, Zhejiang's 8.60% and Jiangsu's 8.02%. Then we treated the regional weather indicators as comprehensive weather indicators:

$$\begin{aligned} \text{weather} = {} & 36.66\% * \text{weather}_{SH} + 26.06\% * \text{weather}_{GD} + 20.66\% * \text{weather}_{BJ} \\ & + 8.60\% * \text{weather}_{ZJ} + 8.02\% * \text{weather}_{JS} \end{aligned} \tag{7}$$

3.4 Model Setting

Model 1 is to explore the influence of weather on SSE Composite Index:

$$\begin{aligned} y_t = {} & C + \alpha_1 FW_t + \alpha_2 JAN_t + \alpha_3 MON_t + \alpha_4 SAD_t + \alpha_5 TEM_t \\ & + \alpha_6 HUM_t + \alpha_7 PRE_t + \varepsilon_t \end{aligned} \tag{8}$$

Model 2 introduces weather volatility to explore the impact of weather and weather changes on SSE Composite Index:

$$\begin{aligned} y_t = {} & C + \alpha_1 FW_t + \alpha_2 JAN_t + \alpha_3 MON_t \\ & + \alpha_4 SAD_t + \alpha_5 TEM_t + \alpha_6 HUM_t + \alpha_7 PRE_t \\ & + \alpha_8 TEMV_t + \alpha_9 HUMV_t + \varepsilon_t \end{aligned} \tag{9}$$

The explained variable Y includes the return rate, turnover rate and volatility of SSE Composite Index.

4 Analysis of Empirical Results

4.1 Descriptive Statistical Results

The descriptive statistics of the sample data are shown in Table 3. The average and median of the sample turnover rates of SSE Composite Index are 1.308% and 0.73% respectively, and the maximum and minimum values are 8.76% and 0.2% respectively. The mean and median of volatility were 1.898% and 1.504%, respectively, with the highest volatility reaching 10.632% and the lowest volatility 0.252%. The

Table 3 Descriptive statistical results

	Mean	Median	Max	Min	STD	SKW	Kurt	JB
TURN	1.3080	0.7300	8.7600	0.2000	1.2816	2.0503	7.5968	4607.1490
VOL	1.8981	1.5048	10.6323	0.2521	1.3077	2.1056	9.2783	7841.5090
R	0.0273	0.0877	9.0345	9.2561	1.6596	0.6199	7.3527	2810.4470
TEM	19.6858	20.6291	33.6944	0.7978	7.8406	0.2767	1.9550	191.8514
A HUM	71.3852	72.3369	92.3556	24.9423	10.3893	0.5151	3.0880	146.6892
The PRE	29.9145	29.9095	30.5370	29.2630	0.2429	0.1145	2.0203	138.8795
TEMV	6.5894	5.2171	71.2593	0.3272	5.7710	3.7589	30.8424	114,118.5
HUMV	42.6430	35.6583	192.3306	2.8630	30.1393	1.3234	4.8829	1447.7440
MON	0.1947	0.0000	1.0000	0.0000	0.3960	1.5424	3.3790	1325.3770
JAN	0.0820	0.0000	1.0000	0.0000	0.2744	3.0472	10.2856	12,379.29
FW	0.4774	0.0000	1.0000	0.0000	0.4996	0.0906	1.0082	548.8246
SAD	0.5692	0.0000	1.9337	0.0000	0.7337	0.7821	1.9114	498.3219

mean and median of return rate are 0.027% and 0.087%, respectively. The highest return rate in the sample is 9.035% and the lowest yield is −9.256%. The JB test statistics of the three are large, so they do not obey the normal distribution. Moreover, the peak coefficient of the three is greater than 3, indicating that they are steeper than the normal distribution. The skewness coefficient of turnover rate and volatility is greater than 0, belonging to the right-skew distribution, and the average value of samples is greater than the median. The skewness coefficient of the return rate is less than 0, belonging to the left-skewed distribution, and the average value of the sample is less than the median.

The mean and median temperatures in the samples were 19.69 and 20.63 °C, respectively; the highest temperature and lowest temperature were 33.69 and −0.8 °C, respectively; the kurtosis was less than the kurtosis of normal distribution, and it was left-skewed distribution. The average humidity of the samples was 71.39 and 72.34%, and the humidity was as high as 92.36% and as low as 24.94%. The kurtosis distribution was close to the normal distribution, and both of them belonged to the left-skewed distribution. The mean and median of the pressure were both 29.91, the difference between the highest value and the lowest value was not big, the skewness was close to 0, and the kurtosis was 2, which was lower than the kurtosis of normal distribution.

4.2 Correlation Analysis

Through correlation analysis of explanatory variables and explained variables in the model, we found that among the three explained variables, turnover rate, volatility and return rate were positively correlated, and the correlation coefficient between turnover rate and volatility was higher, which was 0.4129. However, there is an inverse correlation between volatility and return, with a correlation coefficient of −0.1249. From the correlation analysis, we can see the influence of explanatory variables on explained variables as follows: Temperature has a negative influence on the turnover rate, volatility and return rate of SSE Composite Index; Humidity is inversely correlated with turnover rate and return rate, but positively correlated with volatility. Air pressure has a negative effect on turnover rate, but a positive effect on return rate and volatility. The influence of temperature change and humidity change on the three explained variables is positive. The Monday effect and The January effect have the same influence direction on each explained variable. Only from the correlation analysis, it can be seen that the January effect has a greater influence on SSE Composite index than the Monday effect. In autumn and winter, the turnover rate and volatility of SSE Composite Index are significantly lower than those of other seasons, and the yield rate is higher than that of other seasons (Table 4).

4.3 Regression Analysis

From the results of the least square regression, except that temperature has a significant and slight influence on turnover rate, other weather and weather changes have no significant influence on turnover rate, return rate and volatility. Therefore, we need to analyze whether there is conditional heteroskedasticity in the error term of this regression (Fig. 1; Table 5).

The "cluster" phenomenon of residuals fluctuation can be observed by regression of residual timing diagram with return rate: the fluctuation is small in some time periods (such as the observed values from 2009 to 2014) and very large in some time periods (such as the observed values from 2006 to 2008 and 2015–2016). The arch-LM test and Ljung-box test were used to analyze the square term of residual error, and the null hypothesis that there was no ARCH effect was rejected. Therefore, the sequence has a significant correlation at the significance level of 5%, so this model has the ARCH effect.

In view of the high-order ARCH effect in the residual series of explained variables, the EGARCH(1,1) model can be used to continue to explore the influence of weather and weather change on turnover rate, return rate and volatility of Shanghai composite index (Table 6).

Firstly, we analysis the influence of weather and the weather changes on the turnover rate of SSE Shanghai Composite Index: the effects of humidity on turnover rate is negative at the 1% level. It indicates that the higher the humidity is, the higher

Table 4 Correlation analysis results

	VOL	R	TEM	HUM	PRE	TEMV	HUMV	MON	JAN	FW	SAD
TURN	0.4129	0.0926	−0.0086	−0.0015	−0.0201	0.0239	0.0280	0.0071	0.0262	−0.0668	−0.0526
VOL		−0.1249	−0.0187	0.0061	0.0050	0.0011	0.0093	0.0464	0.0615	−0.0013	0.0175
R			−0.0107	−0.0134	0.0159	0.0242	0.0157	0.0357	−0.0273	0.0198	0.0140
TEM				0.2744	−0.8894	−0.1757	−0.1659	−0.0027	−0.4272	−0.7125	−0.7264
HUM					−0.4092	−0.4665	−0.6070	−0.0033	−0.0538	−0.1849	−0.2160
PRE						0.1748	0.2295	0.0023	0.3674	0.7553	0.7465
TEMV							0.5639	−0.0207	0.0800	0.0825	0.0977
HUMV								0.0246	0.0108	0.0813	0.0750
MON									0.0001	0.0136	0.0086
JAN										0.3007	0.4287
FW											0.8098

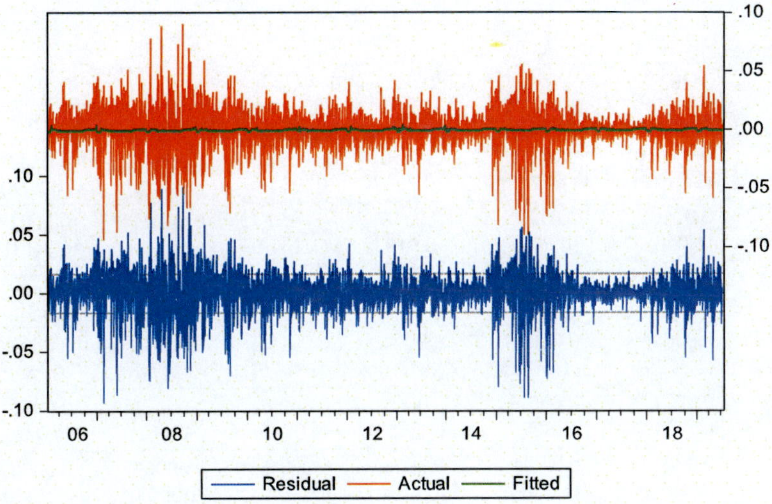

Fig. 1 Time sequence diagram of residual

Table 5 OLS regression results

	TURN		R		VOL	
	Model 1	Model 2	Model 1	Model 2	Model 1	Model 2
C	0.1034 (1.3301)	0.0953 (1.2116)	0.0449 (0.4693)	0.0592 (0.6128)	0.1215 (1.6143)	0.1261* (1.6588)
TEM	0.0002*** (3.5564)	0.0002*** (3.3939)	0.0138 (0.2374)	0.0222 (0.3770)	0.0541 (1.1803)	0.0563 (1.2175)
HUM	0.0092 (0.5196)	0.0088 (0.4044)	0.0066 (0.0842)	0.0102 (0.3900)	0.0183 (1.0889)	0.0175 (0.8424)
PRE	0.0028 (1.0882)	0.0025 (1.0074)	0.0015 (0.4670)	0.0019 (0.6026)	0.0033 (1.3494)	0.0035 (1.4014)
MON	0.0003 (0.4698)	0.0003 (0.4315)	0.0016** (2.1224)	0.0016** (2.1509)	0.0016*** (2.8203)	0.0016*** (2.7841)
JAN	0.0017* (1.7135)	0.0017* (1.7102)	0.0018 (1.4963)	0.0019 (1.5346)	0.0016* (1.6433)	0.0016* (1.6883)
FW	0.0025*** (2.8135)	0.0025*** (2.8352)	0.0007 (0.6607)	0.0007 (0.6550)	0.0002 (0.2517)	0.0002 (0.2567)
SAD	0.0011* (1.7767)	0.0010* (1.6639)	0.0555 (0.1092)	0.0598 (0.1174)	0.0002 (0.3677)	0.0002 (0.3963)
TEMV		0.0089 (0.2611)		0.0451 (1.0536)		0.0193 (0.5728)
HUMV		0.0086 (1.1719)		0.0094 (0.2837)		0.0133 (0.6085)

Note In brackets are the T values of each variable. In the table, ***represents the significance level of 1%, **represents the significance level of 5%, and *represents the significance level of 10%

Table 6 EGARCH fitting results

	TURN		R		VOL	
	Model 1	Model 2	Model 1	Model 2	Model 1	Model 2
C	0.0133** (2.0127)	0.0130* (1.7243)	0.0456 (0.7337)	0.0598 (0.9341)	0.1206*** (2.9618)	0.0282 (0.7990)
TEM	0.0160 (1.050)	0.0129 (0.7793)	0.0348 (0.9193)	0.0451 (1.1694)	0.0002*** (4.9122)	0.0001*** (3.4131)
A HUM	0.0231*** (3.7515)	0.0075*** (3.4274)	0.0180 (1.3067)	0.0243 (1.4378)	0.0299*** (3.1937)	0.0236** (2.0180)
The PRE	0.0003 (1.3729)	0.0003 (1.1943)	0.0014 (0.7001)	0.0019 (0.8876)	0.0034** (2.5168)	0.0003 (0.2850)
MON	0.0001** (2.1038)	0.0001* (1.8875)	0.0008* (1.9534)	0.0009** (2.0615)	0.0012*** (3.7161)	0.0013*** (4.0221)
JAN	0.0002 (1.1939)	0.0002 (1.0560)	0.0004 (0.4620)	0.0004 (0.4755)	0.0006 (1.0803)	0.0006 (1.1230)
FW	0.0149 (0.2254)	0.0087 (0.0096)	0.0009 (1.2167)	0.0009 (1.1402)	0.0004 (0.7740)	0.0002 (0.3741)
SAD	0.0002*** (3.0188)	0.0002*** (2.7371)	0.0006 (1.1513)	0.0005 (1.0547)	0.0101 (0.0448)	0.0522 (0.2273)
TEMV		0.0142 (1.1594)		0.0263 (0.9124)		0.0350* (2.5920)
HUMV		0.0039 (0.3221)		0.0010 (0.0129)		0.0209 (1.2679)

Note In brackets are the Z values of each explanatory variable. In the table, ***represents the significance level of 1%, **represents the significance level of 5%, and *represents the significance level of 10%

the turnover rate will be. Indeed, the increase of humidity will bring discomfort to investors physically, increase the anxiety of investors, increase the frequency of trading, and thus cause the increase of turnover rate; We note that the seasonal affective disorder index has a positive impact on the turnover rate of the SSE Composite Index regardless of whether the weather change variable is introduced and is significant at the 99% confidence level. The index of seasonal affective disorder reflects the relative length of the days in autumn and winter. The shortened days will make investors feel depressed and the trading volume will also decline. However, with the increase of daylight, the mood of investors tends to be cheerful and the investment confidence increases, thus improving the stock market liquidity and increasing the turnover rate correspondingly. In addition, the Monday variables in Model 1 and Model 2 are significantly positive at 95% and 90% confidence levels respectively, and the coefficient is 0.0001, indicating that the turnover rate of SSE Composite Index on Monday is slightly higher than that on other trading days. After introducing the volatility of temperature and humidity in model 2, it has no significant influence on the turnover rate of SSE Composite index.

Secondly, we focus on the influence of weather and weather change on the return rate of SSE Composite Index. In the two regression models of the return rate, only Monday variable has a significant and positive influence on the return rate indicating that the return rate of SSE Composite Index is slightly higher than that of other trading days on Monday. The volatility of temperature and humidity has no significant influence on the return rate of SSE Composite index.

Finally, we examine the impact of weather and weather changes on the volatility of SSE Composite Index. In model 1, temperature, humidity, pressure and Monday variables all have significant influences on the volatility of SSE Composite Index. The influence of the temperature on volatility of SSE composite index is negative, shows that the higher the temperature is, the smaller the volatility is. It is worth noting that the maximum temperature indicators in our sample is 33.7 °C, between minus 1 to 33.7 °C, with the increase of temperature to remove above 30 °C range, body feeling comfort for investors is increase, will further stabilize investor sentiment, reduce the volatility of SSE Composite Index; Humidity also has a negative impact on the volatility of SSE Composite Index. An increase in humidity will reduce the volatility of SSE Composite Index. Different from the turnover rate, the volatility essentially reflects the degree of bullish or bearish divergence among investors in the market, thus indicating that the increase in humidity will actually narrow the divergence. Air pressure is significantly negative in the model without the introduction of weather change index. The volatility of SSE Composite Index will increase with the decrease of air pressure, because the decrease of air pressure will cause the anxiety of investors, thus the difference between many investors on bullish or bearish will increase, which will increase the volatility of SSE Composite index. The volatility of SSE Composite Index on Monday is also higher than that of other trading days. In model 2, the introduction of weather change index, we find that the volatility of the temperature is positive under 10% significance level, means that the temperature increase in volatility will significantly increase the volatility of the stock market, this is due to the significant change of temperature in a day will make the investor sentiment is not stable, so as to increase the volatility of SSE Composite Index.

5 Summary and Prospect

5.1 Summary

With the rise of behavioral finance, some recent psychological studies have found that the climate environment has an indispensable influence on the change of human emotions. Investors may be susceptible to fluctuations due to the change of weather and environmental factors, thus affecting investors' rationality or risk ideology in the decision-making process. In this study, with return rate, turnover rate and volatility of SSE Composite Index as the dependent variable and "air temperature, air pressure, light, humidity and volatility" as the independent variable, the correlation between weather indicators and SSE Composite Index was discussed through EGARCH fitting regression. In view of the few domestic literature studies on the impact of weather changes on the return rate, turnover rate and return rate of the stock market, this paper not only discusses the impact of weather indicators on SSE Composite Index, but also deeply discusses the impact of weather changes on SSE Composite Index, in order to find out a more comprehensive conclusion. This paper summarizes the influence of weather variables on return rate, turnover rate and volatility.

Firstly, through empirical analysis, we find that weather and weather change variables have no significant impact on the return rate of Shanghai Composite Index.

Secondly, among the weather variables, the variables affecting SSE Composite Index include humidity and seasonal affective disorder index. The influencing mechanism lies in that the increase of humidity will increase people's body discomfort, thus causing emotional anxiety, further affecting the trading frequency and increasing the turnover rate. The seasonal affective disorder index reflects that with the increase of the length of the day, the mood of investors tends to be cheerful and the investment confidence increases, thus improving the stock market liquidity and increasing the turnover rate accordingly.

Finally, we find that weather and weather change variables have a significant impact on the volatility of SSE Composite Index. Among them, weather indicators including temperature, humidity and air pressure have a negative impact on the volatility of SSE Composite Index. And the volatility of temperature will have a positive impact on the volatility of SSE Composite Index. From the perspective of the coefficients, the impact of weather changes on the volatility of SSE Composite Index is more significant and strong than the impact of the weather itself.

5.2 *Prospect*

Due to the availability of data, the data used in this paper are the comprehensive data of Shanghai, Beijing, Guangdong, Jiangsu and Zhejiang. Although these five regions account for a large proportion in the trading volume of SSE Composite Index, there are still 30% transactions conducted in other regions, which will also affect the rigor of the conclusion to a certain extent.

Secondly, this article takes investors as a large group and ignores the heterogeneity of investors. So, the conclusion in this paper is only a summary on the whole, and there are large differences between individual investors, such as high degree of education of investors especially have received professional training of investors will be more rational, investors have rich investment experience is more rational than e investors who lack experience. Therefore, we can separate out individual investors in the future research.

Finally, in order to make the research more simple, this paper only considered the emotional changes of individual investors caused by the influence of weather, but did not consider institutional investors. However, Goetzmann, William et al. revealed that the weather-based emotional indicators would affect institutional investors' perception of mispricing and thus affect their trading. Therefore, if institutional investors can be taken into account, it will increase the persuasive power of this study.

References

1. Baker, M., & Jeffery, W. (2006). Investor sentiment and the cross-section of stock returns. *Journal of Finance, 61,* 1645–1680.
2. Yu, J., & Yuan, Y. (2011). Investor sentiment and the mean-variance relation. *Journal of Financial Economics, 100*(2), 367–381.
3. Chi, L. X., & Zhuang, X. T. (2009). Spillover effect of investor sentiment and stock returns volatility. *Journal of System Management, 4,* 367–372.
4. Yan, J. H. (2013). Research on the asymmetry of stock market volatility based on investor sentiment. *Technology and Market, 5,* 333–335.
5. Wang, C. (2014). The impact of investor sentiment on stock market returns and volatility—An empirical study based on net inflows into open-end stock funds. *China Management Science,* (9), 49–56.
6. Saunders, E. M. (1993). Stock price and wall street weather . *American Economic Review, 85*(5), 1337–1345.
7. Keef, S. P., & Roush, M. L. (2005). Influence of weather on New Zealand financial securities. *Accounting & Finance, 45*(3), 415–437.
8. Brown, G. W., & Cliff, M. T. (2005). Investor sentiment and assets valuation. *The Journal of Business, 78*(2), 405–440.
9. Schmeling, M. (2009). Investor sentiment and stock return: Some international evidence. *Journal of Empirical Finance, 16*(3), 394–408.
10. Dragos, S. O., & Laura, B. (2014). Investor sentiment and stock returns: Evidence from Romania. *International Journal of Academic Research in Accounting, Finance and Management Sciences, 4*(2), 19–25.
11. Sayim, M., & Reham, M. (2015). The relationship between individual investor sentiment, stock return and volatility: Evidence from the Turkish market. *International Journal of Emerging Markets, 10*(3), 504–520.
12. Da, Z., Engelberg, J., & Gao, P. J. (2015). The sum of ALL FEARS investor sentiment and asset prices. *Review of Financial Studies, 28*(1), 1–32.
13. Zhang, Q., Yang, S. E., & Yang, H. (2007). An empirical study on investor sentiment and stock returns in Chinese stock market. *Systems Engineering, 7,* 13–17.
14. Zhang, Q., Yang, S. E., & Yang, H. (2008). An empirical study on the characteristics of cross-sectional returns and investor sentiment in China's stock market. *Systems Engineering,* (7), 22–28.
15. Zhang, Q., & Yang, S. E. (2009). Noise trading, investor sentiment swings and stock returns. *Systems Engineering Theory and Practice, 3,* 40–47.
16. Hirshleifer, D., & Shumway, T. (2015). Good day sunshine, stock returns and the weather. *Journal of Finance, 58*(3), 1009–1032.
17. Yi, C. L., & Wang, J. Q. (2009). Weather, seasonal affective disorder and stock returns—Based on the shanghai composite index. *Statistics and Decision Making, 8,* 19–82.
18. Cao, G. X., & Xu, G. F. (2010). An empirical study on the influence of weather change on the volatility of Chinese securities market. *Quantitative Economic and Technological Research, 8,* 848–854.
19. Lu, J. (2011). An empirical study on the weather effect of Chinese stock market. *Chinese Soft Science, 6,* 65–78.

Performance Evaluation of Urban Rail Transit PPP Mode Project

Chang Liu, Xuemeng Guo, and Rong Men

Abstract With the development of China's economy and the continuous acceleration of urbanization, the amount of funds required to construct and operate urban rail transit is huge. Since the introduction of the PPP model, this model has developed in various public transportation areas. However, the application of the PPP financing model in China's urban rail transit construction is still in its infancy, and the application situation is still immature, and many problems still exist. Based on the balanced scorecard and the characteristics of the project itself, this article establishes a project performance evaluation system, revises and determines the evaluation method, and establishes an index system and performance evaluation model.

Keywords PPP model · Urban rail transit · Stakeholders · Performance evaluation

1 Introduction

Up to now, there have been nearly 700 public service projects developed in the UK using the PPP model, of which more than 450 have been put into operation. The PPP model was first started in China by Shenzhen Shajiao B Power Plant, and it was initially developed due to the subsequent construction of Changsha Power Plant and Chengdu No. 6 Water Plant.

In terms of urban rail transit, the traditional financing model mainly based on government investment can no longer meet the development needs of funds required for urban rail transit, and market-oriented and diversified forms of financing for public goods will become a new development trend. The PPP mode has been applied to the Netherlands High Speed Rail South Line, London Underground, Beijing Subway

C. Liu · X. Guo (✉) · R. Men
School of Economics and Management, Beijing Jiaotong University, Beijing, China
e-mail: 379515516@qq.com

C. Liu
e-mail: 18113073@bjtu.edu.cn

R. Men
e-mail: 939307168@qq.com

M. Li et al. (eds.), *IEIS 2020*, https://doi.org/10.1007/978-981-33-4363-4_30

Line 4 and Shenzhen Metro Line 4. However, the introduction of the PPP model into China's urban rail transit projects still creates many problems. In reality, some private investors only pay attention to their own interests and ignore the cost, benefit, and quality assessments in the specific implementation of the project, which creates a lot of uncertainty.

Therefore, it is particularly important to establish a performance evaluation system that can appropriately reflect the urban rail transit projects under the PPP model, and to identify problems and improve them in a timely manner through the evaluation system to improve management efficiency. Through the performance evaluation, it is possible to obtain the completion of the project in a timely manner, find problems, and establish improvement methods, because the reasonable and effective operation of urban rail transit must be achieved in the end, and public and private interests are taken into account in the evaluation process, thereby promoting urban rail transit. Improved performance of transportation projects under the PPP model.

2 Literature Review

According to the stakeholder theory, the performance evaluation of urban rail transit PPP projects refers to the implementation of the PPP model of urban rail transit projects, and from the needs of stakeholders such as the government, private enterprises, and the public, the overall management of the project and the degree of target completion Input–output measurement.

At present, the research on the performance of urban rail transit PPP projects is more based on the overall perspective of PPP projects, and the construction of the entire life cycle performance index system of urban rail transit PPP projects. Some scholars [1–4] use balanced scoring (BSC), key performance indicators (KPI), stakeholder performance goals, Various theories and methods, such as principal component analysis and DPSIR, are used to identify the key performance and system construction of the urban rail transit PPP project itself.

With the in-depth research and discussion on the performance evaluation of urban rail transit PPP projects, scholars at home and abroad have put forward their views on some key performance indicators of urban rail transit PPP projects. Phang [5] discussions on different urban rail transit PPP projects in cities such as Hong Kong argued that reasonable micro-operational risks and the sharing of macroeconomic and financial risks, the degree of government departments' understanding of the PPP model and its management details, government departments and social capital Factors such as project long-term and short-term cost and demand prediction, the bidding process, and whether the final bidder has a strong competitive ability may play a key role in the successful operation of the urban rail transit PPP model [5]. Willoughby [6] and Yuan et al. [7] proceeded from the goals of the urban rail transit PPP project, arguing that the key goal of implementing the urban rail transit PPP project was to relax the financial constraints of government departments, compared to traditional government construction methods. Urban rail transit and other urban infrastructure service operational efficiency and innovation, efficiency improvement [6, 7]. In addition, in the process of establishing and evaluating urban rail PPP projects, fully consider the sustainable performance of environmental benefits brought by urban rail transit projects [8, 9] and manifestation The social fairness and justice performance of the project's social benefits [10] is also an indispensable part of the construction process of key indicators for evaluating the performance of urban rail transit PPP projects.

Existing research is limited to a single perspective. It does not build a unified performance evaluation guide based on the evaluation goals of the overall PPP project company and each stakeholder. It cannot effectively solve the technical evaluation, economic evaluation, external evaluation and other aspects of the PPP model. The issue of mutual compatibility makes the parties involved in the PPP model unable to obtain the specific basis of their effective participation path, which needs further research.

3 Design of Performance Evaluation Model for Urban Rail Transit Project Under PPP Mode

3.1 Model Design Ideas

The design concept of the performance evaluation system of urban rail transit projects under the PPP mode is based on the interests of all stakeholders as the starting point, the basic idea of the balanced scorecard as the core, and the original four dimensions as the basis. Satisfaction of all stakeholders, and increased consideration of project environment, divided the "internal business process" dimension into "environment characteristics" and "management benefit" dimensions, and revised the "innovation and learning" dimension to "sustainability and growth" dimension, which constitutes a five-dimensional project performance evaluation system that affects each other during the project implementation process and reflects the key factors that affect such project performance from different aspects. The specific relationship is shown in Fig. 1.

In 2006, the scholar Zhang Xueqing summarized the best influencing factors on PPP projects, and made reference to relevant domestic and foreign policies and literature, such as the key performance indicators formulated by the British KPI Working Group, and China's "Government Investment Project Procurement and Performance Evaluation Study", "Public Engineering Project Performance Evaluation Study", "Metro Operation Safety Evaluation Standard", "Urban Public Transport Classification Standard", "Metro and Light Rail System Operation Management Standard", "Urban Rail Transit Passenger Service", International Metro Association (CoMET) Based on the "Key Performance Index" system and MOPES of urban rail transit operation performance evaluation system, the domestic and foreign public–private cooperation projects and urban rail transit performance evaluation theory are summarized. With reference to the research conclusions of scholars at home and abroad,

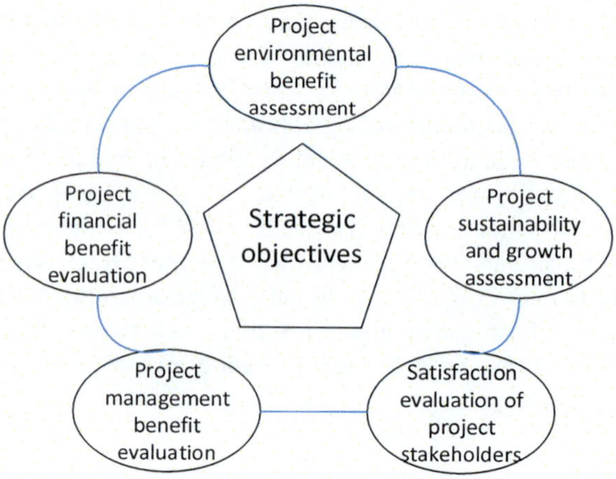

Fig. 1 Performance evaluation model of urban rail transit project under PPP mode

the four dimensions are modified into five dimensions: environmental characteristics, financial benefits, management benefits, satisfaction of various stakeholders, sustainability and growth. Establish a performance evaluation index system for urban rail transit under the PPP model.

The indicators of the environmental characteristic dimension evaluate the impact of the external and internal environment on the project during its operation. The external environment includes the political environment, economic environment, legal environment, and social environment; the internal environment includes the characteristics of the project itself. With reference to the financial indicators of enterprise performance evaluation, according to the factors that the private sector pays more attention to operating efficiency, the financial dimension indicators are divided into four secondary dimensions of debt repayment ability, profitability, operating ability and development ability. The management benefit performance evaluation dimension of urban rail transit projects under the PPP model is modified from the "internal business process" in the balanced scorecard theory, which not only includes the management level, project completion and operation aspects of the transportation project implementation process, It also includes safety services and equipment management unique to urban rail transit projects. Because the PPP model is used, there are many stakeholders in urban rail transit projects. The public, the private sector, and the public sector are all stakeholders in such projects, so there must be indicators that can measure the satisfaction of multiple stakeholders to ensure the overall The unity of interests. Because it is an urban rail transit project and uses the PPP model, it will bring higher risks and uncertainties. And as a transportation project, sustainable development factors, future development potential and development stability also need to be considered.

3.2 Using the AHP Method to Determine the Weight of Project Performance Evaluation Indicators

The multi-level analysis method can divide each factor into ordered levels, and then simplify and handle complex problems. By comparing the relative importance of each element, each pair is quantified according to the relative importance level 1–9, then the consistency is tested, and finally the relative importance of each factor is ranked.

The basic method of the AHP method is: according to the characteristics and needs of the target, it is refined into different possible influence factors; the hierarchical analysis structure model is established according to the attribute relationship and mutual influence of each factor; Determine weights; comprehensively calculate the combined weight of relative importance as the basis for evaluation.

According to the expert's score of the coefficient, 12 expert questionnaires were distributed and 10 were recovered. The effective recovery rate was 83.33%. Expert Choice software is used to calculate the weights of each level of elements, and the weight summary table is summarized in Table 1.

4 Empirical Research

4.1 Empirical Case Selection

After determining the weight of each indicator, the experts will score each indicator of the urban rail transit project under the specific PPP mode, and establish a relevant comment set $V = \{v_1, v_2, v_3, \ldots v_5\}$ = {Excellent, Good, Medium, Poor, Inferior}. Take the number of expert evaluations as the fuzzy evaluation matrix, and then multiply the weights corresponding to different levels, and then carry out the fuzzy comprehensive evaluation, starting from the first-level fuzzy comprehensive evaluation, that is, starting from the comment set of the three-level measurement indicators, and calculating the corresponding weight. The results of the first-level fuzzy comprehensive evaluation; the results of the first-level fuzzy comprehensive evaluation are combined into a second-level fuzzy relationship membership matrix, and the corresponding second-level weights are used to calculate the second-level fuzzy comprehensive evaluation results; the third-level fuzzy evaluation results are calculated by the same method, The evaluation result $B = \max(B_1, B_2, \ldots, B_5)$. Among them, if the value of B_1 is the largest, it indicates that the overall benefit is "excellent", and if the value of B_2 is the largest, it indicates that the overall benefit is "good". The specific indicators can also be judged by the fuzzy evaluation results of each level, and the requirements of different levels are different at different stages, and the standards can also be dynamically adjusted.

Take Beijing Metro Line 4 as an example to conduct research and analysis on the performance evaluation of the project.

Table 1 Summary table of weights for performance evaluation indicators of urban rail transit projects in PPP mode

First-level dimension	Second-level index	Third-level index
Project environmental characteristics. U_1 (0.0410)	Internal environment. U_{11} (0.2500)	Both sides reached a consensus on benefit and risk sharing. U_{111} (0.2473)
		Complexity of project design and construction. U_{112} (0.2473)
		Familiarity with PPP models and urban rail transit projects. U_{113} (0.5054)
	External environment. U_{12} (0.7500)	Political environment stability. U_{121} (0.1500)
		Economic environment stability. U_{122} (0.2000)
		Legal environment support degree. U_{123} (0.4000)
		Social environment support degree. U_{124} (0.1500)
Financial benefit. U_2 (0.1722)	Solvency. U_{21} (0.1200)	Current ratio. U_{211} (0.1500)
		Quick ratio. U_{212} (0.1500)
		Asset-liability ratio. U_{213} (0.5500)
		Interest multiple. U_{214} (0.1500)
	Profitability. U_{22} (0.4874)	Return on total assets. U_{221} (0.2135)
		Return on net assets. U_{222} (0.6283)
		Capital appreciation rate. U_{223} (0.1582)
	Operating capacity. U_{23} (0.1200)	Total asset turnover rate. U_{231} (0.6000)
		Operating index. U_{232} (0.4000)
	Development capacity. U_{24} (0.2726)	Project income growth rate. U_{241} (0.2500)
		Capital accumulation rate. U_{242} (0.7500)
Management benefits. U_3 (0.4424)	Effective communication management. U_{31} (0.0367)	Number of effective communications. U_{311} (0.1000)
		Effective communication. U_{312} (0.9000)

(continued)

Table 1 (continued)

First-level dimension	Second-level index	Third-level index
	Cost management. U_{32} (0.2144)	Cost control rate. U_{321} (0.7500)
		Cost reduction rate. U_{322} (0.2500)
	Progress control management. U_{33} (0.0367)	Number of rework stoppages. U_{331} (0.2500)
		Project progress completion rate. U_{332} (0.7500)
	Risk management. U_{34} (0.2282)	Risk allocation ratio. U_{341} (0.6283)
		Risk aversion ability. U_{342} (0.1582)
		Risk transfer ability. U_{343} (0.2135)
	Quality management. U_{35} (0.2282)	Quality of service. U_{351} (0.2000)
		Device management. U352 (0.2000)
		Traffic operation capacity. U_{353} (0.6000)
	Security management. U_{36} (0.2144)	Number of safety accidents. U_{361} (1.0000)
	Government administration. U_{37} (0.0414)	Government subsidy ratio. U_{371} (0.2500)
		Public complaint resolution degree. U_{372} (0.1500)
		Supervision of project strength. U_{373} (0.6000)
Satisfaction of various stakeholders. U_4 (0.1722)	Satisfaction with the public. U_{41} (0.600)	Satisfaction with transportation price. U_{411} (0.3000)
		Satisfaction with transportation services. U_{412} (0.3000)
		Public satisfaction with environmental benefits. U_{413} (0.1500)
		Public satisfaction with social benefits. U_{414} (0.1500)
	Private sector satisfaction. U_{42} (0.200)	Satisfaction with obtaining reasonable profit. U_{421} (0.6283)

(continued)

Table 1 (continued)

First-level dimension	Second-level index	Third-level index
		Private sector satisfaction with risk allocation. U_{422} (0.2135)
		Private sector satisfaction with partnership. U_{423} (0.1582)
	Government department satisfaction. U_{43} (0.200)	Satisfaction of government departments on environmental benefits. U_{431} (0.1425)
		Satisfaction of government departments on social benefits. U_{432} (0.2137)
		Satisfaction of government departments on shared risk. U_{433} (0.1023)
		Satisfaction of government departments on quality levels. U_{434} (0.5024)
		Satisfaction of government departments on cooperative relationships. U_{435} (0.0391)
Sustainability and growth. U_5 (0.1722)	Resource use efficiency. U_{51} (0.3626)	Adaptability of urban rail transit and regional economic development.U_{511} (0.3781)
		Reasonable allocation and utilization of resources. U_{512} (0.2321)
		Pollution emission control. U_{513} (0.0734)
		Possibility of resource utilization. U_{514} (0.2321)
		Energy saving. U_{515} (0.0843)
	Social benefits. U_{52} (0.3626)	Regional income level improvement. U_{521} (0.5000)
		Regional employment level improvement. U_{522} (0.5000)
	Operational innovation. U_{53} (0.0762)	Service innovation level. U_{531} (0.0750)
		Design innovation degree. U_{532} (0.0750)
		Implementation technology innovation degree. U_{533} (0.4250)

(continued)

Table 1 (continued)

First-level dimension	Second-level index	Third-level index
		Hardware facilities innovation degree. U_{534} (0.0750)
		Management technology innovation degree. U_{535} (0.3500)
	Staff training. U_{54} (0.1986)	Staff training days. U_{541} (0.0500)
		Training employee input ratio. U_{542} (0.7500)
		Knowledge employee absorption rate. U_{543} (0.2000)

Data source It is obtained by using Expert Choice software to process and summarize expert scoring results

Beijing Metro Line 4 is a public–private joint venture established by Beijing Infrastructure Investment Co., Ltd. and Hong Kong Metro Company and Beijing Capital Venture Group Co., Ltd., the investment and construction of franchised trains and mechanical and electrical equipment and the operation of Line 4 became the first domestic Metro projects operating in PPP mode.

All investment construction tasks of the project are divided into two parts, A and B. Part A accounts for 70% of the total investment, about 10.7 billion yuan, including land acquisition and demolition, civil engineering (including subway stations, caves, car depots and parking lots), tracks, and civil defense projects. Be responsible for. After completion, this part of the assets will be provided to the company for use in two ways: capital contribution and lease. After completion, this part of the asset will be provided to the PPP company in two ways: right-of-use capital contribution and lease. The asset part of the capital contribution of the right-of-use is referred to as Al and the leased asset part is referred to as A2. Part B accounts for 30% of the total investment, about 5 billion yuan, including the purchase and sale of electromechanical equipment such as vehicles, automatic ticket inspection systems, signaling and communication, air conditioning and ventilation, water supply and drainage, and fire fighting, escalators and elevators, control equipment, and power supply facility installation. Hong Kong Metro holds 49%, while government investors (CBN and Infrastructure Corporation) hold 51%. The Beijing Municipal Government granted a 30-year franchise right to Metro Line 4 of the PPP company through a franchise agreement, and stipulated the line construction standards, operating standards, and project facility transfer standards after the end of the concession period. The concession rights cannot be transferred during the franchise period. The infrastructure company and the PPP company signed a lease agreement to lease A2 to the PPP company. The infrastructure company signed a contract agreement with the relevant engineering contracting company to transfer the completion risk of the construction of Asset A to the engineering contracting company in the form of a guarantee of completion.

4.2 Project Performance Evaluation

4.2.1 Performance Evaluation Process

The experts involved in the performance evaluation of this PPP project include both theoretical research scholars and actual staff of PPP projects. 12 expert questionnaires were distributed and 10 were recovered, with an effective recovery rate of 83.33%.

The fuzzy evaluation is carried out layer by layer according to the fuzzy evaluation method.

The first-level fuzzy evaluation of the internal environment is performed, and the evaluation weight is assigned as $A_{11} = (0.2473, 0.2473, 0.5054)$.

The fuzzy evaluation matrix is:

$$R_{11} = \begin{bmatrix} 0.3 & 0.7 & 0 & 0 & 0 \\ 0.8 & 0.2 & 0 & 0 & 0 \\ 0.6 & 0.3 & 0.1 & 0 & 0 \end{bmatrix}$$

The evaluation of the internal environment is:

$$\begin{aligned} B_{11} &= A_{11}R_{11} \\ &= (0.2473, 0.2473, 0.5054) \cdot \begin{bmatrix} 0.3 & 0.7 & 0 & 0 & 0 \\ 0.8 & 0.2 & 0 & 0 & 0 \\ 0.6 & 0.3 & 0.1 & 0 & 0 \end{bmatrix} \\ &= (0.5250\ 0.4750\ 0.0000\ 0.0000\ 0.0000) \end{aligned}$$

According to the principle of maximum membership, the largest number in B_{11} is 0.5250, which corresponds to "excellent", indicating that experts evaluate the internal environment of the project environmental characteristics as "excellent".

All other levels are calculated according to this process, and the first-level fuzzy evaluation results are summarized as follows:

$B_{11} = (0.5250\ 0.4750\ 0.0000\ 0.0000\ 0.0000)$;
$B_{12} = (0.6468\ 0.2859\ 0.0000\ 0.0000\ 0.0000)$;
$B_{21} = (0.3468\ 0.6428\ 0.0105\ 0.0000\ 0.0000)$;
$B_{22} = (0.5000\ 0.4400\ 0.0600\ 0.0000\ 0.0000)$;
$B_{23} = (0.3250\ 0.6000\ 0.0750\ 0.0000\ 0.0000)$;
$B_{24} = (0.5350\ 0.4750\ 0.0000\ 0.0000\ 0.0000)$;
$B_{31} = (0.6000\ 0.3000\ 0.1000\ 0.0000\ 0.0000)$;
$B_{32} = (0.6000\ 0.4000\ 0.0000\ 0.0000\ 0.0000)$;
$B_{33} = (0.2180\ 0.4363\ 0.2757\ 0.0637\ 0.0000)$;
$B_{34} = (0.3500\ 0.5000\ 0.1500\ 0.0000\ 0.0000)$;
$B_{35} = (0.4000\ 0.3500\ 0.2000\ 0.0500\ 0.0000)$;
$B_{36} = (0.4846\ 0.3637\ 0.1301\ 0.0258\ 0.0000)$;
$B_{37} = (0.9000\ 0.1000\ 0.0000\ 0.0000\ 0.0000)$;

B_{41} = (0.4617 0.3598 0.1448 0.0046 0.0000);
B_{42} = (0.5428 0.3258 0.1105 0.0209 0.0000);
B_{43} = (0.4000 0.3200 0.2800 0.0000 0.0000);
B_{51} = (0.2600 0.4200 0.2400 0.0800 0.0000);
B_{52} = (0.3250 0.4750 0.2000 0.0000 0.0000);
B_{53} = (0.3500 0.5750 0.0750 0.0000 0.0000);
B_{54} = (0.3000 0.4572 0.2286 0.0143 0.0000).

The above evaluation results are then combined into a second-level membership matrix, and the corresponding fuzzy vector is used to perform a second-level fuzzy evaluation through the same calculation steps. The evaluation results obtained are summarized as follows:

B_1 = (0.6164 0.3332 0.0478 0.0000 0.0000);
B_2 = (0.4121 0.5665 0.0214 0.0000 0.0000);
B_3 = (0.4268 0.3984 0.1517 0.0271 0.0000);
B_4 = (0.4409 0.3291 0.3212 0.0051 0.0000);
B_5 = (0.3007 0.4893 0.1936 0.0164 0.0000).

Finally, the above evaluation results are composed into a third-layer membership matrix, and the corresponding weight vectors are used to perform a three-level fuzzy evaluation through the same calculation steps. The evaluation results are obtained:

B = (0.4126 0.4285 0.1615 0.0157 0.0000).

4.2.2 Analysis of Performance Evaluation Results

In the actual evaluation process, care needs to be taken to ensure the authenticity of the acquired data and the accuracy of the data processing process. Because the calculation result depends to a large extent on the authenticity of the acquired data, and in the calculation process, the calculation process is more complicated, manual calculation is prone to errors, and can be processed with the help of statistical software.

According to the evaluation criteria, performance evaluation is performed on each indicator of the project, and the evaluation results are finally obtained. The evaluation results are analyzed as follows:

1. The project has good environmental characteristics during the process. With the support of relevant laws and policies, the project construction and operation can be carried out well.
2. In terms of financial evaluation, the financial capacity of the private sector is generally good, but the profit of the project is not high. This is because the project is controlled by the government, the subway fare adjustment should not be too large, and government supervision has effectively suppressed the profitability of social capital.
3. The project management benefit is good, and the project reaches high-quality projects. Under the public–private partnership model, subways effectively reduce costs by optimizing resource allocation. At the same time, the government has

given higher credit to the private sector, which has effectively reduced the policy risk of investment and the investment cost of projects.

4. The satisfaction of the government, private and public sectors are high. Although the project may cause some inconvenience to the surrounding residents during the implementation of the project, it will bring great convenience to the surrounding environment after the completion of the project and improve it to a certain extent Economic level.
5. In terms of sustainability, the project has achieved a sustainable use of environmentally friendly resources, and innovations in technology and management can also ensure the continued operation and competitiveness of the project.

To a certain extent, the project considered both efficiency and fairness, and the performance evaluation was "good".

5 Conclusion

For the time being, China has not formed a performance evaluation system for urban rail transit projects under the PPP mode, so problems such as uneven risk sharing, inadequate benefit sharing mechanisms, and low operation quality may occur in actual project implementation. This article takes the performance evaluation system of urban rail transit projects under the PPP model as the research object. After analyzing the relevant research results at home and abroad, it is based on the stakeholder theory, key performance indicator methods, and balanced scorecard theory. The card theory sets up five dimensions of environmental characteristics, financial benefits, management benefits, satisfaction of various stakeholders, sustainability and growth to evaluate the project's performance level. Then use the fuzzy comprehensive evaluation method and AHP method to build performance evaluation system. The conclusions are mainly as follows:

1. Summarize and analyze the results of performance evaluation studies on PPP mode projects and urban rail transit projects at home and abroad, and put forward the importance of establishing a performance evaluation system for urban rail transit projects under the PPP mode in China. Each stakeholder, urban rail transit project has the characteristics of quasi-public goods and so on. Instead of evaluating the performance of the project from a single stakeholder, it takes into account the needs of multiple parties.
2. Describes the connotation of urban rail transit projects under the PPP model. Based on relevant theories, the environmental characteristics, financial benefits, management benefits, satisfaction of various stakeholders, and sustainability are established based on the balanced scorecard theory and project characteristics. And growth five dimensions of evaluation, clearly evaluate the subject, and set the key performance indicators from the five dimensions as the starting point.
3. Determined the method and the weight ratio of such project performance evaluation, constructed the performance evaluation system of urban rail transit projects

under the PPP mode, and provided a detection tool for evaluating performance levels for the use of the PPP mode in the urban rail transit field. To a certain extent, it promotes the improvement of application maturity.

However, there are still many deficiencies in the research of this paper. First, the degree of innovation and depth of this paper is limited. The PPP model is a relatively complex subject and requires a knowledge base in urban rail transit, because the related theoretical research on project management is insufficient. As the knowledge of urban rail transit projects is limited, the data is difficult to obtain, which leads to fewer innovative theories, and there are still many areas that need to be improved. Secondly, the evaluation system constructed needs to be further improved, because the theoretical understanding is still insufficient, the urban rail transit industry is not fully understood, and the indicators and standard settings of the performance evaluation system have uncertain elements. Corrections can ultimately reflect the project's performance level scientifically. And in terms of indicator design, because in specific projects, many indicators are difficult to carry out quantitative research, such as environmental characteristics, the reasonable degree of risk sharing, and the satisfaction of various stakeholders, etc. In the process of obtaining qualitative indicator evaluation, the data is vulnerable to human Subjective impact, so designing these qualitative indicators to be more reasonable and measurable is also a problem to be solved.

Acknowledgements Funding for National Natural Science Foundation "Research on Performance Evaluation System of urban rail transit PPP mode based on resource "passenger-value flow"" (71973009).

Funding for Fundamental Scientific Research Business Funds of Central Universities Funds "Does proprietary technology transfer cause local governments in various regions to choose PPP mode financing differences?—Based on empirical research in infrastructure field" (2019YJS075).

References

1. Yi, X. (2015). Performance evaluation of public-private partnerships in urban rail transit from the perspective of knowledge transfer. *Urban Rail Transit Research, 18*(04), 27–33.
2. Wang, J., Liu, F., You, W., Zhang, S., & Peng, L. (2018). Life cycle performance evaluation of urban rail transit PPP projects. *Journal of Civil Engineering and Management, 35*(06), 23–29.
3. Wang, J., Liu, F., You, W., Zhang, W., & Yang, D. (2018). Performance evaluation of urban rail transit PPP projects based on balanced scorecard. *Engineering Economy, 28*(08), 71–75.
4. Wang, J., Liu, F., You, W., Yang, D., & Zhang, W. (2018). Performance evaluation of urban rail transit PPP projects based on DPSIR and matter element analysis. *Tunnel Construction (Chinese and English), 38*(07), 1100–1107.
5. Phang, S. Y. (2007). Urban rail transit PPPs: Survey and risk assessment of recent strategies. *Transport Policy, 14*(3), 0–231.
6. Willoughby, C. (2013). How much can public private partnership really do for urban transport in developing countries? *Research in Transportation Economics, 40*(1), 34–55.
7. Yuan, J.-F., Skibniewski, M, J., Li, Q., et al. (2010). The driving factors of China's public-private partnership projects in metropolitan transportation systems: Public sector's viewpoint. *Journal of Civil Engineering and Management, 16*(1), 5–18.

8. Grasman, S. E., Faulin, J., & Lera-López, F. (2014). Integrating environmental outcomes into transport public–private partnerships. *International Journal of Sustainable Transportation, 8*(6), 399–422.
9. Qin, T., Zhang, Y., & Ouyang, H. (2018). Research on the influencing factors of sustainable performance of urban rail transit projects . *Journal of Wuhan University of Technology (Information and Management Engineering Edition), 40*(06), 644–648.
10. Tang, Z., & Chen, S. (2016). Social justice performance evaluation of urban rail transit network distribution in downtown Shanghai. *Shanghai Urban Planning, 02,* 102–108.